AF389227

Intramolecular Hydrogen Bonding 2021

Intramolecular Hydrogen Bonding 2021

Editor

Mirosław Jabłoński

MDPI • Basel • Beijing • Wuhan • Barcelona • Belgrade • Manchester • Tokyo • Cluj • Tianjin

Editor
Mirosław Jabłoński
Nicolaus Copernicus University
in Toruń
Poland

Editorial Office
MDPI
St. Alban-Anlage 66
4052 Basel, Switzerland

This is a reprint of articles from the Special Issue published online in the open access journal *Molecules* (ISSN 1420-3049) (available at: https://www.mdpi.com/journal/molecules/special_issues/Intramolecular_Hydrogen_Bonding_2021).

For citation purposes, cite each article independently as indicated on the article page online and as indicated below:

LastName, A.A.; LastName, B.B.; LastName, C.C. Article Title. *Journal Name* **Year**, *Volume Number*, Page Range.

ISBN 978-3-0365-2518-1 (Hbk)
ISBN 978-3-0365-2519-8 (PDF)

Contents

About the Editor

Mirosław Jabłoński

The entire scientific career of Dr. Mirosław Jabłoński is associated with the Nicolaus Copernicus University in Toruń. Immediately after graduating from Chemical Studies in 2000, he was employed at this university, first as an assistant, then as an assistant professor, and now as a professor. His research interests relate to intra- and intermolecular interactions including, for example, hydrogen, halogen, triel and agostic bonds. However, most of his scientific research is also concerned with the so-called charge-inverted hydrogen bonds (CIHB), which he proposed in 2009. His research also concerns the issues of estimating the energy of intramolecular interactions and the interpretation of the so-called bond paths defined within QTAIM. He is also interested in the steric and substituent effects, and more recently in the chemistry of carbenes, especially the carbene-metal bond.

Preface to "Intramolecular Hydrogen Bonding 2021"

Studied for over a hundred years, hydrogen bonds are one of the most described phenomena in chemistry. This is because they occupy an important position in the world of inter- and intramolecular interactions, which in turn is related to their position on the scale of the strength of interactions and chemical bonds. Particularly, on this scale, hydrogen bonds are between the weaker van der Waals interactions and the stronger covalent bonds. For this reason, intermolecular hydrogen bonds act as a glue that binds individual molecules into dimers, trimers, or larger molecular clusters. However, the network of hydrogen bonds is dynamic, old hydrogen bonds are broken and new ones are formed. This ability is at least in part due to the small size of the hydrogen atom, which allows the X and Y atoms in the X-H$\cdots$Y bridge to come closer to each other without major steric effects. The positive atomic charge of H is also important here, which, being contrary to the atomic charges on X and Y, enables this contact. The most frequently mentioned symptom of the presence of intermolecular hydrogen bonds is the curiously high boiling point of liquid water compared to slightly heavier hydrogen sulfide. Another manifestation of the presence of intermolecular hydrogen bonds is the lower density of ice than that of liquid water, thanks to which it floats on its surface, making it easier for fish to survive during cold winters. The good solubility of polar and ionic substances in water is also due to the formation of hydrogen bonds.

However, hydrogen bonds are not all about water. Their role is equally important for many much larger molecules, including macromolecules. For example, the presence of intermolecular hydrogen bonds between complementary nitrogen base pairs (cytosine–guanine and adenine–thymine) binds two strands of DNA together, giving them a double-helical structure, which is fundamental in the replication of genetic information. Intermolecular hydrogen bonds also affect the structure of proteins and maintenance of cellulose or polymer chains.

It seems that intramolecular hydrogen bonds are not as often described as their intermolecular counterparts. This is due to the fact that their detection is generally based on indirect methods, and this is often associated with some problems, for example in the form of finding an appropriately reliable reference system. In spectroscopic methods, the presence of intramolecular hydrogen bond is ascertained, for example, on the basis of a change in a certain parameter (e.g., the stretching vibration frequency of X-H) describing the proton-donor group X-H, but determining the value of this change obviously requires establishing a certain reference value characterizing the unperturbed X-H group. Like their intermolecular counterparts, intramolecular hydrogen bonds are, of course, also of great importance. First of all, they can significantly affect the conformational equilibrium, giving preference to certain conformers over others. A known case is the keto-enol equilibrium, which is related to the alpha hydrogen atom movement. Such equilibria can, in turn, affect the crystallographic structure of the compound in the solid. A very important effect related to the movement of the hydrogen atom in the X-H$\cdots$Y bridge is the proton transfer effect, which occurs especially in the excited state of a molecule.

In my opinion, the great advantage of this book is that it contains the results of both theoretical and experimental research, and with the use of many different research methods. Therefore, it is an excellent review of these methods, while showing their applicability to the current scientific issues regarding intramolecular hydrogen bonds. The experimental techniques used include X-ray diffraction, infrared and Raman spectroscopy (IR), nuclear magnetic resonance spectroscopy (NMR), nuclear quadrupole resonance spectroscopy (NQR), incoherent inelastic neutron scattering (IINS), and differential scanning calorimetry (DSC). The solvatochromic and luminescent studies are also described. On the other hand, theoretical research is based on ab initio calculations and the Car–Parrinello Molecular Dynamics (CPMD). In the latter case, a description of nuclear quantum effects (NQE) is also possible. This book also demonstrates the use of theoretical methods such as Quantum Theory of Atoms in Molecules (QTAIM), Interacting Quantum Atoms (IQA), Natural Bond Orbital (NBO), Non-Covalent Interactions (NCI) index, Molecular Tailoring Approach (MTA), and many others.

Mirosław Jabłoński
Editor

Editorial

Intramolecular Hydrogen Bonding 2021

Mirosław Jabłoński

Faculty of Chemistry, Nicolaus Copernicus University in Toruń, Gagarina 7, 87-100 Toruń, Poland; teojab@chem.umk.pl; Tel.: +48-056-611-4695

Citation: Jabłoński, M. Intramolecular Hydrogen Bonding 2021. *Molecules* **2021**, *26*, 6319. https://doi.org/10.3390/molecules26206319

Received: 5 October 2021
Accepted: 15 October 2021
Published: 19 October 2021

Publisher's Note: MDPI stays neutral with regard to jurisdictional claims in published maps and institutional affiliations.

Undoubtedly, hydrogen bonds occupy a leading place in the rich world of intermolecular interactions. This fact results from their intermediate strength, being between stronger covalent bonds and weaker van der Waals interactions. For this reason, intermolecular hydrogen bonds act as a glue that binds individual molecules into dimers or larger molecular clusters, but at the same time allows for dynamic changes consisting in breaking old bonds and possible formation of new ones. This ability is partly due to the small size of the hydrogen atom, which, being poor in electron density, allows the X and Y units in the X-H...Y contact to come closer to each other with relatively small steric effects. The most frequently quoted academic symptoms of the presence of intermolecular hydrogen bonding are the unexpectedly high boiling point of water, lower density of ice compared to liquid water, and the good solubility of polar and ionic substances in water. Nevertheless, equally important is the presence of hydrogen bonds between the complementary nitrogen bases (C-G, A-T) in the nucleotides of two DNA strands, because it leads to a double-helical structure that is fundamental in the replication of genetic information. Intermolecular hydrogen bonds also have a huge impact on the formation of the structure of proteins or the maintenance of cellulose and polymer chains.

It seems that intramolecular hydrogen bonds are described somewhat less frequently. The reason may be the problem of direct detection of intramolecular hydrogen bonds, which in turn entails the need to use some indirect methods, both experimental and theoretical. This generally requires finding a reasonable reference system, which is often not so trivial. Like their intermolecular counterparts, intramolecular hydrogen bonds are also often of great importance. They significantly influence the stability and physico-chemical properties of the conformers in which they occur.

The year 2021 will be known historically as the year of the SARS-CoV-2 coronavirus, which has made scientific research difficult for many. However, despite these difficulties, this Special Issue has collected as many as 12 articles, including 4 reviews. Thus, this has to be seen as a success, all the more so as these articles deal with many issues of intramolecular hydrogen bonding. Moreover, these articles gather both experimental and theoretical results. At the same time, the rich subject matter of the articles of this Special Issue proves that intramolecular hydrogen bonds are, despite over a hundred years of investigations in this field, still an important object of scientific research.

Two of the four review articles are devoted to theoretical methods for estimating the energy of intramolecular interactions, including, of course, hydrogen bonds [1,2]. In the first of them, written by Jabłoński [1], the main goal was to present the theoretical rationale of a given method and to show that often many variants of a given method are possible. This generally leads to a significant range of the estimate values obtained. The limitations of the methods used are also discussed. This review is then brilliantly complemented by Deshmukh and Gadre in their review on the Molecular Tailoring Approach [2]. Importantly, this fragmentation method allows the energy of an individual intramolecular hydrogen bond to be estimated in systems containing many such interactions. The authors present many examples of the use of this method.

Hansen [3] reviews intramolecular NH...X (X = O, S, N) bonds in various systems and describes the usefulness of spectroscopic parameters (mainly based on NMR) in assessing their strength. He points out that the NH chemical shift should be corrected for ring current effects when the substituent at the nitrogen is an aromatic ring. The importance of the two-bond deuterium isotope effects (TBDIE) on ^{13}C chemical shifts is then underlined.

Studies on excited states are usually regarded as more problematic than ground-state studies and are therefore much rarer. For this reason, the review by Jankowska and Sobolewski on the excited-state intramolecular proton transfer (ESIPT) is particularly important [4]. These authors discuss the modern theoretical methods of the ESIPT description, paying attention to their range of applications and limitations.

The remaining articles are good evidence, showing that in order to study the phenomenon of hydrogen bonding, many different research techniques, both experimental and theoretical, are used. For example, Filarowski and collaborators [5] use infrared (IR) and Raman spectroscopy, incoherent inelastic neutron scattering (IINS), X-ray diffraction, nuclear quadrupole resonance spectroscopy (NQR), differential scanning calorimetry (DSC) along with DFT calculations to explain the influence of the O-H...O intramolecular hydrogen bond on the polymorphic states and isomerization of 5-chloro-3-nitro-2-hydroxyacetophenone. In a similar way, Tolstoy, Filarowski and collaborators [6] also investigate the intramolecular O-H...N hydrogen bond in 2-[(E)-(phenylimino)methyl]phenol, which is one of the most photo-thermochromic compounds. Their spectroscopic studies confirm that this bond can be classified as a resonance-assisted hydrogen bond (RAHB).

Alkhimova, Babashkina and Safin [7] investigate solvatochromic and luminescent properties of four N-salicylidene aniline derivatives derived from the ethyl ester of glycine, whose photophysical properties are dictated by the intramolecular proton transfer in the O-H...N bridge. It is shown that the physicochemical properties of these compounds, related to a subtle keto-enamine-enol-imine equilibria, can be tuned by the nature of the solvent used (non-polar vs. polar aprotic vs. polar protic).

Many methods (including Car-Parrinello Molecular Dynamics (CPMD) for the gas and crystalline phases) are used by Jezierska, Panek and collaborators in their theoretical studies of 2,3-dimethylnaphthazarin and 2,3-dimethoxy-6-methylnaphthazarin [8]. These authors show that the proton transfer phenomenon takes place in both the compounds as well as in both phases. Importantly, the nuclear quantum effects (NQE) are shown to localize the proton closer to the half of the O...O contact. Nevertheless, NQE should have no qualitative impact on the properties of the investigated molecules. The strong mobility of the bridged protons is also confirmed by spectroscopic data.

Alkorta, Elguero and Del Bene [9] investigate intramolecular O-H...O hydrogen bond in 1-oxo-3-hydroxy-2-propene (i.e., 3-hydroxy-2-propenal or simply the enol form of malondialdehyde) and the change of its characteristics during interaction of this molecule with Lewis acids LiH, LiF, BeH_2, and BeF_2. They found that the binding of these acids to the -C=O group is more preferred than to -OH and that $^{2h}J(O-O)$, $^{1h}J(O-H)$, and $^{1h}J(H-O)$ spin–spin coupling constants exhibit a second-order dependence on the O...O, O-H, and H...O distances, respectively.

Lamsabhi, Mó, and Yáñez [10] utilize the high-level ab initio G4 theory to study the O-H...N intramolecular hydrogen bond in a series of the most stable conformers of $HOCHX(CH_2)_nCH_2NH_2$ and $HOCH_2(CH_2)_nCHXNH_2$ ($n = 0$–5) where X is H, F, Cl, or Br substituted in position α with respect to either -OH or -NH_2. The strongest hydrogen bond occurs when $n = 2$ as shown by shortest H...N distance, isodesmic reaction-based largest interaction energy, largest red-shift of ν_{OH}, and NBO, QTAIM, and NCI theoretical methods. In the group of substituents X, Br gives the greatest influence on OH...N, but interestingly, it is the opposite depending on whether this substituent is in position α with respect to -OH or with respect to -NH_2. This article [10] also investigates the effect of interaction with the BeF_2 molecule.

Intramolecular O-H...O hydrogen bond in malonaldehyde is also theoretically investigated by Pendás and collaborators [11]. Additionally, the influence of eight substituents (both electron-withdrawing and electron-donating) at each of the three skeletal carbon atoms is investigated, and then the OH...O energy is determined using the proprietary IQA method and compared with their equivalents obtained using the OCM and EM methods (see also [1]). While in general the O-H...O bond can either be weakened or strengthened depending on the substituent and the site of substitution, the substitution next to -OH always significantly strengthens this bond (see also [10]). It turns out that for the tested RAHB systems, IQA energies correlate well with EM energies, while there is no such correlation with OCM.

Noticeably, using Local Mode Analysis (and QTAIM and NCI), Altun, Bleda and Trindle [12] order the various intramolecular hydrogen bonds present in tautomers and isomers of 3-hydroxy-2-butenamide according to their strength as follows: the strongest O-H...O=C > N-H...O=C > O-H...N, intermediate N-H...O=C ≥ N-H...O ≈ C-H...O=C, the weakest C-H...N > C-H...O.

Funding: This research received no external funding.

Acknowledgments: I would like to thank all the authors for their valuable contributions to the Special Issue "Intramolecular Hydrogen Bonding 2021", all the reviewers for their responsible effort in evaluating the submitted manuscripts, and the editorial staff (especially Lucy Chai) of Molecules for their kind and professional assistance.

Conflicts of Interest: The author declares no conflict of interest.

References

1. Jabłoński, M. A Critical Overview of Current Theoretical Methods of Estimating the Energy of Intramolecular Interactions. *Molecules* **2020**, *25*, 5512. [CrossRef] [PubMed]
2. Deshmukh, M.M.; Gadre, S.R. Molecular Tailoring Approach for the Estimation of Intramolecular Hydrogen Bond Energy. *Molecules* **2021**, *26*, 2928. [CrossRef] [PubMed]
3. Hansen, P.E. A Spectroscopic Overview of Intramolecular Hydrogen Bonds of NH...O, S, N Type. *Molecules* **2021**, *26*, 2409. [CrossRef] [PubMed]
4. Jankowska, J.; Sobolewski, A.L. Modern Theoretical Approaches to Modeling the Excited-State Intramolecular Proton Transfer: An Overview. *Molecules* **2021**, *26*, 5140. [CrossRef] [PubMed]
5. Hetmańczyk, Ł.; Szklarz, P.; Kwocz, A.; Wierzejewska, M.; Pagacz-Kostrzewa, M.; Melnikov, M.Y.; Tolstoy, P.M.; Filarowski, A. Polymorphism and Conformational Equilibrium of Nitro-Acetophenone in Solid State and under Matrix Conditions. *Molecules* **2021**, *26*, 3109. [CrossRef] [PubMed]
6. Hetmańczyk, Ł.; Goremychkin, E.A.; Waliszewski, J.; Vener, M.V.; Lipkowski, P.; Tolstoy, P.M.; Filarowski, A. Spectroscopic Identification of Hydrogen Bond Vibrations and Quasi-Isostructural Polymorphism in N-Salicylideneaniline. *Molecules* **2021**, *26*, 5043. [CrossRef] [PubMed]
7. Alkhimova, L.E.; Babashkina, M.G.; Safin, D.A. A Family of Ethyl *N*-Salicylideneglycinate Dyes Stabilized by Intramolecular Hydrogen Bonding: Photophysical Properties and Computational Study. *Molecules* **2021**, *26*, 3112. [CrossRef] [PubMed]
8. Kułacz, K.; Pocheć, M.; Jezierska, A.; Panek, J.J. Naphthazarin Derivatives in the Light of Intra- and Intermolecular Forces. *Molecules* **2021**, *26*, 5642. [CrossRef] [PubMed]
9. Alkorta, I.; Elguero, J.; Del Bene, J.E. Perturbing the O–H···O Hydrogen Bond in 1-oxo-3-hydroxy-2-propene. *Molecules* **2021**, *26*, 3086. [CrossRef] [PubMed]
10. Lamsabhi, A.M.; Mó, O.; Yáñez, M. Perturbating Intramolecular Hydrogen Bonds through Substituent Effects or Non-Covalent Interactions. *Molecules* **2021**, *26*, 3556. [CrossRef] [PubMed]
11. Guevara-Vela, J.M.; Gallegos, M.; Valentín-Rodríguez, M.A.; Costales, A.; Rocha-Rinza, T.; Pendás, A.M. On the Relationship between Hydrogen Bond Strength and the Formation Energy in Resonance-Assisted Hydrogen Bonds. *Molecules* **2021**, *26*, 4196. [CrossRef] [PubMed]
12. Altun, Z.; Bleda, E.A.; Trindle, C. Focal Point Evaluation of Energies for Tautomers and Isomers for 3-hydroxy-2-butenamide: Evaluation of Competing Internal Hydrogen Bonds of Types -OH...O=, -OH... N, -NH...O=, and CH...X (X=O and N). *Molecules* **2021**, *26*, 2623. [CrossRef] [PubMed]

 *molecules*

Review

A Critical Overview of Current Theoretical Methods of Estimating the Energy of Intramolecular Interactions

Mirosław Jabłoński [†]

Faculty of Chemistry, Nicolaus Copernicus University in Toruń, 87-100 Toruń, Poland; teojab@chem.umk.pl;
Tel.: +48-056-611-4695
† Current address: 7-Gagarina St., 87-100 Toruń, Poland.

Academic Editor: Maxim L. Kuznetsov
Received: 4 November 2020; Accepted: 23 November 2020; Published: 25 November 2020

Abstract: This article is probably the first such comprehensive review of theoretical methods for estimating the energy of intramolecular hydrogen bonds or other interactions that are frequently the subject of scientific research. Rather than on a plethora of numerical data, the main focus is on discussing the theoretical rationale of each method. Additionally, attention is paid to the fact that it is very often possible to use several variants of a particular method. Both of the methods themselves and their variants often give wide ranges of the obtained estimates. Attention is drawn to the fact that the applicability of a particular method may be significantly limited by various factors that disturb the reliability of the estimation, such as considerable structural changes or new important interactions in the reference system.

Keywords: intramolecular interaction; interaction energy; hydrogen bond

---

1. Introduction

Undoubtedly, intermolecular hydrogen bonds [1–16] occupy the main place among various intermolecular interactions. This is largely due to their intermediate strength, between weaker van der Waals interactions [7,11] and much stronger chemical bonds [1,17]. It is this intermediate strength of intermolecular hydrogen bonds that allows for them to act as a glue that binds various molecules into dimers or larger molecular aggregates. On the other hand, their relative weakness allows for the full dynamics of the bonding motif; the hydrogen bond can be broken relatively easily and a new one can be formed in its place.

It is already visible at this point that the knowledge of the strength of intermolecular hydrogen bonding is a very important element in the full description of the characteristics of this bond. Such knowledge would allow, for example, to classify them according to the strength found and study the impact of various internal and external factors on it. Because of the fact that the total energy of a molecule is a fundamental quantity available to quantum mechanics [18], the appropriate balance of total energies can be successfully used to write a strict definition of the energy of interaction (i.e., the interaction energy) between A and B systems in AB dimer:

$$E_{\text{int}} = E(\text{AB}) - [E(\text{A}) + E(\text{B})] \tag{1}$$

Therefore, it is clear that the reference system for the bound AB dimer is that of the isolated monomers A and B. Another thing that I will leave behind is that these monomers may have their own, i.e., fully optimized geometries or geometries taken from the dimer. Anyway, the energy that is described by Equation (1) is strictly defined. Not quite rightly, E_{int} obtained by Equation (1) is commonly taken as the interaction energy associated with the closest contact between A and B, e.g., an intermolecular hydrogen bond. Therefore, this equation has also become the main source of information regarding the strength of intermolecular hydrogen bonds in the so-called supermolecular method.

It is quite natural that one would like to have such an important quantity also in the case of intramolecular interactions, including intramolecular hydrogen bonds. However, there is a fundamental problem here. Namely, unlike in its intermolecular counterpart, breaking the intramolecular interaction is impossible without disturbing the structure of the molecule. Because of this fact, not only it is impossible to find a strict definition of the intramolecular interaction energy, but what is more, this energy is not even strictly definable (see, however, the further discussion on the QTAIM-based methods).

Nevertheless, one can try to introduce a method that results in a number that is treated (in this method) as the energy of a given intramolecular interaction. It is obvious that, in the general case, the energies obtained will differ (perhaps even significantly) among the adopted methods. For this reason, an important aspect of the proposed method is the evaluation of the reliability of the energy obtained. It would rather be a worthless result to obtain for intramolecular hydrogen bond of e.g., the OH$\cdots$O type an energy of the order of, say -50 kcal/mol, if the intermolecular equivalent in the case of a similar configuration of O and H atoms gives energy from about -4 to about -8 kcal/mol. One of the possible ways of assessing the reliability of the obtained energy value is thus comparing it to the appropriate intermolecular interaction, in which not only the type of X and Y atoms (from the XH$\cdots$Y contact), but also their spatial configuration (e.g., the key distance H$\cdots$Y) is largely preserved. Another sensible possibility is to check the fulfillment of various correlations between the found energy values and other parameters describing the strength of the H$\cdots$Y bond. One should also compare the obtained estimates for structurally closely related molecules.

This article reviews the current theoretical methods introducing the concept of the XH$\cdots$Y intramolecular hydrogen bond interaction energy (or more generally the intramolecular X$\cdots$Y interaction) and allowing for the computational generation of these energies. The main emphasis will be on the problems associated with these methods, which may naturally lead to their different, more or less reliable, variants. On the other hand, due to the multitude of numerical data concerning the interaction energy values that were determined with these methods, this issue will necessarily be of minor importance. Rather, I will limit myself to a few examples that illustrate how a given method works.

2. Theoretical Methods of Estimating the Energy of Intramolecular Interactions

2.1. Conformational Methods

As noted in the Introduction, it is impossible (see, however, the further discussion on the QTAIM-based methods) to precisely define the energy of the intramolecular hydrogen bond XH$\cdots$Y (or more generally of the intramolecular interaction X$\cdots$Y), because it is impossible to create a reference system in which there would be no such interaction, but in which the configuration of all atoms would be preserved. In such a reference system, the interaction of interest would be simply "switched off". Crucially, this approach is eqivalent to the following partition of the total energy of the system (the so-called closed or chelate form) containing the interaction of interest (e.g., a hydrogen bond)

$$E(\text{closed}) = E^f(\text{closed}) + E_{\text{HB}} \tag{2}$$

in which $E(\text{closed})$, $E^f(\text{closed})$, and E_{HB} correspond successively to the total energy of the closed form, the total energy of a fictitious closed form with the interaction switched off, and the hydrogen bond interaction energy. Of course, such an exclusion is impossible, but, nevertheless, one may be tempted to find another system being very similar to the fictitious closed one. Due to the fact that total energy depends on the number and type of particles making up a given molecule, the phrase "another but very similar system" should be understood as a different conformer of the closed form of a molecule. This leads to so-called conformational methods, i.e., methods which use total energies of at least two conformers of a molecule having the intramolecular interaction.

2.1.1. The Open-Closed Method (OCM)

The simplest and the most frequently used method of estimating the energy of intramolecular interactions, including intramolecular hydrogen bonds, is the so-called open-closed method (OCM) [5,19]. Apart from the molecule that contains a given interaction (i.e., the closed or chelate form), OCM requires the use of one more reference form (the so-called open), in which this interaction is absent [20–45]. It is then assumed that

$$E^f(\text{closed}) \approx E(\text{open}) \tag{3}$$

which means that the total energy of another conformer, i.e., the open form, can be used instead of the impossible to obtain total energy of the fictitious closed system. Substituting expression (3) to (2) leads to a simple expression for the intramolecular hydrogen bond energy in OCM:

$$E_{\text{HB}}^{\text{OCM}} = E(\text{closed}) - E(\text{open}) < 0 \tag{4}$$

It should be emphasized that this article adopts the convention according to which a negative value of the obtained interaction energy means local stabilization that results from H$\cdots$Y, while on the contrary, a positive value means local destabilization. Thus, of course, as being stabilizing interactions, hydrogen bonds should be characterized by negative values.

Equation (3) requires that the open form does not differ much from the (fictitious) closed form. Therefore, the open form is most often obtained by rotating the donor or acceptor group by 180°, as shown in Figure 1.

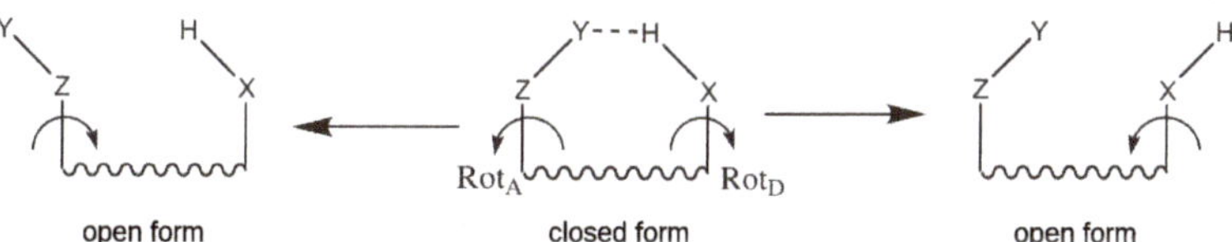

Figure 1. Scheme showing two open forms obtained by rotation of either the hydrogen-acceptor (lhs) or the hydrogen-donor (rhs) group.

It is understood that, in general, these reference open forms give different values of $E_{\text{HB}}^{\text{OCM}}$ [33]. In principle, one can also try to use a different open form. However, I will come back to this issue further. Although the expression (3) suggests that the open form should be fully optimized, i.e., it should correspond to a local minimum on the potential energy hypesurface, another possibility is to use an open form having the geometry (more precisely, geometrical parameters) of the closed form [5,19,33,38,39,42,44,45]

$$E^f(\text{closed}) \approx E^{\text{closed}}(\text{open}) \tag{5}$$

Of course, this leads to a different energy value

$$E_{\text{HB}}^{\text{OCM}} = E(\text{closed}) - E^{\text{closed}}(\text{open}) \tag{6}$$

since $E^{\text{closed}}(\text{open}) \neq E(\text{open})$. In fact, Schuster advocated this option, suggesting that the reference open form should have "the least changes in molecular geometry besides a cleavage of the H-bond" and proclaiming that it "need not be a local minimum of the energy surface" [5]. Moreover, in his opinion, the full optimization of the open form geometry is even inadvisable, because this approach mixes the energy of isomerization (resulting from the change of the conformer) into the determined energy value [5]. In fact, both of these approaches introduce different definitions of the intramolecular interaction energy (cf. Equations (4) and (6)). This situation is somewhat similar to the one that occurs when determining the interaction energy from Equation (1). Namely, the use of the monomers A and B with their geometries taken from the AB dimer defines the interaction energy, while their full optimization leads to the binding energy. The latter quantity also takes into account the correction for geometry change that takes place during the transition from the isolated form to the bound form in the dimer. Because of the fact that, in OCM, the fictitious closed form is replaced by the open form obtained by some conformational change, Schuster stressed that any splitting of the energetical difference between both forms is artificial [5]. However, it seems that this opinion may be slightly weakened by some corrective approaches [39,44]. It is valuable to present both variants of the partition of the total energy of the closed form in one scheme, as shown in Figure 2, where more concise notations are used for the respective energies.

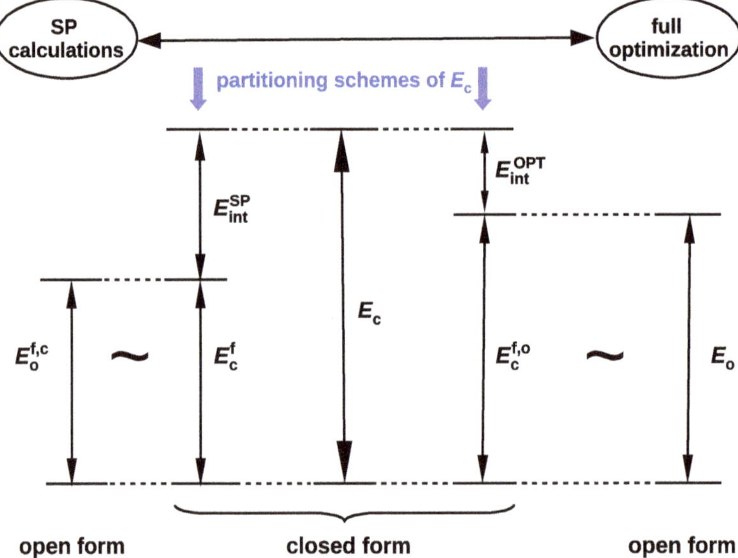

Figure 2. Scheme showing two variants of the closed form total energy partition (E_c) to the interaction energy (E_{int}) and the total energy of the fictitious closed form (E_c^f) obtained after 'excluding' this interaction.

It is worth noting that $|E_o| > |E_o^{f,c}|$, which, in principle, should lead to the relationship $|E_{int}^{OPT}| < |E_{int}^{SP}|$. It seems that at present the variant based on full geometry optimization of the open form (leading to E_o and then E_{int}^{OPT}) is much more popular [37] than the variant based on single point calculations (leading to $E_o^{f,c}$ and then E_{int}^{SP}). In this variant, the isomerization energy mentioned by Schuster [5] is 'absorbed' into the interaction energy. In other words, this variant assumes that the changes in geometrical parameters that take place during the open form → closed form transition are related to the continuous process of creating the interaction (e.g., an intramolecular hydrogen bond) in the closed form [42,45].

Although OCM seems to be the most frequently [20–45] used theoretical method of estimating the energy of an intramolecular interaction, it is not free from further problems. The rotation of the proton-donor or the proton-acceptor group quite often leads to a new, significant interaction (either repulsive or attractive) in an open form [24,27–29,31,33,39–46]. Unfortunately, this possibility is quite often ignored. Moreover, sometimes, one or even both of the open forms cannot be used due to symmetry of these groups. Some simpe examples representing both cases are shown in Figure 3. Of course, similar examples can be easily invented endlessly.

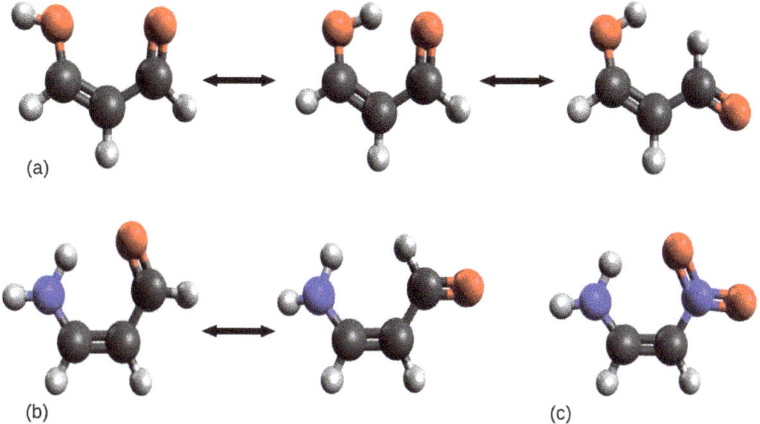

Figure 3. Examples of problematic cases in the open-closed method: (**a**) malondialdehyde, (**b**) 3-aminoacrolein, (**c**) 1-amino-2-nitroethylene.

In the case of (a) relating to the intramolecular hydrogen bond O-H$\cdots$O in malondialdehyde, the rotation of the proton-donor group -OH leads to a new rather significant interaction O$\cdots$O, while the rotation of the proton-acceptor group -CHO leads to also rather significant new interaction H$\cdots$H. In the case of (b) (3-aminoacrolein), due to the symmetry of the amino group, its rotation leads to practically the same system, while the rotation of the aldehyde group leads to a new significant H$\cdots$H interaction, similar to the case of (a). The closed form of 1-amino-2-nitroethylene does not have any such simple open forms due to the symmetry of both groups, i.e., -NH$_2$ and -NO$_2$.

Another, but important, question is whether these new interactions can be completely ignored [24,27–29,33,39–46]. For example, in the case of malondialdehyde, geometry optimizations (B3LYP/aug-cc-pVTZ) of the open form shown on the left-hand side of Figure 3 gives 2.89 Å for the O$\cdots$O distance and 2.02 Å for H$\cdots$H in the open form shown in the right-hand side of this figure. In the case of the open form of 3-aminoacrolein, the distance H$\cdots$H is 2.18 Å. Therefore, it would seem that these distances are too large for the interaction energy to be uncertain. However, on the other hand, the comparison of the CCC angle values in both forms (119.6° vs. 126.8° and 125.2° in malondialdehyde and 122.0° vs. 125.1° in 3-aminoacrolein) shows that the closed form $\rightarrow$ open form transition leads to an opening of the molecular skeleton, which may suggest significant repulsive actions of both these interactions. It seems that the O$\cdots$O contact, in particular, cannot be completely ignored here. It is worth mentioning that both forms, i.e., closed and open, may differ in some structural aspects, e.g., the amino group in 3-aminoacrolein (b) is flat in the closed form, whereas slightly pyramidal in the optimized open form. In this case, one would have to decide whether the pyramidalization energy of the amino group should be shelled out or included in the hydrogen bond energy value [47].

In such and similar cases, it may be tempting to find other reasonable open forms, obtained after the rotation of one of the groups around the CC double bond. On the one hand, such new interactions will be avoided, but on the other hand, the configuration of the carbon skeleton of the molecule will be changed. For example, Buemi et al. [33] rebuked the use of the most extended enol and enethiol tautomers of thiomalondialdehyde [48,49] as reference structures [24,50], because, in their opinion, the *trans* configuration of double bonds seems to be too different that the *cis* arrangement in the closed form (Figure 4).

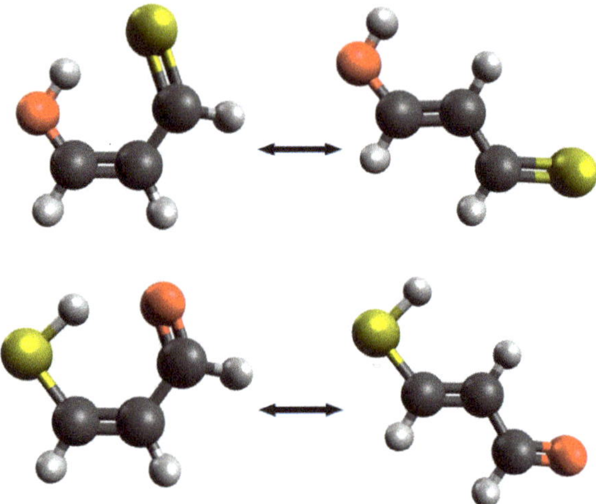

Figure 4. Closed and the most extended enol and enethiol forms of thiomalondialdehyde.

It is also worth adding that the most extended conformers are very often the global minima of a given molecule. On the other hand, open systems with a changed configuration of backbone atoms can be more reasonable in many cases. In fact, the selection of the most reasonable reference system is an individual matter for the closed form of the molecule under consideration. Therefore, this issue should be carefully analyzed before starting the appropriate calculations while using OCM.

The fundamental issue for OCM is that the presence of a new significant interaction in the reference open form leads to either an overestimation or underestimation of the determined value of the interaction energy in the closed form [42,45]. Both of the situations are shown in Figure 5.

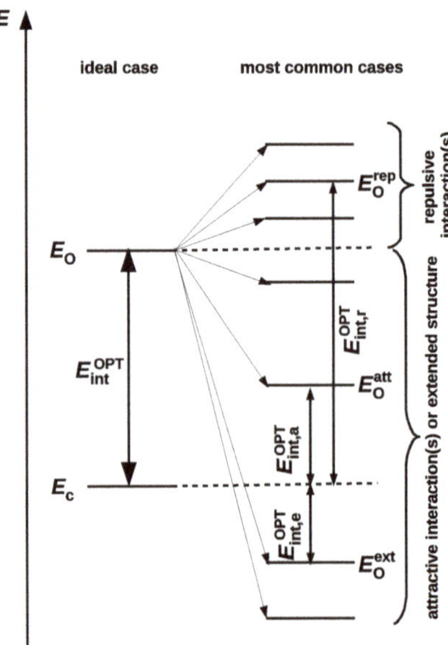

Figure 5. Scheme showing the presence of a new either repulsive or attractive interaction as a cause of either overestimating or underestimating the determined value of the intramolecular interaction energy.

The presence of a new significantly repulsive interaction in the reference open form leads to a less negative total energy of this form (E_o^{rep}), and thus to an overestimation of the determined value of the interaction energy ($E_{int,r}^{OPT} > E_{int}^{OPT}$). Conversely, the presence of a significant attractive (stabilizing) interaction, e.g., a new hydrogen bond, results in underestimating the determined energy value ($E_{int,a}^{OPT} < E_{int}^{OPT}$). Moreover, because the most extended forms are often the most stable (as already mentioned), $E_o^{ext} < E_c$, their frivolous use can underestimate the value of the interaction energy so much that this value can even change the sign ($E_{int,e}^{OPT}$) [42]. As open forms with presence of new locally repulsive interactions $X \cdots Y$ (e.g., $O \cdots O$, $O \cdots S$, $S \cdots S$, etc.) and, in particular, $H \cdots H$ are often treated favorably, the resulting energies may often be overestimated. This, in turn, may lead to overinterpretations of the considerable strength of some intramolecular hydrogen bonds [39].

Given the fact that the full geometry optimization of the open form can lead to a new significant interaction (repulsive or attractive) or to a significant change in structure as compared to the closed form, a solution may be to perform a partial (i.e., constrained) geometry optimization [42]. In many cases, it is enough to 'freeze' one or two dihedral angles that define the spatial orientation of the proton-donor or proton-acceptor group, the optimization of which would lead to the previously mentioned undesirable effects. However, sometimes, it is also necessary to freeze other geometric parameters [42]. The approach that is based on partial geometry optimization of the open form is, in fact, another variant of OCM, leading to the interaction energy value between these described by Formulas (4) and (6).

This variant was first proposed [42] to estimate the energy of Si-H$\cdots$Al intramolecular charge-inverted hydrogen bonds [51,52] in ten model systems. The energy values of Si-H$\cdots$Al in these systems were determined while using seven variants of OCM. In addition to either the full optimization (OPT) or complete freeze (SP) of the open form geometry, five variants of the constrained optimizations of the open form geometry were also used: (P1) only bonds optimized, (P2) only bonds and plane angles optimized, (P3) all geometric parameters optimized but dihedral angles governing the positions of the Si atom and the -AlH$_2$ group in relation to the carbon skeleton of the reference form, (P4) all geometric parameters optimized but dihedral angles governing the positions of the Si atom and both hydrogen atoms from the -AlH$_2$ group, and (P5) all geometric parameters optimized, but dihedral angles governing the positions of both hydrogen atoms from -AlH$_2$. Of course, the values of the non-optimized geometric parameters in the variants SP and P1–P5 were taken from the closed form. Therefore, it can be seen that the P1–P2 variants in a controlled manner increase the number of optimized parameters (degrees of freedom), which increases the flexibility of the approach. Because the obtained results [42] very well reflect the mentioned problems related to the use of OCM, these results are shown for three molecules (Figure 6) in Table 1.

Table 1. Determined (B3LYP/aug-cc-pVTZ) energy values (in kcal/mol) of Si-H$\cdots$Al interactions in 1_c, 2_c and 3_c (see Figure 6).

System	Rotated Group	SP	P1	P2	P3	P4	P5	OPT
1	$-SiH_2$	−5.31	−4.88	−4.41	−3.36	−1.50	−0.99	−1.08
	$-AlH_2$	−7.01	−6.61	−5.79	−4.41	−3.24	−1.75	
2	$-SiH_2$	−5.23	−4.75	−3.93	−2.46	−2.42	−2.42	$p_\pi \to Al$
	$-AlH_2$	−8.09	−7.37	−6.27	−4.85	−4.73	−4.72	
3	$-SiH_2$	−6.04	−5.03	−3.47	−1.41	−1.40	−1.40	−4.97/−0.75
	$-AlH_2$	−10.61	−10.05	−8.44	−6.23	−6.19	−6.19	

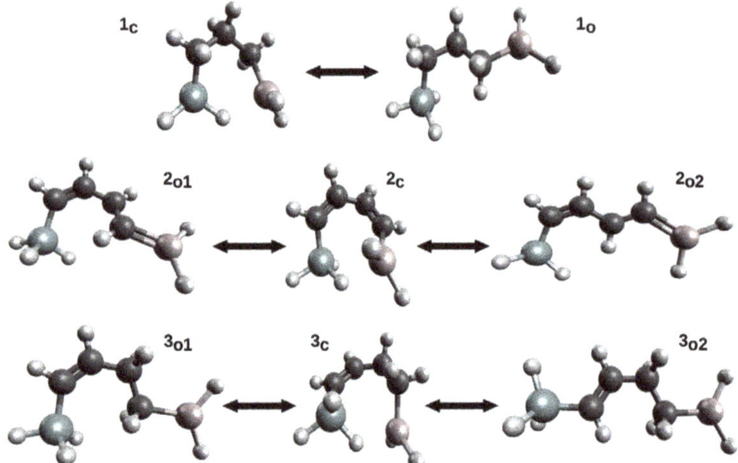

Figure 6. Closed and some open forms of (**1**) H_3Si-CH_2-CH_2-CH_2-AlH_2, (**2**) H_3Si-CH=CH-CH=CH-AlH_2 and (**3**) H_3Si-CH=CH-CH_2-CH_2-AlH_2.

First of all, it can be seen that the determined values of the interaction energy vary widely, depending on the variant of the open-closed method used in the calculations. In the case of molecule **1**, it is from -7 to -1 kcal/mol and, in the case of **3**, from -10.6 to about -0.8 kcal/mol. The values decrease (i.e., become less negative) with an increased degree of flexibility regarding the geometric parameters optimized in a given variant. It can be seen that especially even a partial optimization of dihedral angles has a large influence on the determined interaction energy values. Moreover, the rotation of the $-SiH_3$ group in general gives significantly different values from that when the $-AlH_2$ group is rotated. This is especially visible for the least flexible variant SP, while on going from P1 to P5 these differences become smaller and smaller. It is instructive to analyze the results from the last column of Table 1, i.e., regarding the variant with full geometry optimization of the proposed open form. While in case of **1** one reasonable value was found (-1.08 kcal/mol), in the case of **6** two significantly different values were obtained (-4.97 and -0.75 kcal/mol). The latter results from the fact that two open reference forms (see **3$_{o1}$** and **3$_{o2}$** in Figure 6) with quite different characteristics were obtained. Despite the fact that both forms have identical carbon frame configuration (*cis*), the **3$_{o2}$** form has two new $H^{\delta+} \cdots H^{\delta+}$ interactions. On the other hand, the **3$_{o1}$** form has two pairs of probably less important $H^{\delta-} \cdots H^{\delta+}$ interactions. Case **2**, on the other hand, is an important example illustratating the significant impact of the presence of a completely new type of interaction in an open form on the quality of the estimation of the interaction energy is a closed form. Namely, in both open forms (**2$_{o1}$** and **2$_{o2}$**), the $-AlH_2$ group (Al has an empty p orbital) takes a coplanar position to the CH=CH fragment with a formal C=C double bond. This arrangement allows for the $p_\pi \rightarrow$Al coupling (highlighted in Figure 6 by drawing a C=Al double bond), which significantly lowers total energies of these forms. Consequently, the estimates of the interaction energy of Si-H$\cdots$Al in **2$_c$** are highly unreliable.

The variant of OCM with partial geometry optimization of the open form was then used [45] to estimate energies of Si-H$\cdots$B contacts in some 1-silacyclopent-2-enes and 1- silacyclohex-2-enes and helped to successfully support the earlier Wrackmeyer's suggestion based on NMR spectroscopic data [53] that this contact is considerably stronger in the latter system than in the former one. Additionally, the energies of Ge-H$\cdots$Al and Ge-H$\cdots$H-N interactions in some alkenylhydrogermanes were estimated [46] in a similar way (see Figure 7).

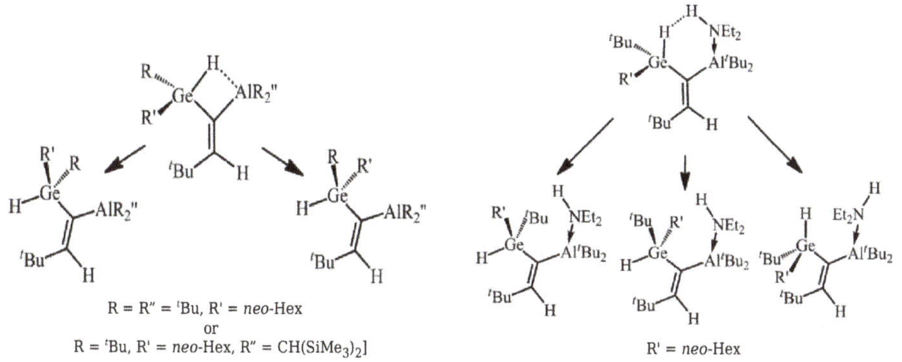

Figure 7. Closed and open forms investigated in ref. [46].

The variant of OCM with partial geometry optimization of the open form should rather be treated as a certain, but probably not the only, possible solution when the full geometry optimization of this form gives (for the reasons discussed earlier) highly unreliable estimates of the interaction energy [42,45].

The results that are presented here are enough to show that OCM in which only one reference system is utilized must be used with great caution so as not to write with reserve. It should be so especially when its—nevertheless the most popular—variant with the full geometry optimization of the reference open form is used. Not as rare as it may seem at first, the occurrence of new interactions (whatever attractive or repulsive) or significant structural changes (e.g., changing the skeleton of a molecule) can lead to highly unreliable estimates of the energy value of the intramolecular interaction of interest in a closed form. Indeed, Rozas et al. [32] went so far as to say that the energy value obtained from OCM should scarcely be taken as the value of the energy of the interaction in a closed form. Simply, it should rather be treated as the energetical difference between the respective forms of a molecule. On the other hand, this criticism seems a bit exaggerated. If the open form is very similar to the closed form both in terms of structure and the interactions occurring in these forms, then it seems that OCM is a worthy method of choice. The substantial similarity that is referred to herein can be provided by the presence of some rigid part of the molecule to which both the donor and acceptor groups are attached. This is the case, for example, with a benzene ring, leading to the variant of OCM, described as the *ortho-para* method [38,54,55], which is described in more detail in the next subsection.

2.1.2. Ortho-Para Method (opM)

The *ortho–para* method (opM) was most likely used for the first time by Estácio et al. [38] for estimating the energy of intramolecular hydrogen bonds in four 1,2-disubstituted benzene derivatives: 1,2-dihydroxybenzene (catechol), 1,2-benzenedithiol, benzene-1,2-diamine, and 2-methoxyphenol (guaiacol). To describe opM, it is enough to refer to the O-H$\cdots$O hydrogen bond in 1,2-dihydroxybenzene, i.e., catechol (Figure 8).

The use of the open form that was simply obtained by rotating the hydroxyl group around the C-O bond resulted in hydrogen bond energy estimates of −3.7 or −4.0 kcal/mol at the MPW1PW91/aug-cc-pVDZ and CBS-QMPW1 levels of theory, respectively. These values were considered to be unreliable and significantly overestimated as a result of the presence of new repulsive interactions between oxygen atoms as well as the O-H dipole–dipole interactions [38]. As a consequence, it was concluded that the energetic difference between the open and closed forms cannot be regarded as the energy of the O-H$\cdots$O hydrogen bond in the latter form. However, in this and similar cases, the *para* form is a very reliable reference form. The comparison of total energy of this form with the total energy of the closed form of the *ortho* configuration gives opM, which can be seen as a variant of OCM. Based on this approach, the respective hydrogen bond energies were -2.1 and -2.3 kcal/mol [38].

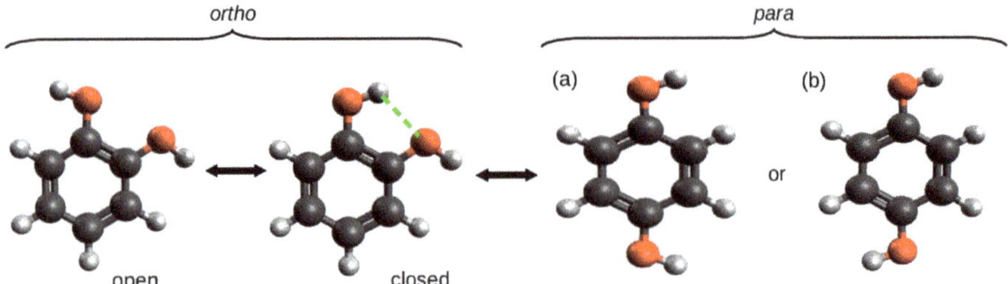

Figure 8. Various forms of catechol (the subfigures (**a**) and (**b**) represent different forms of *para*-catechol).

It is worth emphasizing here that the high reliability of the estimate obtained by means of opM results from the high stiffness of the main part of the molecule, i.e., the benzene ring and, hence, the significant transferability of the related geometric parameter values. In other words, the stiffness of the molecular framework and its high preservation when going to the *para*-substituted reference system allowed for avoiding the typical problems that are faced by the standard version of OCM which were mentioned earlier. On the other hand, it should be noted that this method assumes that the substituent electronic effects in the *ortho* and *para* forms are similar. However, this is in line with the general knowledge on substituent effects [56–59]. Nevertheless, another question, which is completely not addressed by Estácio et al., is which form of the *para* conformer (see (a) and (b) in Figure 8) to use. While this rather purely theoretical issue seems to be insignificant for catechol due to the negligible difference in total energies between the two forms (e.g., 0.1 kcal/mol at the B3LYP/aug-cc-pVTZ level of theory), the difference may become slightly larger for other substituents or molecular frameworks.

It should be mentioned that Estácio et al. described the O-H$\cdots$O hydrogen bond in the closed form of catechol by means of a simple model that is based on the description of interacting dipoles of the O-H bonds. This model resulted in the following formula

$$E_{\mathrm{HB}} = -[E_{\mathrm{LJ}}^{\mathrm{O}\cdots\mathrm{O}} + E_{\mathrm{dd}}] \tag{7}$$

where $E_{\mathrm{LJ}}^{\mathrm{O}\cdots\mathrm{O}}$ is the Lennard–Jones interaction energy for the relevant pair of oxygen atoms and E_{dd} is the dipole-dipole interaction energy for the closed form [38]. The energy value that was determined using this formula was -2.0 kcal/mol (MPW1PW91/aug-cc-pVDZ) and it was very closed to the one determined while using opM (-2.1 kcal/mol). Estácio et al. considered this result to be significant, because it shows that opM correctly describes both interactions, i.e., the O$\cdots$O repulsion and the interaction between the dipoles of both O-H bonds in the closed form of catechol.

2.1.3. Related Rotamers Method (RRM)

As we have seen, the choice of a reasonable open form in OCM is often problematic and even sometimes impossible. This is due to the requirement that this form should be as close to the closed form as possible. This means that the conformer change should not lead to significant changes in the values of geometric parameters. In order to overcome any inaccuracies, another approach is to use more than just two conformers of a given molecule [47,60–63]. This idea will be shown on the example of 3-aminopropenal (3-aminoacrolein), which has four conformers. The N-H$\cdots$O hydrogen bond energy in the ZZ conformer of 3-aminoacrolein was quite often estimated [47,61,63–65], but the methods used did not take into account changes in the values of geometric parameters when switching from the bound system (ZZ-3-aminoacrolein) to reference forms (in particular, to ZE-3-aminoacrolein) [64,65]. The specific system of conjugated double bonds and, hence, the presence of four conformers (see Figure 9), allowed for proposing a method that was derived from the analysis of the mutual energy relations between the four conformers of 3-aminoacrolein (Figure 9) [47].

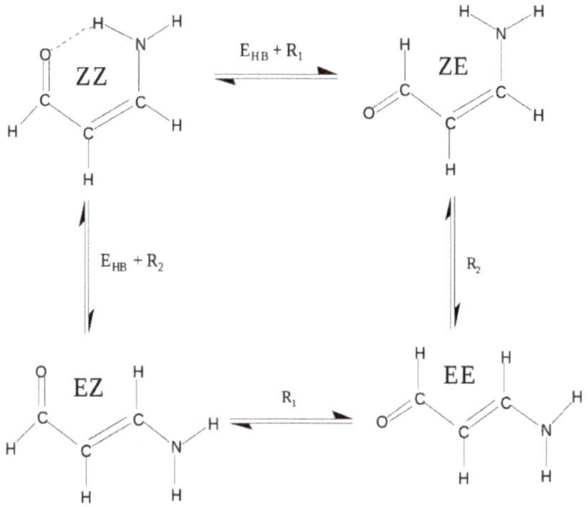

Figure 9. Four conformers of 3-aminoacrolein and energetic relationships between them.

This method takes use of approximations

$$E_{HB} + R_1 = E^{ZZ} - E^{ZE}, \quad R_1 \approx E^{EZ} - E^{EE} \tag{8}$$

$$E_{HB} + R_2 = E^{ZZ} - E^{EZ}, \quad R_2 \approx E^{ZE} - E^{EE} \tag{9}$$

that lead to the following formula for the hydrogen bond energy in the ZZ form of 3-aminoacrolein

$$E_{HB}^{RRM} = \left(E^{ZZ} - E^{ZE}\right) + \left(E^{EE} - E^{EZ}\right) \tag{10}$$

Calculations that are based on MP2(Full)/6-31G** and MP2(FC)/6-311+G** level of theory gave values of −8.2 and −7.5 kcal/mol, respectively [47]. Later, B3LYP/6-311++G** (however, most likely Nowroozi et al. [61] used a smaller 6-31G** basis set, as evidenced by the number of 100 basis functions mentioned by them and the obtained value of −8.4 kcal/mol, which is close enough to the value of −8.2 kcal/mol obtained [47] at the MP2(Full)/6-31G** level of theory) computation by Nowroozi et al. [61] gave value of −8.4 kcal/mol.

The term in the first bracket of Equation (10) is equivalent to the energy that is obtained from the most commonly used variant of OCM in which the open reference form is obtained by the rotation of the proton-acceptor group. Hence, the relationship between RRM and OCM can be expressed by the following relationship between the total energies of the conformers EE and EZ [44]:

$$E_{HB}^{RRM} - E_{HB}^{OCM} = E^{EE} - E^{EZ} \tag{11}$$

Because, in most cases, the extended EE conformer is more stable than the EZ conformer, the difference defined by the above equation is negative. For this reason, as compared to OCM, RRM gives greater stabilizations of interactions. It may even happen that interactions that are weakly destabilizing based on OCM are weakly stabilizing if RRM is considered instead, and this result is only due to "different zeros" in both of these methods [44].

It should be mentioned that RRM [47,60] has been readily adopted by Nowroozi et al. [61–63], who called it the Related Rotamers Method (RRM) and in this review it functions under that name. However, even a little earlier, practically the same method was used by Lipkowski et al. [60] to estimate the energy of O-H$\cdots$N intramolecular hydrogen bonds in some chloro-derivatives of 2-(N-dimethylaminomethyl)-phenols, but they used the term "thermodynamic cycle". Therefore, as one can see, not only are different methods used, but even the same method can function under different names [47,60–63].

2.1.4. Geometry-Corrected Method (GCM)

All of the methods for estimating the energy of intramolecular interactions (e.g., hydrogen bonds) discussed so far do not take into account changes in the values of geometric parameters upon considering an open reference form of a molecule. However, the presence of a conjugated system of double bonds, which is characteristic for 3-aminoacrolein and, thus, the existence of its four conformers (Figure 9) allowed for proposing a method to estimate the energy of the N-H$\cdots$O intramolecular hydrogen bond in the ZZ conformer with simultaneous partial consideration of geometric factors [39]. This method initially functioned under the name "Scheme A" [39–41,43], but later its meaningless name was changed to the Geometry-Corrected Method (GCM) [44].

Very helpful in understanding the idea of GCM and how to derive it are the diagrams presented in Figure 10, which show the energy relationships between the respective forms of 3-aminoacrolein.

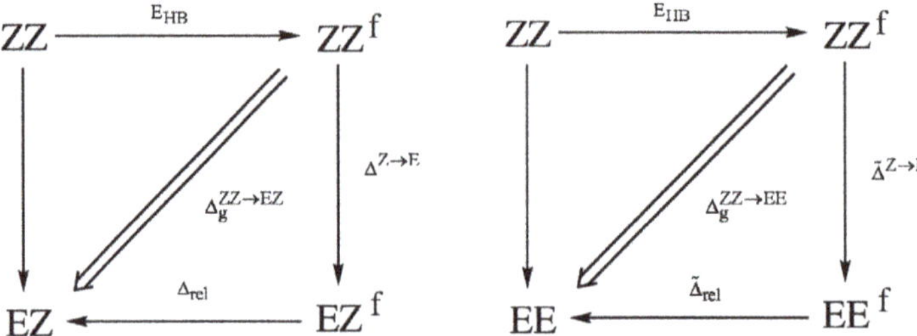

Figure 10. Energy dependencies between the respective forms of 3-aminoacrolein used in deriving the formula for the energy of the intramolecular N-H$\cdots$O hydrogen bond in the ZZ form, according to Geometry-Corrected Method (GCM).

As in Equation (2), in the first step, it is assumed that the hydrogen bond in the ZZ form of 3-aminoacrolein can be simply 'turned off' without any changes in the electron density distribution of the system, therefore also without any changes in the geometrical parameters of this form. By introducing an approximation of the energy additivity, we obtain:

$$E_{HB} = E^{ZZ} - E^{ZZ,f} < 0 \tag{12}$$

where $E^{ZZ,f}$ is simply the total energy of the fictitious form of ZZ with the hydrogen bond just 'turned off'. The rotation of the aldehyde group around the C=C double bond, i.e., the transition ZZ$\to$EZ leads not only to breaking the hydrogen bond, but also to some changes in the geometrical parameters. If the energy associated with these changes in geometric parameters is $\Delta_g^{ZZ\to EZ}$, then

$$E^{EZ} \approx E^{ZZ} + E_{HB} + \Delta_g^{ZZ\to EZ} = E^{ZZ,f} + \Delta_g^{ZZ\to EZ} \tag{13}$$

and quite similarly for the ZZ$\to$EE transition

$$E^{EE} \approx E^{ZZ} + E_{HB} + \Delta_g^{ZZ\to EE} = E^{ZZ,f} + \Delta_g^{ZZ\to EE} \tag{14}$$

Dividing the sum of Equations (13) and (14) by two, one obtains an expression that can be interpreted as the averaged energy that is related to the configuration change Z→E:

$$\Delta_g^{av} = \frac{1}{2}(\Delta_g^{ZZ\rightarrow EZ} + \Delta_g^{ZZ\rightarrow EE}) = \frac{1}{2}(E^{EZ} + E^{EE}) - E^{ZZ,f} \tag{15}$$

Combining this equation with (12) gives the expression for the hydrogen bond energy in the conformer ZZ

$$E_{HB} = E^{ZZ} - \frac{1}{2}(E^{EZ} + E^{EE}) + \Delta_g^{av} \tag{16}$$

but in which there is (so far) the unknown quantity Δ_g^{av}. In fact, the hydrogen bond energy, E_{HB}, and the averaged contribution to the configuration change Z→E, Δ_g^{av}, are formally non-separable quantities. However, the existence of conformers allowed for determining the unknown contribution Δ_g^{av} from yet another source. Let us introduce the fictitious equivalents of the conformers EZ and EE (EZ^f and EE^f, respectively), having the same values of all (of course, except the dihedral angle(s) changing the conformation) geometrical parameters as the conformer ZZ (see Figure 10). The energy that is associated with the transition ZZ^f →EZ can then be assumed in the form

$$\Delta_g^{ZZ\rightarrow EZ} = \Delta^{Z\rightarrow E} + \Delta_{rel} \tag{17}$$

where $\Delta^{Z\rightarrow E}$ is the energy resulting from the change of the Z→E configuration while maintaining the constant values of all geometrical parameters, while Δ_{rel} is the relaxation energy of the fictitious EZ^f form to its fully relaxed equivalent obtained after the full geometry optimization. The energy associated with the transition ZZ^f →EE can be presented quite similarly

$$\Delta_g^{ZZ\rightarrow EE} = \tilde{\Delta}^{Z\rightarrow E} + \tilde{\Delta}_{rel} \tag{18}$$

Changing the conformation from ZZ^f to either EZ^f or EE^f (i.e., maintaining the same values of bond lengths and angles) should not have a significant influence on the energy change. With the neglect of changing the interactions between 'unbound' atoms, it can therefore be assumed that $\Delta^{Z\rightarrow E} \approx \tilde{\Delta}^{Z\rightarrow E} \approx 0$. This approximation gives, after adding Equations (17) and (18) to each other, another expression for Δ_g^{av}

$$\Delta_g^{av} = \frac{1}{2}(\Delta_g^{ZZ\rightarrow EZ} + \Delta_g^{ZZ\rightarrow EE}) \approx \frac{1}{2}[(E^{EZ,f} - E^{EZ}) + (E^{EE,f} - E^{EE})] \tag{19}$$

which, inserted into Equation (16), gives the final formula for the value of the hydrogen bond energy in ZZ-3-aminoacrolein [39]

$$E_{HB}^{GCM} = E^{ZZ} - \frac{1}{2}(E^{EZ,f} + E^{EE,f}) < 0 \tag{20}$$

Thus, it can be seen that, to determine E_{HB}^{GCM}, only the total energy of the fully optimized ZZ conformer and the total energies of the fictitious EZ and EE conformers with the values of geometric parameters (except for the dihedral angle O=C–C=O) from the ZZ conformer are needed. It is worth repeating at this point that GCM, i.e., formula (20), to some extent takes into account the changes in geometric parameters when moving from the ZZ form to the reference forms.

At this point, it is instructive to compare GCM with the OCM variant, in which the open reference form is the ZE conformer, i.e., $E_{HB}^{OCM} = E^{ZZ} - E^{ZE}$ (cf. with Equation (4)). As already discussed, the assumption of OCM is that the reference open system does not differ significantly from the closed form, whereas, in the case of ZZ-3-aminoacrolein, the rotation of the aldehyde group around the C–C bond leading to the ZE conformer introduces a new rather significant interaction of the H$\cdots$H type (see Figure 9). This interaction is practically not present in the closed form ZZ. Any estimate that refers to the ZE conformer as the reference form should take this change into account. Suppose (similar to the hydrogen bond in the ZZ conformer) that this H$\cdots$H interaction in the fictitious ZE^f form can be 'turned off', which gives ZE$^{f'}$ (note the prime sign in the superscript). The energy associated with the rotation of the aldehyde group ($\Delta_s^{Z\to E}$) at the transition ZZ$^f \to$ ZE$^{f'}$ can be assumed to be negligible due to both the conservation of the same geometric parameters as in the conformer ZZ and also due to the neglect of additional H$\cdots$H repulsion at this stage (additionally, the negligible influence of changes in the interactions of unbound atoms other than H$\cdots$H is also assumed). This repulsion leads to an energy increase of Δ_{rep}^{ZE} and to the form ZE^f, which still maintains the geometry of ZZ. Only full relaxation of the ZE^f geometry leads to the optimized ZE conformer. The energy that is associated with this relaxation has been designated as $\tilde{\tilde{\Delta}}_{rel}$ (see Figure 11).

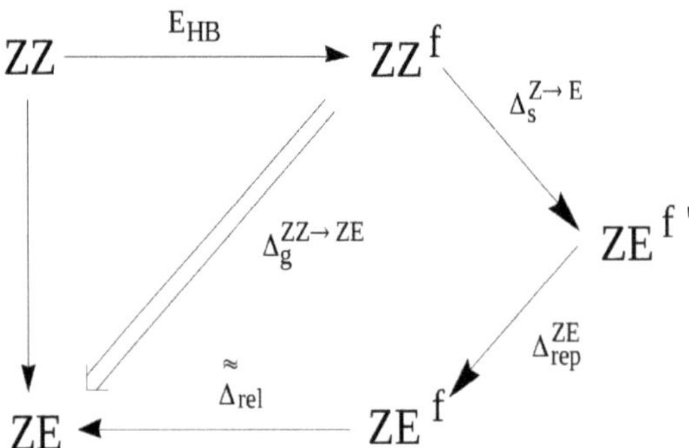

Figure 11. Diagram showing the way of obtaining the ZE conformer from the ZZ one through various fictitious forms.

Therefore, the energy that is associated with the transition from the ZZ^f conformer to the ZE conformer can be expressed as:

$$\Delta_g^{ZZ\to ZE} \approx \Delta_{rep}^{ZE} + \tilde{\tilde{\Delta}}_{rel} \tag{21}$$

Given the assumption (12) and by the similarity to the previously defined changes in energies $\Delta_g^{ZZ\to EZ}$ (13) and $\Delta_g^{ZZ\to EE}$ (14), one gets

$$\Delta_g^{ZZ\to ZE} = E^{ZE} - E^{ZZ,f} \tag{22}$$

Inserting this expression together with Equation (21) into Equation (12) gives a relationship between the estimation of the hydrogen bond energy in ZZ-3-aminoacrolein that is obtained by GCM and that obtained by OCM with the ZE conformer as the reference open form

$$E_{HB}^{GCM} = E^{ZZ} - E^{ZE} + (\Delta_{rep}^{ZE} + \tilde{\tilde{\Delta}}_{rel}) = E_{HB}^{OCM} + (\Delta_{rep}^{ZE} + \tilde{\tilde{\Delta}}_{rel}) \tag{23}$$

Equation (23) shows that, when compared to OCM, the estimation that is based on GCM takes into account two terms with opposite signs. The repulsive term Δ_{rep}^{ZE} is positive, whereas the relaxation term $\tilde{\tilde{\Delta}}_{rel}$ is negative. The mutual weights of these two terms cause that the value of the intramolecular hydrogen bond energy determined by GCM is either below or above the value obtained by OCM. Strong hydrogen bonds should cause significant changes within the X-H$\cdots$Y bridge and, thus, both a small distance H$\cdots$Y in the conformer ZZ and a small distance H$\cdots$H in the fictitious form ZE$^{f'}$ (or ZE^f) obtained after rotation of the proton-acceptor group while maintaining the geometrical parameters from the conformer ZZ (except for the dihedral angle O=C–C=C). As a consequence, in molecules with a strong intramolecular hydrogen bond, the role of H$\cdots$H repulsion at the ZE$^{f'} \to$ ZE^f should be significant. On the other hand, the significance of the relaxation term $\tilde{\tilde{\Delta}}_{rel}$ should be dominant in the case of relatively small distances $H \cdots H$ in the ZE^f form ($H \cdots Y$ in ZZ) and, which seems more important, in the case of bulky proton-acceptors.

At this point, it is worth comparing the hydrogen bond energy values that were obtained with GCM with those obtained with the traditional variant of OCM. The first comparison of this type was made for the ZZ-3-aminopropenal (ZZ-3-aminoacrolein) [47] discussed here and for the related ZZ-3-aminopropential [39], where sulfur atom replaces the oxygen atom. The energy values of hydrogen bonds E_{HB}^{GCM} and E_{HB}^{OCM} are shown in Table 2. Additionally, this table also shows the relative energies (in relation to ZZ) of the respective conformers, the $H \cdots H$ distances in ZE^f and ZE forms, and the values of $\tilde{\Delta}_{rel}$, which will be used in the current discussion. All of these values are limited to the best method used (MP2/6-311++G**), so as not to increase the amount of numerical data [39].

Table 2. Some energetic (in kcal/mol) and geometric (in Å) parameters computed (MP2/6-311++G**) for different forms of 3-aminoacrolein (Y = O) and 3-aminopropential (Y = S).

Y	E^{EZ}	E^{EE}	E^{ZE}	E_{HB}^{OCM}	E_{HB}^{GCM}	$d_{H \cdots H}^{ZE,f}$	$d_{H \cdots H}^{ZE}$	$\Delta d_{H \cdots H}^{ZE,f \to ZE}$	$\tilde{\Delta}_{rel}$
O	4.77	3.77	6.50	−6.50	−5.28	1.840	2.141	−0.301	−1.87
S	6.07	4.09	6.02	−6.02	−6.96	1.968	2.127	−0.159	−1.86

In the case of 3-aminoacrolein, the following order of relative energies of conformers was obtained: $E^{ZZ} < E^{EE} < E^{EZ} < E^{ZE}$. It suggests a significant relaxation of the most extended EE conformer and a significant role of the $H \cdots H$ repulsion in the ZE conformer. In the case of 3-aminopropential (Y = S), the relative energy of the EZ conformer, on the other hand, is significantly lifted up, so that it equals that of the ZE conformer. In turn, this result suggests a greater role of $S \cdots H$ valence repulsion in 3-aminopropential than $O \cdots H$ in the EZ conformer (a complementary explanation may also be the greater role of the attractive component in the $O \cdots H$ interaction than $S \cdots H$, which lowers the relative energy of EZ-3-aminoacrolein in relation to EZ-3-aminopropential). It is also manifested by the values of the angle CCY, which is 128.3° and only 125.4° for Y = S and O, respectively. As for the estimated values of the hydrogen bond energy, interestingly, OCM suggests a somewhat stronger N-H$\cdots$O hydrogen bond in ZZ-3-aminoacrolein (−6.50 kcal/mol) than N-H$\cdots$S in ZZ-3-aminopropential (−6.02 kcal/mol), whereas, in the case of GCM, the opposite is obtained, i.e., this method suggests that the latter bond is stronger (−6.96 kcal/mol) than the former one (−5.28 kcal/mol). Table 2 also presents the values of the distances $H \cdots H$ in the ZE^f and ZE forms of both molecules, as well as the changes of these distances at the $ZE^f \to ZE$ transition, i.e., upon relaxation of this conformer. The much higher $\Delta d_{H \cdots H}^{ZE,f \to ZE}$ value for the ZE-3-aminoacrolein (−0.301 Å) than for ZE-3-aminopropential (−0.159 Å) suggests a much stronger $H \cdots H$ repulsion in the former of these systems, which is most likely due to the much shorter initial distance (1.840 Å vs. 1.968 Å). This suggestion is actually confirmed by the obtained results. Namely, as can be seen from the last column of Table 2, 3-aminoacrolein and 3-aminopropential are characterized by the same value (−1.87 kcal/mol) of $\tilde{\Delta}_{rel}$, i.e., the relaxation term $ZE^f \to ZE$. Therefore, the greater change in the $H \cdots H$ distance at the transition $ZE^f \to ZE$ for the former of these molecules must result primarily from the greater repulsion Δ_{rep}^{ZE}, which, as a consequence, should significantly exceed the relaxation component $\tilde{\Delta}_{rel}$. In turn, this should lead to a significantly lower E_{HB}^{GCM} when compared to E_{HB}^{OCM}. As can be seen from Table 2, such a relationship for ZZ-3-aminoacrolein does indeed take place since E_{HB}^{GCM} and E_{HB}^{OCM} amount to −5.28 and −6.50 kcal/mol, respectively. This result shows that the hydrogen bond energies obtained within GCM are consistent with the observable geometric changes.

In addition to the case of 3-aminoacrolein [47] and 3-aminopropential [39] discussed here, GCM was later used to estimate the energy of intramolecular C-H$\cdots$O/S interactions in few systems featuring a similar *quasi*-ring structure (Figure 12) [40,41].

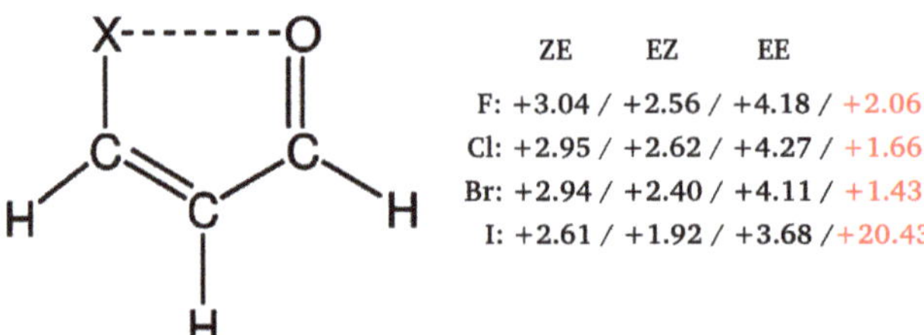

Figure 12. Energy values (in kcal/mol) of intramolecular C-H···O/S interactions obtained [41] (B3LYP/aug-cc-pVTZ) by either OCM (**black**) or GCM (**red**).

Importantly, contrary to popular belief, these calculations showed that the C-H···O/S contacts in these systems are actually destabilizing. Therefore, no hydrogen bond in the usual sense is formed between the proton-donating C-H bond and proton-acceptor O or S atoms. This result was interpreted [40,41] in terms of the steric compression, which leads to the dominance of the valence repulsion contribution in the C-H···O contact and it was further supported by observing both the increase in contact destabilization and the corresponding geometric changes during the flattening of some systems. Further detailed studies on an even larger group of systems (*vide infra*) showed, however, that intramolecular C-H···O interactions may be destabilizing in some systems, while stabilizing in others [44]. The fact that the large number of X···O (X = F, Cl, Br, I), O···O and F···F interactions, which some consider stabilizing due to the presence of a bond path tracing these contacts are, in fact, destabilizing in many molecules was also shown [43] by means of the energy values obtained, *inter alia*, by GCM and OCM. An example is shown in Figure 13.

Figure 13. Interaction energies (in kcal/mol) of the X···O (X = F, Cl, Br, I) contact obtained [43] by either OCM (**black**) or GCM (**red**). The MP2/aug-cc-pVTZ level of theory was used for all systems but that with Y = I, for which MP2/aug-cc-pVTZ-PP was used instead.

Theoretical studies [39–41,43,44] show that GCM can be considered to be a reliable method of estimating the energy of both intramolecular hydrogen bonds as well as intramolecular non-bonding interactions. As this method takes into account changes in geometric parameters that occur when passing to reference systems, it is a more reliable approach than the standard OCM, which does not take into account these changes at all. Of course, the applicability of GCM, like most other methods, is limited. For example, the presence of bulky substituents can significantly reduce the reliability of this method. Moreover, of course, the analyzed molecule must have appropriate conformers, which is not always the case. However, OCM also has to deal with similar requirements. Nevertheless, OCM is less tricky.

It is obvious that obtaining the individual conformers needed while using conformational methods requires a great deal of care and attention. Unfortunately, this is not always the case. In their study of the N-H$\cdots$O and N-H$\cdots$S intramolecular hydrogen bonds in β-aminoacrolein, β-thioaminoacrolein, and their halogenated derivatives, Nowroozi and Masumian claimed that GCM performs worse than RBM and RRM, in particular [63]. However, it is enough to look at their Scheme 3 to realize that they used wrong conformers labeled as EZ and EE. Briefly, both of these conformers should have H and R_3 at reversed positions! (Starting with the ZZ conformer, rotation of the -NHR$_3$ group around the C=C double bond obviously leaves the H atom rotated with this group on the "inside" of the molecule, i.e., at the R_3 site and close to R_1.) Because EZ and EE conformers (either real or fictitious) are used in RRM and GCM, it is obvious that the results that are presented by Nowroozi and Masumian [63] are completely wrong (as evidenced, e.g., by low R^2 values). Moreover, these authors ignored the fact that some of the conformers they used experience new significant interactions, such as O$\cdots$Br, which, of course, significantly affect the total energy of a given conformer.

2.1.5. Geometry-Corrected Related Rotamer Method (GCRRM)

It is worth noting that, when compared to GCM, RRM should give too negative values of interaction energy, because the total energy of the ZE conformer, E^{ZE}, which is not present in the formula for E_{HB}^{GCM} (Equation (20)), appears with a negative sign. At first glance, it would seem difficult to further directly compare the two methods, as they do not use the same EZ and EE conformer structures; GCM overlays them with the values of the geometric parameters from the closed ZZ form, while RRM uses fully relaxed geometries. Nevertheless, the difference in estimations of the two methods can be written, as follows [44]:

$$E_{HB}^{GCM} - E_{HB}^{RRM} = \frac{1}{2}(E^{EZ} - E^{EZ,f}) + \frac{1}{2}(E^{EE} - E^{EE,f}) + \frac{1}{2}(E^{EZ} - E^{EE}) + (E^{ZE} - E^{EE}) \tag{24}$$

This expression shows that the difference between the estimates that were obtained with GCM and RRM results from the balance of the relaxation terms (first two terms) and the conformational changes (EZ→EE and ZE→EE). Importantly, both of these contributions have opposite signs and the latter ones are larger than the former. Moreover, the last term contributes without the factor $1/2$. As a consequence, the difference (24) is positive. In the case of 13 molecules containing intramolecular C-H$\cdots$O contacts considered in the reference [44], the energies of the consecutive terms were, as follows: -1.3 ± 0.3, -2.0 ± 0.6, 2.5 ± 0.8, and 2.7 ± 0.3 kcal/mol, so that the difference (24) was 2.3 ± 0.3 kcal/mol. Because the first three values almost cancel themselves (-0.4 kcal/mol), it can be assumed that the difference (24) comes mainly from the configurational change EE→ZE. This configurational change can then be considered as a two-step process: EE→EE$^{f'}$→ZE, where EE$^{f'}$ is a fictitious conformer EE having the geometric parameters of the ZE conformer. Hence, the energy of the EE→ZE process can be written as the sum of the preparation energy of the ZE conformer and the E→Z isomerization energy:

$$E^{ZE} - E^{EE} = (E^{EE,f'} - E^{EE}) + (E^{ZE} - E^{EE,f'}) \tag{25}$$

In the considered systems with the intramolecular C-H$\cdots$O interactions, the first term was 1.0 ± 0.2 kcal/mol. The second term is related to the H$\cdots$H repulsion that appears in the conformer ZE, the value of which was estimated at 0.60 ± 0.17 kcal/mol (median value) [44]. Together with the preparation energy, this energy suggests that the EE→ZE process is affected by close H$\cdots$H contact by roughly 1.6 kcal/mol, which is close to the actual value of 2.7 kcal/mol as well as the E→Z isomerization energy in 2-butene (1.04 kcal/mol). This fairly good agreement led to the proposition of a corrected RRM known as the Geometry Corrected Related Rotamers Method (GCRRM) [44]. According to GCRRM, the estimated value of an intramolecular hydrogen bond (or other interaction) can be obtained from the following formula

$$E_{HB}^{GCRRM} = E_{HB}^{RRM} + (E^{EE,f'} - E^{EE}) + E_{HH} \tag{26}$$

where the E^{HH} value is 0.6 kcal/mol. The values obtained with GCRRM are between the values obtained with GCM and RRM and closer to the former, as shown in Figure 14.

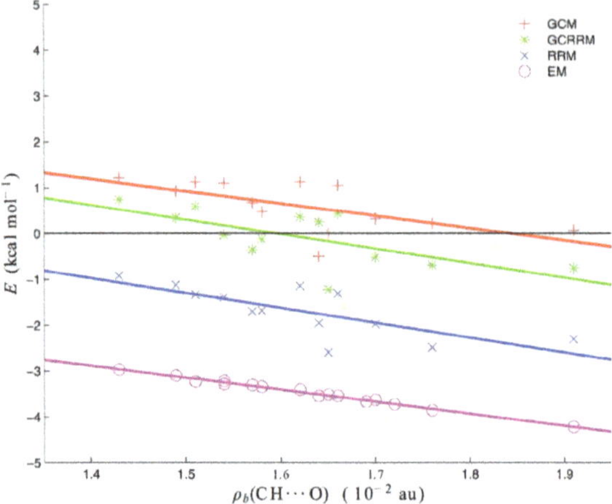

Figure 14. Interaction energies (in kcal/mol) of the intramolecular C-H$\cdots$O contacts in the molecules investigated in ref. [44].

It is noteworthy that all four lines that are shown in Figure 14 have similar slopes; therefore, the methods differ by their intercepts that can be seen as "zeros of the interaction energy" [44]. Indeed, $E_{\text{HB}}^{\text{RRM}} = E_{\text{HB}}^{\text{EM}} + 1.7\,\text{kcal/mol}$, $E_{\text{HB}}^{\text{GCRRM}} = E_{\text{HB}}^{\text{EM}} + 3.4\,\text{kcal/mol}$ and $E_{\text{HB}}^{\text{GCM}} = E_{\text{HB}}^{\text{EM}} + 4.0\,\text{kcal/mol}$. At the same time, Figure 14 is a wonderful illustration displaying that a given intramolecular interaction in a certain system may be suggested to be much or less stabilizing according to one estimating method while another method may suggest its rather repulsive nature.

2.2. Rotation Barriers Method (RBM)

A strong alternative to OCM with its various variants is the Rotation Barriers Method (RBM) [33,65–72] first used by Buemi et al. in order to estimate the energy of the O-H$\cdots$O intramolecular hydrogen bond in malonaldehyde [66] and a bit later of N-H$\cdots$N in formazan [67]. Quite rightly, this method assumes that an intramolecular hydrogen bond in the closed (chelate) form raises the height of the energy barrier that is associated with either the proton-donor or proton-acceptor group rotation by 180° to form an open form. Hence, when assuming the additivity of the respective energy terms, it can be written that

$$E_{\text{RB}} = E_{\text{HB}}^{\text{RBM}} + E_{\text{ARB}} \tag{27}$$

where E_{RB} is the rotation barrier and E_{ARB} is (to use Buemi's terminology) the actual rotation barrier of the considered group [33]. The actual rotation barrier introduced as a result of the above additivity scheme is obviously related to a fictitious equivalent of a closed system in which the intramolecular hydrogen bond is 'turned off', and, therefore, it is not possible to calculate its value exactly. Nevertheless, E_{ARB} can be estimated while using a certain reference system (see Figure 15).

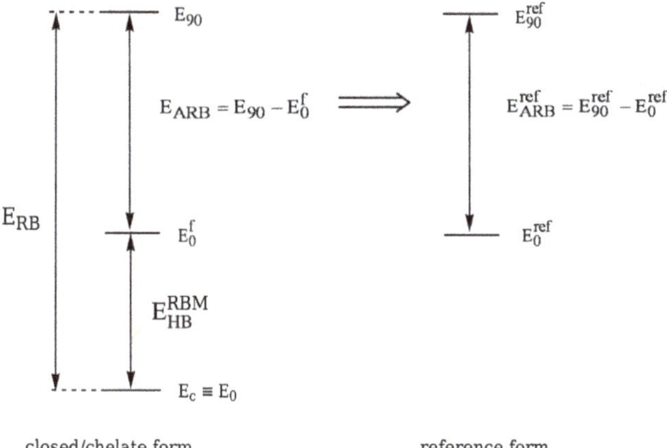

Figure 15. Scheme showing the way of estimating the energy of an intramolecular hydrogen bond according to RBM.

Hence,

$$E_{HB}^{RBM} = E_{ARB}^{ref} - E_{RB} = \left(E_{90}^{ref} - E_0^{ref}\right) - \left(E_{90} - E_c\right) < 0 \tag{28}$$

where the expressions in the former and in the latter brackets are rotation energy barriers for either the proton-donor or the proton-acceptor group in the reference and the closed form, respectively. In fact, the transition states for the rotations in both systems do not have to correspond exactly to the perpendicular orientation of the group. Nevertheless, the symbols denoting total energies of the transition states are given the subscript 90 in order to emphasize that often the transition state, that is associated with the rotation of a given group, roughly corresponds to its perpendicular orientation with respect to the molecular framework. Importantly, just like in the case of OCM, in RBM it is assumed that the reference system retains the earlier described significant similarity to the bound, i.e., closed form. This condition is not always easy to meet. On the other hand, the use of RBM is a reasonable method of choice in many of those cases where the energy estimate based on OCM is unreliable due to the presence of some bulky or highly electronegative substituents leading to new important interactions in the open reference form [33].

As already mentioned, this method was first used by Buemi et al. [67] in order to estimate the energy of the N-H$\cdots$N intramolecular hydrogen bond in one of the conformers of formazan (Figure 16).

Figure 16. The open and closed forms of formazan and the two reference systems (**A1** and **A2**) used by Buemi et al. [67] in RBM.

The abandonment of the traditional OCM and the need to use a different method, which led to RBM, resulted from the inability to find a reliable reference form. Because of the symmetry of the amino group, its rotation is useless, and the rotation of the N=N–H group leads to a close H$\cdots$H contact.

On the other hand, the use of other conformers was considered [67] impractical, because it led to too large structural change. As a consequence of these problems, Buemi et al. proposed using, in RBM, two reference systems shown in Figure 16 as A1 and A2. Buemi et al. emphasized that they had previously successfully used this method to determine the interaction energy of the $O\text{-}H\cdots O$ intramolecular hydrogen bond in malonaldehyde (using vinyl alcohol as a reference), obtaining (MP2/6-31G**) a value similar to that of the traditional open-closed method (-14.07 and -14.01 kcal/mol, respectively) [66]. Depending on the reference molecule A1 or A2 and on more subtle conditions concerning the structure of the amino group (planar vs. pyramidal), Buemi et al. obtained energies that ranged from -9.38 to -4.85 kcal/mol. Subsequently, however, the value close to the middle, i.e., -7.17 kcal/mol, was considered as the most reliable. Nevertheless, quite reasonably, Buemi and Zuccarello pointed out that such wide range of the obtained estimates does not allow for stating that the estimate of the $N\text{-}H\cdots N$ hydrogen bond energy in formazan is as good as $O\text{-}H\cdots O$ in malonaldehyde [66].

Buemi and Zuccarello then used RBM to estimate interaction energies of various intramolecular hydrogen bonds ($O\text{-}H\cdots O$, $O\text{-}H\cdots$ halogen, $O\text{-}H\cdots N$, $N\text{-}H\cdots O$, $N\text{-}H\cdots N$, $S\text{-}H\cdots O$, $O\text{-}H\cdots S$, and $S\text{-}H\cdots S$) in many molecules (e.g., malondialdehyde, acetylacetone, and their variously substituted derivatives, formazan, 3-aminoacrolein, some β-thioxo- and β-dithioketones, 2-halophenols, 2-nitrophenol) [33]. From the many data shown there, I will only mention those obtained for malondialdehyde, acetylacetone, and 3-aminoacrolein. The closed form, the two open forms, and the two reference molecules used in RBM are shown in Figure 17, and the quoted values of the respective estimates are listed in Table 3.

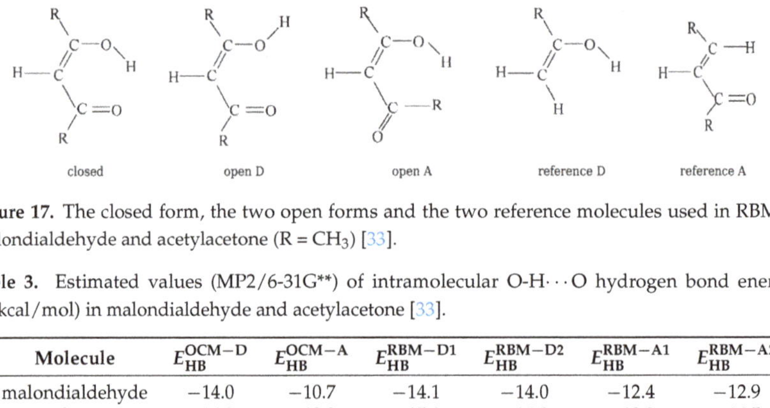

Figure 17. The closed form, the two open forms and the two reference molecules used in RBM for malondialdehyde and acetylacetone ($R = CH_3$) [33].

Table 3. Estimated values (MP2/6-31G**) of intramolecular $O\text{-}H\cdots O$ hydrogen bond energies (in kcal/mol) in malondialdehyde and acetylacetone [33].

Molecule	E_{HB}^{OCM-D}	E_{HB}^{OCM-A}	E_{HB}^{RBM-D1}	E_{HB}^{RBM-D2}	E_{HB}^{RBM-A1}	E_{HB}^{RBM-A2}
malondialdehyde	-14.0	-10.7	-14.1	-14.0	-12.4	-12.9
acetylacetone	-16.2	-13.3	-15.1	-16.9	-12.3	-14.5

As can be seen from Table 3, in the RBM calculations, Buemi and Zuccarello used four reference structures, two for the ARB for the -OH proton-donor group, and two for the ARB for the proton-acceptor -CHO group. In the former case, these systems were the open form A and the reference D obtained by replacing the group -CHO by the H atom. In the latter case, these were the open form D and the reference A obtained by replacing the OH group by H. Buemi and Zuccarello emphasized the very good agreement of the estimates based on OCM and RBM, whenever these methods use the proton-donor group rotation [33]. This result is especially obvious in the case of malondialdehyde (ca. -14 kcal/mol), whereas slightly less in the case of acetylacetone (from ca. -17 kcal/mol to ca. -15 kcal/mol), which was attributed to the new, probably quite significant, interaction between the methyl group and the hydrogen atom from the hydroxyl group in the open form A. On the contrary, worse agreement of the OCM and RBM results was noted for the estimates that are based on the rotation of the proton-acceptor group. However, it is noted that, in general, the estimates that are based on RBM (no matter whether it is a rotation of the proton-donor or the proton-acceptor group) are closer to OCM estimates based on the proton-donor group rotation than the corresponding OCM based on the proton-acceptor group rotation [33].

As already mentioned, for ZZ-3-aminoacrolein (see Figure 9), it seems that the most reasonable reference form is EZ (although Buemi and Zuccarello also admitted the conformer ZE, this form experiences a new significant H$\cdots$H interaction). In the case of malonaldehyde, the EZ conformer gives a value of −9.7 kcal/mol, thus approximately 5.3 kcal/mol lower than the classic value of the O-H$\cdots$O hydrogen bond energy in malondialdehyde. Assuming that a similar underestimation would also act for 3-aminoacrolein, Buemi and Zuccarello renormalized the obtained value (−5.2 kcal/mol), finally obtaining a value of about −10.5 kcal/mol [33]. The main model problem in the estimation of the N-H$\cdots$O hydrogen bond energy in ZZ-3-aminoacrolein using RBM is the change in the degree of amino group pyramidalization during rotation [33,47]. Because of the presence of the hydrogen bond, this group is planar in the ZZ conformer, whereas slightly pyramidal when rotating around the C-N bond. Depending on the constraint put on the rotating amino group and the reference system utilized (Figure 18), the estimated value of the N-H$\cdots$O hydrogen bond energy in ZZ-3-aminoacrolein is between −11.7 and −8.4 kcal/mol (MP2/6-31G**).

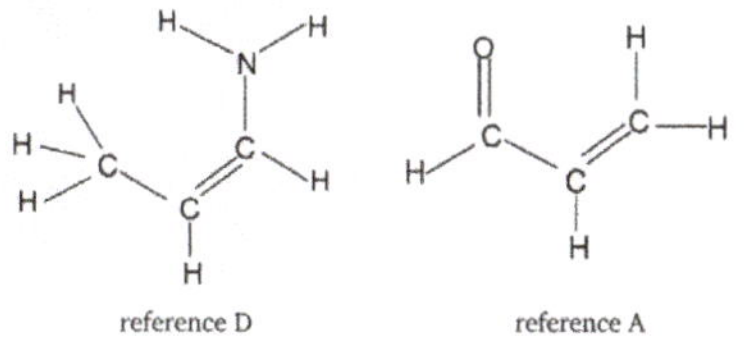

Figure 18. Two reference molecules used in RBM for 3-aminoacrolein [33].

Unfortunately, this example shows quite a lot of freedom in terms of the possible choice of reference systems. On the one hand, the reference molecule A was obtained for ZZ-3-aminoacrolein by replacing the amino group with a hydrogen atom, whereas molecule D by replacing the aldehyde group with a methyl group (and not only with hydrogen). On the other hand, both of these reference molecules have the same number of heavy framework atoms. However, unlike D, molecule A features the presence of a conjugated system of two double bonds. Hence, it should be expected that the π-electron structure in both of these reference molecules is quite different.

In summary, RBM is a reasonable approach for estimating intramolecular hydrogen bond energy in many simple molecules and it can be successfully used as a replacement or supplement to the estimation based on OCM. However, like OCM, this method should also be used with great caution, because the presence of new interactions during rotation of a group in the parent or reference molecule may significantly reduce the reliability of the estimation. Moreover, in this method, both the problem of choosing a reasonable reference form and a certain freedom of this choice are noticeable. As noted by Buemi and Zuccarello [33], RBM is much more computationally expensive than OCM, as it requires calculating the rotation barriers for two systems, the bound, i.e., the closed one, and the reference molecule (Equation (28)). It is worth reminding here that the maxima of these barriers do not have to correspond exactly to the perpendicular arrangement of the rotated groups.

2.3. Dimer Model (DM)

Importantly, in the context of the present considerations, all of the methods of estimating the energy of intramolecular hydrogen bond (or more generally, interaction) in the closed form that were discussed so far, were based on the assumed model of energy additivity, which leads to quite a lot of freedom in choosing a reasonable reference system. This, in turn, leads to the known problem that the resulting hydrogen bond energy value can be quite dependent on this reference system. Moreover, even within the adopted estimation method there are often many possible variants (e.g., in OCM with only partial optimization of the open reference form). Hence, the idea was born to abandon the assumed total energy partition of the closed form and refer to the strictly defined interaction energy of the *inter*molecular contact (Equation (1)). This idea leads to the Dimer Model (DM) [73,74]. This model was most likely used for the first time by Palusiak and Krygowski in order to estimate the interaction energy of the intramolecular $\pi \cdots \pi$ contact in 1,3,5,7-cyclooctatetraene [73]. Subsequently, DM was used by Jabłoński and Palusiak [74] to support the previously obtained result [43] that the intramolecular Cl$\cdots$O interaction in 3-chloropropenal is, in fact, destabilizing (repulsive) and not stabilizing [75].

To relate to the results that were obtained for the 3-halogenopropenal previously presented in Figure 13, details on the ideas of DM will be discussed on the basis of this molecule [74]. The first step in this model is to build a reasonable dimer in which the fragment of the considered interaction from the bound molecule, i.e., its closed form, is preserved. In the case of 3-halogenopropenal, this is obviously the C–X$\cdots$O=CH fragment in the ZZ conformer (see Figure 13). In order to test the reliability of the model, two dimers, namely ZE$\cdots$EZ and ZE$\cdots$EE, were constructed where, fundamentally, the C–X$\cdots$O=CH fragment was taken from the ZZ-3-halogenopropenal and then built into these dimers, as shown by green ovals in Figure 19.

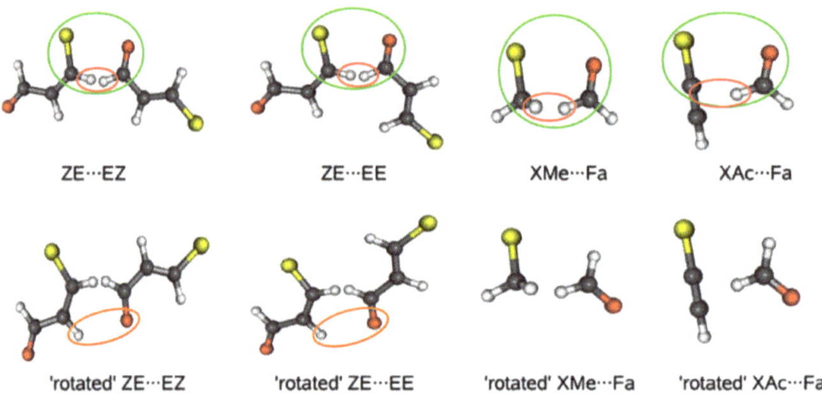

Figure 19. Spatial arrangement of reference dimers and their 'open' forms used in estimating the interaction energies of intramolecular X$\cdots$O contacts in ZZ-3-halogenopropenal [74].

Unfortunately, as clearly seen (red ovals) in Figure 19, the new very short C-H$\cdots$H-C contacts appear in the dimers thus constructed. However, they can be accounted for (and 'subtracted') by using appropriate rotated (inverted) forms of the proposed dimers, in which, importantly, the relative arrangement of all atoms in the C-H$\cdots$H-C fragment is conserved. Consequently, the formula for the interaction energy of the intramolecular X$\cdots$O contact in ZZ-3-halogenopropenal has the following form:

$$E^{DM}_{X\cdots O} = E_{int}(\text{dimer}) - E_{int}(\text{rotated dimer}) \tag{29}$$

Table 4 presents the results obtained using this formula.

Table 4. Interaction energies (kcal/mol) of the X$\cdots$O intramolecular contact in ZZ-3-halogenopropenal estimated by means of several dimers utilized in Dimer Model (DM) [74].

X	$E^{ZE\cdots EZ}_{X\cdots O}$	$E^{ZE\cdots EE}_{X\cdots O}$	$E^{XMe\cdots Fa}_{X\cdots O}$	$E^{XAc\cdots Fa}_{X\cdots O}$
F	1.82	1.97	0.52	0.44
Cl	2.98	3.34	1.06	0.38
Br	3.49	3.95	1.08	0.19

First, it should be noted that the thus obtained estimates are positive, not negative. This result confirmed the previously [43] obtained conclusion that the intramolecular $X \cdots O$ interactions in ZZ-3-halogenopropenal are in fact locally destabilizing, i.e., repulsive. As expected, the obtained repulsion values increase in the order $F < Cl < Br$, and those obtained for Cl and Br are similar to each other, whereas the values for F differ from them. Due to a probably slight (2.44 Å) contamination of the rotated dimers with weak $C\text{-}H \cdots O$ hydrogen bond (orange ovals), simplification of DM was then applied. Namely, halogenomethane$\cdots$formaldehyde (XMe$\cdots$Fa) and halogenacetylene$\cdots$formaldehyde (XAc$\cdots$Fa) dimers were then designed (Figure 19) with imposed restrictions on the structural requirements discussed earlier [74]. Although these simplified variants introduce some subtle problems [74] and the resulting estimates are clearly lower, the values are still positive, which supports the earlier conclusion regarding the repulsive nature of the $X \cdots O$ contact in ZZ-3-halogenopropenal.

2.4. Isodesmic Reactions Method (IRM)

In many areas of physical organic and theoretical chemistry the so-called isodesmic reactions are used [63,76–93]. These are more or less hypothetical reactions, in which the same numbers of single and multiple bonds of the same type are present on both sides of this reaction, i.e., of the reagents and of the products. If, in addition, the relevant atoms conserve their hybridization, then these reactions are called the homodesmotic reactions [79–83,87–92]. The conservation of the atomic hybridizations makes the homodesmotic reaction a more reliable description of a given phenomenon than the less demanding isodesmic reaction. The use of isodesmic and homodesmotic reactions allowed for a more detailed theoretical description of many physical processes and effects, such as the extra stability due to cyclic π-electron delocalization [88]. Homodesmotic reactions are also often used in order to estimate the energy of intramolecular hydrogen bonds [32,38,39,44,63,84,86,87,89,91] or some other interactions of interest [42–44,55,87,93].

The reliability of the Isodesmic Reactions Method (IRM) is based on the assumption that the total energy of a molecule I can be partitioned into energies of chemically recognizable fragments, such as bond energies, and that those energies are transferable among various molecules which, however, involve similar chemical units. A general scheme of a simple homodesmotic (also isodesmic) reaction for a model system featuring an intramolecular $X\text{-}H \cdots Y$ hydrogen bond is shown in Figure 20.

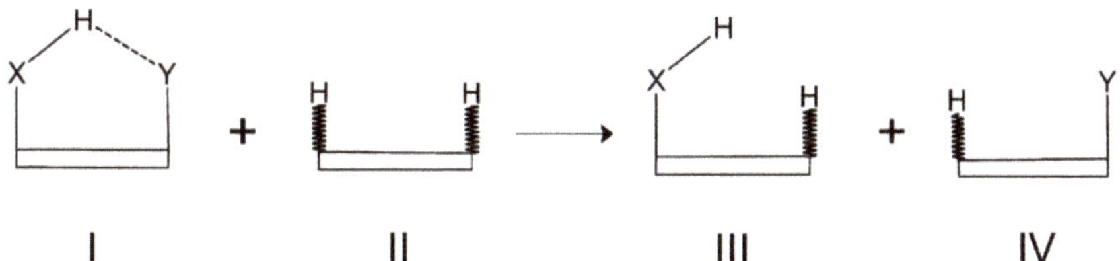

Figure 20. General scheme showing a homodesmotic reaction for a model molecule featuring an intramolecular $X\text{-}H \cdots Y$ hydrogen bond.

In this figure, the molecular framework, which, of course, may vary from molecule to molecule, is drawn as a box for simplicity and, moreover, those C-H bonds in molecules II, III, and IV, which are not present in the parent molecule I are marked by a zigzag bond line. When comparing both sides of the homodesmotic reaction shown in Figure 20, it can be easily seen that all the bonds, except the only one denoting the $H \cdots Y$ contact in the parent molecule I, on the left side of this reaction are also present on the right side of this reaction. Accordingly, the only missing 'bond' on the right side of this reaction equation is the intramolecular $H \cdots Y$ contact in the parent molecule I. Thus, the interaction energy of this contact can be obtained by the following expression

$$E_{HB}^{IRM} = E(I) - E^{f}(I) < 0 \tag{30}$$

where

$$E^{f}(I) = E(III) + E(IV) - E(II) \tag{31}$$

In these two equations, $E(I)$ is the total energy of the fully optimized parent molecule I, whereas $E^{f}(I)$ can be regarded to as the total energy of its fictitious counterpart featuring no $H \cdots Y$ contact.

As with conformational methods, the question now arises as to what geometries to use for the auxiliary molecules II, III, and IV [42–44,91]. The vast majority of calculations that are related to isodesmic or homodesmotic reactions use fully optimized geometries, so that the total energies in Equations (30) and (31) are total energies of fully optimized molecules. For this reason, such an approach can lead to considerable doubts regarding the reliability of E_{HB}^{IRM} if only full geometry optimization of at least one of the molecules leads to new significant interaction(s) or to a significant change in molecular structure compared to I [42]. On the other hand, if the structural fragments in II, III, and IV do not differ significantly from those in I, then IRM can give reasonable estimates of interaction energies of the hydrogen bond (or any other contact of interest) in I. It seems that rigid ring molecules should be privileged here [44]. Another possibility that is very rarely considered [42–44] is that the geometry of the parent molecule I is transferred to the auxiliary molecules II, III, and IV. However, then, the question arises, what to do with the C-H bonds that the molecule I does not have (they are indicated in Figure 20 by a zigzag line). Hence, a field for different IRM variants arises here. For example, these bonds can be optimized, or they can be given the length of either the C-X or C-Y bond of molecule I, or any other reasonable value as, e.g., 1 Å. It is worth mentioning that the use of not fully optimized molecules II, III, and IV leads to an overestimation of the hydrogen bond energy value in I, i.e., E_{HB}^{IRM}.

Despite the fact that, as mentioned above, the method of estimating the interaction energy either of some hydrogen bonds or some other kinds of intramolecular interactions in I that is based on isodesmic/homodesmotic reactions is quite popular [32,38,39,42–44,55,63,84,86,87,89,91,93], its reliability, in general, may raise some doubts. For example, it has been shown that the comparison of the interaction energy in closely related systems disqualifies IRM (Figure 21) [44].

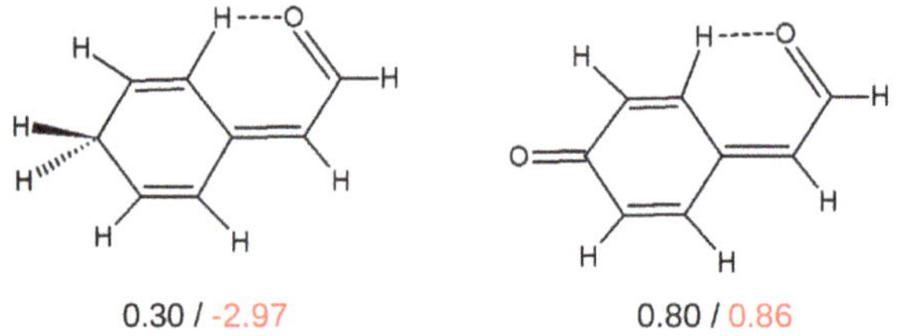

Figure 21. The values (in kcal/mol) of the interaction energies of the C-H$\cdots$O contacts estimated by OCM (black values) and IRM (red values) [44].

As one can see, the values of the interaction energies of C-H$\cdots$O contacts in both of these similar molecules are very close to each other (0.30 and 0.80 kcal/mol) when OCM is used, while IRM gives values very different from each other, also in terms of sign (-2.97 and 0.86 kcal/mol). Moreover, the failure of IRM also manifested itself in the unphysical positive slope of the dependence of E_{HB}^{IRM} on the electron density at the critical point (ρ_b) of the C-H$\cdots$O interaction featuring a sp^3 hybridized carbon atom (see Figure 22) [44].

In contrast, RRM and GCM gave physically justifiable negative slopes.

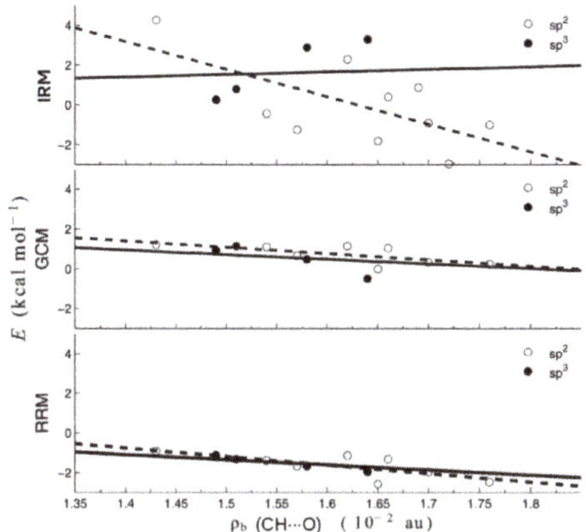

Figure 22. Correlations between RRM-, GCM-, and IRM-based interaction energies of C-H$\cdots$O contacts featuring either sp^2 (open circle) or sp^3 (full circle) hybridized carbon atoms and the electron density at the bond critical points of these contacts [44].

2.5. QTAIM-Based Methods

By operating with the topology of the electron density distribution, the Quantum Theory of Atoms in Molecules (QTAIM) created by Bader [94–96] makes it possible to divide the space of a molecule into closely adjoining, i.e., non-overlapping, three-dimensional subspaces. According to QTAIM, they are understood as individual atoms. These atoms are separarated from each other by surfaces through which there is no electron density gradient flow, and atomic nuclei are most often attractors of this gradient field. Importantly, the possibility of unequivocally defining the space of an atom in a molecule, introduced by QTAIM, enables the determination of many atomic quantities by integration over the volume of a given atom. Yet another useful result of QTAIM are the concepts of the so-called bond path (BP) and bond critical point (BCP) [94–96]. The latter is a saddle point in the electron density distribution that features positive curvature (thus, experiencing the minimum) in the direction of the two neighboring nuclei and negative curvatures (thus, experiencing maxima) in the two perpendicular directions, and the former is a pair of gradient vector paths originating at this BCP and terminating at both nuclei. Thus, a bond path is also a ridge of the highest electron density between a pair of the so linked atoms. For this reason, the pattern of such electron density ridges, i.e., bond paths (the so-called molecular graph) very often corresponds to the pattern of bonds that are drawn by chemists, i.e., the structural formula [97]. The fundamental issue in the theory of inter- and intramolecular hydrogen bonds or other stabilizing interactions is that, according to QTAIM [94–96], the simultaneous presence of a bond path and a bond critical point between any pair of atoms is a necessary and sufficient proof that this bond (interaction) is stabilizing [98]. This view has resulted in many people using the presence of a bond path as a sufficient criterion for the presence of a hydrogen bond (or other bonding interaction) between a pair of atoms. Unfortunately, this happens without even specifying the bonding (interaction) energy value, in a way ignoring the fact that the hydrogen bond must be stabilizing to deserve this name. However, it should be noted that the treatment of both BP and BCP as sufficient evidence for stabilizing interaction has been criticized [99–114], as it turns out that the presence of these topological features between a pair of atoms in general does not determine the stabilizing or destabilizing nature of the interaction between the pair [112,113].

However, in the context of this article, it is more important that QTAIM makes it possible to determine the interatomic interaction energy. The first of these methods was proposed by Espinosa et al. [115] and it is based on a quantity computed at the bond critial point of the interaction, the energy of which is to be determined. The another method is based on the partition of the total energy of a system into individual monoatomic and diatomic contributions and functions under the name of Interacting Quantum Atoms (IQA) [116,117]. These two methods will now be discussed in the next two subsections.

2.5.1. Espinosa's Method (EM)

Based on empirical data for many systems featuring intermolecular hydrogen bonds of the H$\cdots$O type, Espinosa et al. [115] proposed the following relationship between the energy of an intermolecular hydrogen bond and the local electronic potential energy density that is determined at the bond critical point of a given hydrogen bond

$$E_{HB}^{EM} = \frac{1}{2} V_{BCP} \tag{32}$$

It should be noted that both the simplicity of the formula (32) and the easy availability of the V_{BCP} value (QTAIM calculations) resulted in a rather uncritical acceptance of this expression for determining not only the energy of intermolecular hydrogen bonds, but also the energy of intramolecular interactions. The latter, however, must be firmly criticized [44]. Since V_{BCP} is a negatively defined quantity [94], the energy of any hydrogen bond (or other interaction) determined by formula (32) will always give a negative E_{HB}^{EM} value. Therefore, according to EM, any interaction will be stabilizing. However, as shown [44], many C-H$\cdots$O contacts, which many would probably consider weak hydrogen bonds, are, in fact, destabilizing, i.e., repulsive in nature. Indeed, V_{BCP}, which is crucial in formula (32), was interpreted [115] as the pressure exerted by the system on the electrons in the closest vicinity of BCP. Therefore, it is easy to imagine a situation that the short distance H$\cdots$Y is merely forced, e.g., by the stiffness of the molecular skeleton or some steric interactions, which lead to a high value of V_{BCP} and, consequently, to a large value of E_{HB}^{EM}, suggesting strong hydrogen bond, though in reality the interaction may be locally repulsive in nature. For this reason, the use of EM in the cases of intramolecular interactions is not recommended [44].

It should be added that some concerns regarding EM have also been pointed out by Gatti et al. [118] and more recently by Nikolaienko et al. [119]. For example, the latter authors complained that the expression (32) was obtained while using data relating to crystallographic structures in which, as is known, the distances H$\cdots$Y are often much shorter due to lattice forces. Moreover, this expression was obtained for X-H$\cdots$O (X = C, N, O) hydrogen bonds only and its use for hydrogen bonds of other types is unreliable. They also refer to the example of hydrogen bonds of the H$\cdots$F type, for which the Espinosa formula (32) should rather have a factor of 0.31 [120]. To all of these allegations [119], it can be added that the hydrogen bond energies were obtained [115] with a real mixture of theoretical methods. Therefore, it seems necessary to revise the derivation of formula (32). In fact, some modifications to the orginal Espinosa's formula (Equation (32)) have been proposed [44,118,120,121]. For example, Afonin et al. [121] obtained the formula

$$E_{HB} = 0.277 V_{BCP} + 0.45 \tag{33}$$

in which the slope value of 0.277 is very close to the mentioned 0.31 value that was obtained by Mata et al. [120]. Even a little earlier, Jabłoński and Monaco proposed correcting the Espinosa's formula (32) by adding a constant k of 3.4 kcal/mol, thus to use the expression $E_{HB} = E_{HB}^{EM} + 3.4$ [44]. Importantly, this expression was specifically dedicated to intramolecular hydrogen bonds.

2.5.2. Interacting Quantum Atoms (IQA)

As already mentioned, the Interacting Quantum Atoms (IQA) approach [116,117], which is based on QTAIM, allows for the total energy of a system to be divided into mono- and polyatomic components. Among many energy terms available by means of IQA, the most important one in the context of this article is the interatomic interaction energy defined as follows

$$E_{\text{int}}^{E_1 E_2} = V_{\text{nn}}^{E_1 E_2} + V_{\text{en}}^{E_1 E_2} + V_{\text{ne}}^{E_1 E_2} + V_{\text{ee}}^{E_1 E_2} \qquad (E_1 \neq E_2) \tag{34}$$

where $V_{\text{nn}}^{E_1 E_2}$ is the repulsion energy between nuclei of atoms E_1 and E_2, $V_{\text{en}}^{E_1 E_2}$ is the attraction energy between electrons of the atom E_1 and the nucleus of the atom E_2, $V_{\text{ne}}^{E_1 E_2}$ is the attraction energy between the nucleus of the atom E_1 and the electrons of the atom E_2, and $V_{\text{ee}}^{E_1 E_2}$ is the interatomic two-electron repulsion energy. Because the E_1 and E_2 atoms may be e.g., the H and Y atoms from the X-H$\cdots$Y hydrogen bridge, it is evident that IQA via Equation (34) can be a suitable tool for calculating the energy of inter- and, more importantly, intramolecular hydrogen bonds. In this case, Equation (34) takes the following form

$$E_{\text{int}}^{\text{H}\cdots\text{Y}} = E_{\text{HB}}^{\text{IQA}} = V_{\text{nn}}^{\text{HY}} + V_{\text{en}}^{\text{HY}} + V_{\text{ne}}^{\text{HY}} + V_{\text{ee}}^{\text{HY}} \tag{35}$$

It should be emphasized that the determination of the interaction energy of H$\cdots$Y using formula (35) does not require assuming any reference system or referring to empirical data, and from this point of view the IQA-based approach is absolutely unique and, therefore, also worth a wider study of its applicability. It should also be added that E_1 and E_2 of Equation (34) can be any atoms, and therefore the interaction energy of any interatomic contact, not just hydrogen bonding, can be determined in a similar way. Moreover, these atoms do not need to be linked to each other by a bond path, nor do they need be in a close proximity to each other.

Unfortunately, as compared to intermolecular hydrogen bonds [122–126], the IQA-based estimates of the energy of intramolecular hydrogen bonds are relatively rare [127–130]. It is worth noting here that the list of IQA applications given most recently by Guevara-Vela et al. [117] can be easily supplemented with various repulsive interactions [105–108,112–114], which are often related to the presence of an appropriate bond path. Therefore, these cases are important in the very discussion on the interpretation of a bond path and the earlier connection of its presence on molecular graphs with the stabilizing nature of interactions, as stated in the orthodox QTAIM [94].

It should be emphasized that, as it seems, the interatomic interaction energy itself (i.e., $E_{\text{int}}^{E_1 E_2}$) is currently not as popular quantity as the exchange-correlation component ($E_{\text{ee,xc}}^{E_1 E_2}$) of the interelectron interaction energy ($E_{\text{ee}}^{E_1 E_2} = E_{\text{ee,C}}^{E_1 E_2} + E_{\text{ee,xc}}^{E_1 E_2}$). This, in turn, results from the fact that $E_{\text{ee,xc}}^{E_1 E_2}$ was associated with the presence of bond path via the concept of so-called privileged exchange channels [131]. Moreover, even more importantly, at short distances [113] $E_{\text{ee,xc}}^{E_1 E_2}$ is related to the strength of a given bond [123]. Therefore, it turns out that, despite the possibility of determining the interatomic interaction energy and thus also of an intramolecular hydrogen bond, which is significant for the theory of intramolecular interactions, this quantity in IQA has become less important than dimensionally much lower exchange energy.

2.6. Empirically-Based Methods

It is well known that, apart from the energy or ethalphy of hydrogen bond, i.e., the quantities that directly prove its stabilizing nature, there is a whole range of quantities that indicate or rather suggest the presence of a hydrogen bond indirectly [5]. The typical effects that are to prove the presence of (strong) standard (i.e., those where both X and Y atoms in the X-H$\cdots$Y contact are strongly electronegative) hydrogen bonds include an elongation of the proton-donor X-H bond, shifting the frequency of its stretching vibration towards lower values (i.e., the so-called red-shift) [132–135], intensification and broadening of the band associated with this vibration [136,137], and deshilding of the proton participating in the hydrogen bond [138–141] in the magnetic field [142,143] observed in the ^{1}H NMR spectra. While all of these effects can be relatively easily correlated with the hydrogen bond energy in the case of intermolecular hydrogen bonds, which, of course, is due to the relatively simple availability of the hydrogen bond energy (or enthalpy of formation) in such a case, transferring these correlations to the ground of intramolecular hydrogen bonds [144–149] is much more troublesome. Obviously, this, in turn, results from both the lack of an unambiguous definition of the interaction energy of intramolecular hydrogen bonds and the problematic determination of an unperturbed reference value. Moreover, the rationale for such transferability is unclear and certainly deserves to be a hotly debated topic. Nevertheless, some empirical expressions that were basically derived in order to determine the interaction energy (or the enthalpy of formation) in the case of intermolecular hydrogen bonds are also used from time to time in the estimation of the energy of intramolecular hydrogen bonds. In the following subsections, I will discuss the two most common approaches that are based on spectroscopic quantities, namely the red-shift of the X-H proton-donor stretching vibration frequency and the proton downfield shift in the ^{1}H NMR spectrum.

2.6.1. Iogansen's Relationship

Based on the results obtained for various phenol complexes, in 1969 Iogansen and Rassadin proposed an empirical formula for the relationship between intermolecular hydrogen bond energy (the hydrogen bond enthalpy of formation) and the red-shift of the X-H stretching vibration frequency ($\Delta\nu_{XH}$) that takes place upon the hydrogen bond formation [150,151]

$$E_{HB} = 0.33\sqrt{\Delta\nu_{XH} - 40} > 0 \tag{36}$$

The $\Delta\nu_{XH}$ red-shift can be obtained either from experimental measurements or theoretical computations. Unfortunately, although the Iogansen's relationship was derived for intermolecular hydrogen bonds [150,151], it is also used in order to estimate intramolecular hydrogen bond energies [119,152–154], where its applicability is at least unclear. One such use will be discussed in more detail here.

Using the formula (36), Nikolaienko et al. [119] estimated the energies of an impressively large number (However, it is highly doubtful that all of the more than 4000 conformers studied there actually correspond to true minima on the potential energy hypersurface and thus it is unclear what these conformers really mean.) of O-H$\cdots$O, O-H$\cdots$N, N-H$\cdots$O, and O-H$\cdots$C intramolecular hydrogen bonds in some biologically relevant DNA-related molecules. It is obvious that the use of this expression requires the knowledge of the reference vibration frequency ν_{XH}^{free}. Despite the fact that, in the case of an isolated molecule, obtaining such a quantity is, at least in terms of theoretical calculations, a fairly simple process; however, in the case of intramolecular interaction, it is controversial which frequency should be best taken as a reference [40,41]. It is enough to mention here a very common problem with coupling vibrations. Anyway, Nikolaienko et al. [119] stated that "ν_{XH}^{free} has been calculated as the simple average of stretching vibration frequencies for XH groups, such that: (i) their H atom does not participate in any XH$\cdots$Y bonding (i.e., no QTAIM bond path ends on it except for the one corresponding to the XH covalent bond), and (ii) unique normal vibration exists with $c_j^{XH} > c_{th}$.", where $c_j^{XH} = \partial l_{XH}/\partial x_j$ (x_j is the j-th normal coordinate) and c_{th} is the fixed treshold value (=0.92). Subsequently, based on the thus calculated energy values, these authors obtained, for each type of the hydrogen bond under consideration, a relationship with the determined value of the electron density at the bond critical point of a given hydrogen bond, $E_{HB} = A\rho_{BCP} + B$. The linear fit values for A and B thus obtained (the signs have been changed, so that the resulting E_{HB} values are negative) by Nikolaienko et al. [119] for each of the types of hydrogen bonds considered by them are shown in Table 5.

In principle, one could complain that the B values should be exactly zeros, as a zero electron density value should result in no hydrogen bonding and, therefore, also zero energy value. On the other hand, however, a significant portion of the electron density in the BCP is only due to the mutual overlapping of atomic orbitals of various atoms, and not only H and Y [107]. This fact was completely ignored here. Most likely, a correction is required to at least subtract the electron density contributions from H and Y, and perhaps even X. A somewhat similar correction has recently been proposed by Scheiner for calculating corrected NMR chemical shift for a proton involved in an intramolecular hydrogen bonding [141].

Table 5. Linear fit parameters for the linear relation $E_{HB} = A\rho_{BCP} + B$ between the hydrogen bond energy (in kcal/mol) and electron density at the BCP (au) of the indicated hydrogen bond [119].

Type	A	B	R	R^2
O-H$\cdots$O	-239 ± 2.2	3.09 ± 0.07	0.93	0.86
O-H$\cdots$N	-142 ± 2.1	-1.72 ± 0.08	0.97	0.94
N-H$\cdots$O	-225 ± 12	2.03 ± 0.25	0.85	0.72
O-H$\cdots$C	-288 ± 19	0.29 ± 0.22	0.86	0.74
together	-200 ± 2.2	1.70 ± 0.07	0.88	0.77

2.6.2. Chemical Shift—Based Method

As mentioned earlier, one of the characteristic effects accompanying the formation of a hydrogen bond is the ^{1}H NMR signal shift for the donor proton, i.e., the so-called downfield shift or proton magnetic deshilding, $\Delta\delta$ [138–141]. As early as in 1961, Gränacher [155] noticed a linear correlation between the proton chemical shift and the shift of the infrared absorption band, announcing the possibility of obtaining values of intermolecular hydrogen bond energies via the following equation [145]

$$E_{HB} = \Delta\delta + (0.4 \pm 0.2) > 0 \tag{37}$$

Quite recently, this expression was used by Afonin et al. [121] in order to estimate energies of many different intramolecular hydrogen bonds, including improper, blue-shifting [156,157] ones, the energy of which cannot be estimated while using Equation (36). Importantly, their aim was to compare the energies that were obtained in this way with their counterparts obtained using other popular methods of estimating the energy of hydrogen bonds. Moreover, this approach allowed them to obtain new correcting parameters in the formulas combining the hydrogen bond energy with the local potential energy density (V_{BCP}) and the electron density (ρ_{BCP}) at the BCP of these interactions and determine the quality (However, the combination of experimental rather than theoretical ^{1}H NMR data with theoretically determined QTAIM parameters for theoretically obtained geometries of the molecules under consideration is somewhat suspicious.) of the methods that are based on these parameters [121]. In this way, Afonin et al. showed that the estimates based on geometric parameters [158–160] are very poor (therefore, they are not discussed in this review). Nevertheless, even less reliable (2–3 times overestimation) values were obtained while using the uncorrected Espinosa's formula, whereas applying a multiplier of 0.31 significantly improved the results. Based on the values of the hydrogen bond energies that were obtained by Equation (37) and the calculated values of either V_{BCP} or ρ_{BCP}, Afonin et al. [121] obtained Equation (33) for the former parameter and $E_{HB} = -191.4\,\rho_{BCP} + 1.78$ for the latter one, where the coefficients A and B are noticeably close to those that were obtained earlier by Nikolaienko et al. [119] (see last column in Table 5).

3. Summary

This article is a comprehensive critical overview of the currently most commonly used theoretical methods for estimating the energy of intramolecular hydrogen bonds and other intramolecular interactions. All of these methods have been grouped, as follows: conformational methods (The Open-Closed Method, Ortho-Para Method, Related Rotamers Method, Geometry-Corrected Method, Geometry-Corrected Related Rotamers Method), Rotation Barriers Method, Dimer Method, Isodesmic Reactions Method, QTAIM-based methods (Espinosa's Method and IQA-based method) and empirically-based methods (Iogansen's relationship and chemical shift - based method). The main emphasis is placed on two issues, namely the theoretical rationale of a given method and the fact that within the adopted method its diverse variants are often possible. Quite often, the methods themselves and their variants may lead to a wide range of the estimates being obtained by them.

The user should be aware that the applicability of a particular method is quite often very limited. This is especially seen in the case of conformational methods. Urgent attention should be paid that the reference form or forms do not have significant new stabilizing (attractive) or destabilizing (repulsive) interactions and that the structure of such forms is as close as possible to the intramolecularly hydrogen-bonded form. Because of the fact that ensuring such conditions is not always a simple task, the emerging methods of a non-invasive estimation of the intramolecular hydrogen bond interaction energy, i.e., not requiring any open form with the intramolecular hydrogen bond (or any other interaction) of interest being broken, are particularly important.

The performed energy estimates should be completed with an evaluation of their credibility. This can be done by either showing quite good correlations with various types of parameters used for indirect assessment of the bond strength or by an in-depth comparison of the obtained estimates for structurally similar systems. It is also recommended to use several estimation methods simultaneously. It should also be remembered that most of the problems that are encountered result from the fact that just trying to estimate the energy of intramolecular hydrogen bonding is also an attempt to introduce an indefinable quantity.

Funding: This research received no external funding.

Conflicts of Interest: The author declares no conflict of interest.

Abbreviations

The following abbreviations are used in this manuscript:

OCM	Open-Closed Method
opM	*Ortho-Para* Method
RRM	Related Rotamers Method
GCM	Geometry-Corrected Method
GCRRM	Geometry-Corrected Related Rotamers Method
RBM	Rotation Barriers Method
DM	Dimer Model
IRM	Isodesmic Reactions Method
EM	Espinosa's Method
QTAIM	Quantum Theory of Atoms in Molecules
IQA	Interacting Quantum Atoms
BP	bond path
BCP	bond critical point

References

1. Pauling, L. *The Nature of the Chemical Bond*; Cornell University Press: New York, NY, USA, 1960.
2. Pimentel, G.C.; McClellan, A.L. *The Hydrogen Bond*; W.H. Freeman & Co.: San Francisco, CA, USA, 1960.
3. Hamilton, W.C.; Ibers, J.A. *Hydrogen Bonding in Solids*; W. A. Benjamin: New York, NY, USA, 1968.
4. Vinogradov S.N.; Linnell, R.H. *Hydrogen Bonding*; Van Nostrand-Reinhold: Princeton, NJ, USA, 1971.
5. Schuster, P.; Zundel, G.; Sandorfy, C. (Eds.) *The Hydrogen Bond. Recent Developments in Theory and Experiments*; North Holland: Amsterdam, The Netherlands, 1976; Volumes I–III.
6. Schuster, P. *Intermolecular Interactions: From Diatomics to Biopolymers*; Pullman, B., Ed.; Wiley-Interscience: New York, NY, USA, 1978.
7. Hobza, P.; Zahradník, R. *Weak Intermolecular Interactions in Chemistry and Biology*; Academia: Prague, Czech Republic, 1980.
8. Jeffrey, G.A.; Saenger, W. *Hydrogen Bonding in Biological Structures*; Springer: Berlin, Germany, 1991.
9. Hadži, D. (Ed.) *Theoretical Treatments of Hydrogen Bonding*; Wiley: Chichester, UK, 1997.
10. Jeffrey, G.A. *An Introduction to Hydrogen Bonding*; Oxford University Press: New York, NY, USA, 1997.
11. Scheiner, S. (Ed.) *Molecular Interactions. From van der Waals to Strongly Bound Complexes*; Wiley: Chichester, UK, 1997.
12. Scheiner, S. *Hydrogen Bonding: A Theoretical Perspective*; Oxford University Press: New York, NY, USA, 1997.

13. Desiraju, G.R.; Steiner, T. *The Weak Hydrogen Bond in Structural Chemistry and Biology*; Oxford University Press: New York, NY, USA, 1999.

14. Grabowski, S.J. *Hydrogen Bonding—New Insights. Challenges and Advances in Computational Chemistry and Physics*; Springer: Dordrecht, The Netherlands, 2006.

15. Maréchal, Y. *The Hydrogen Bond and the Water Molecule*; Elsevier: Amsterdam, The Netherlands, 2007.

16. Gilli, G.; Gilli, P. *The Nature of the Hydrogen Bond. Outline of a Comprehensive Hydrogen Bond Theory*; Oxford University Press: Oxford, UK, 2009.

17. Frenking, G.; Shaik, S. (Eds.) *The Chemical Bond*; Wiley-VCH Verlag GmbH & Co.: Weinheim, Germany, 2014.

18. Piela, L. *Ideas of Quantum Chemistry*; Elsevier: Amsterdam, The Netherlands, 2020.

19. Schuster, P. LCAO-MO-Beschreibung intramolekularer Wasserstoffbrücken. *Monatshefte Chem.* **1969**, *100*, 2084–2095. [CrossRef]

20. Dietrich, S.W.; Jorgensen, E.C.; Kollman, P.A.; Rothenberg, S. A Theoretical Study of Intramolecular Hydrogen Bonding in Ortho-Substituted Phenols and Thiophenols. *J. Am. Chem. Soc.* **1976**, *98*, 8310–8324. [CrossRef]

21. Emsley, J. The Composition, Structure and Hydrogen Bonding of the β-Diketones. *Struct. Bond.* **1984**, *57*, 147–191.

22. Buemi, G.; Gandolfo, C. Malondialdehyde and Acetylacetone: An AM1 Study of their Molecular Structures and Keto-Enol Tautomerism. *J. Chem. Soc. Faraday Trans.* **1989**, *85*, 215–227. [CrossRef]

23. Millefiori, S.; Di Bella, S. Hydrogen Bonding and Tautomerism in 3-Substituted β-Thioxoketones: An Ab Initio Molecular Orbital Study. *J. Chem. Soc. Faraday Trans.* **1991**, *87*, 1297–1302. [CrossRef]

24. Craw, J.S.; Bacskay, G.B. Quantum-chemical Studies of Hydrogen Bonding involving Thioxoketones, Thienols, Thioformaldehyde and Hydrogen Sulfide with Specific Reference to the Strength of Intramolecular Hydrogen Bonds. *J. Chem. Soc. Faraday Trans.* **1992**, *88*, 2315–2321. [CrossRef]

25. Luth, K.; Scheiner, S. Excited-State Energetics and Proton-Transfer Barriers in Malonaldehyde. *J. Phys. Chem.* **1994**, *98*, 3582–3587. [CrossRef]

26. Lampert, H.; Mikenda, W.; Karpfen, A. Intramolecular Hydrogen Bonding in 2-Hydroxybenzoyl Compounds: Infrared Spectra and Quantum Chemical Calculations. *J. Phys. Chem.* **1996**, *100*, 7418–7425. [CrossRef]

27. Scheiner, S.; Kar, T.; Čuma, M. Excited State Intramolecular Proton Transfer in Anionic Analogues of Malonaldehyde. *J. Phys. Chem. A* **1997**, *101*, 5901–5909. [CrossRef]

28. Chung, G.; Kwon, O.; Kwon, Y. Theoretical Study on 1,2-Dihydroxybenzene and 2-Hydroxythiophenol: Intramolecular Hydrogen Bonding. *J. Phys. Chem. A* **1997**, *101*, 9415–9420. [CrossRef]

29. Cuma, M.; Scheiner, S.; Kar, T. Effect of adjoining aromatic ring upon excited state proton transfer, *o*-hydroxybenzaldehyde. *J. Mol. Struct.* **1999**, *467*, 37–49. [CrossRef]

30. Forés, M.; Scheiner, S. Effects of chemical substitution upon excited state proton transfer. Fluoroderivatives of salicylaldimine. *Chem. Phys.* **1999**, *246*, 65–74. [CrossRef]

31. Koll, A.; Rospenk, M.; Jagodzińska, E.; Dziembowska, T. Dipole moments and conformation of Schiff bases with intramolecular hydrogen bonds. *J. Mol. Struct.* **2000**, *552*, 193–204. [CrossRef]

32. Rozas, I.; Alkorta, I.; Elguero, J. Intramolecular Hydrogen Bonds in *ortho*-Substituted Hydroxybenzenes and in 8-Susbtituted 1-Hydroxynaphthalenes: Can a Methyl Group Be an Acceptor of Hydrogen Bonds? *J. Phys. Chem. A* **2001**, *105*, 10462–10467. [CrossRef]

33. Buemi, G.; Zuccarello, F. Is the intramolecular hydrogen bond energy valuable from internal rotation barriers? *J. Mol. Struct.* **2002**, *581*, 71–85. [CrossRef]

34. Kovács, A.; Szabó, A.; Hargittai, I. Structural Characteristics of Intramolecular Hydrogen Bonding in Benzene Derivatives. *Acc. Chem. Res.* **2002**, *35*, 887–894. [CrossRef]

35. Korth, H.-G.; de Heer, M.I.; Mulder, P. A DFT Study on Intramolecular Hydrogen Bonding in 2-Substituted Phenols: Conformations, Enthalpies, and Correlation with Solute Parameters. *J. Phys. Chem. A* **2002**, *106*, 8779–8789. [CrossRef]

36. Grabowski, S.J. π-Electron delocalisation for intramolecular resonance assisted hydrogen bonds. *J. Phys. Org. Chem.* **2003**, *16*, 797–802. [CrossRef]

37. Grabowski, S.J. Hydrogen bonding strength–measures based on geometric and topological parameters. *J. Phys. Org. Chem.* **2004**, *17*, 18–31. [CrossRef]

38. Estácio, S.G.; do Couto, P.C.; Costa Cabral, B.J.; Minas da Piedade, M.E.; Martinho Simões, J.A. Energetics of Intramolecular Hydrogen Bonding in Di-substituted Benzenes by the ortho–para Method. *J. Phys. Chem. A* **2004**, *108*, 10834–10843. [CrossRef]

39. Jabłoński, M.; Kaczmarek, A.; Sadlej, A.J. Estimates of the Energy of Intramolecular Hydrogen Bonds. *J. Phys. Chem. A* **2006**, *110*, 10890–10898. [CrossRef] [PubMed]

40. Jabłoński, M.; Sadlej, A.J. Blue-Shifting Intramolecular C–H$\cdots$O Interactions. *J. Phys. Chem. A* **2007**, *111*, 3423–3431. [CrossRef] [PubMed]

41. Jabłoński, M. Blue-shifting intramolecular C–H$\cdots$O(S) contacts in sterically strained systems. *J. Mol. Struct.* **2007**, *820*, 118–127.

42. Jabłoński, M. Full vs. constrain geometry optimization in the open–closed method in estimating the energy of intramolecular charge-inverted hydrogen bonds. *Chem. Phys.* **2010**, *376*, 76–83. [CrossRef]

43. Jabłoński, M. Energetic and Geometrical Evidence of Nonbonding Character of Some Intramolecular Halogen$\cdots$Oxygen and Other Y$\cdots$Y Interactions. *J. Phys. Chem. A* **2012**, *116*, 3753–3764. [CrossRef]

44. Jabłoński, M.; Monaco, G. Different Zeroes of Interaction Energies As the Cause of Opposite Results on the Stabilizing Nature of C–H$\cdots$O Intramolecular Interactions. *J. Chem. Inf. Model.* **2013**, *53*, 1661–1675. [CrossRef]

45. Jabłoński, M. Strength of Si–H$\cdots$B charge-inverted hydrogen bonds in 1-silacyclopent-2-enes and 1-silacyclohex-2-enes. *Struct. Chem.* **2017**, *28*, 1697–1706. [CrossRef]

46. Honacker, C.; Kappelt, B.; Jabłoński, M.; Hepp, A.; Layh, M.; Rogel, F.; Uhl, W. Aluminium Functionalized Germanes: Intramolecular Activation of Ge-H Bonds, Formation of a Dihydrogen Bond and Facile Hydrogermylation of Unsaturated Substrates. *Eur. J. Inorg. Chem.* **2019**, 3287–3300. [CrossRef]

47. Jabłoński, M. Enaminoketony: Teoretyczne Badania Dynamiki Wewnątrzmolekularnej w Układzie Modelowym 3-Aminoakroleiny. Master's Thesis, Nicolaus Copernicus University, Toruń, Poland, 2000.

48. Buemi, G. AM1 Study of Tautomerism and Intramolecular Hydrogen Bonding in Thiomalondialdehyde and Thioacetylacetone. *J. Chem. Soc. Faraday Trans.* **1990**, *86*, 2813–2818. [CrossRef]

49. Buemi, G. Intramolecular Hydrogen Bonding of Malondialdehyde and its Monothio and Dithio Analogues Studied by the PM3 Method. *J. Mol. Struct.* **1990**, *208*, 253–260. [CrossRef]

50. Gonzales, L.; Mó, O.; Yáñez, M. High-Level ab Initio Calculations on the Intramolecular Hydrogen Bond in Thiomalonaldehyde. *J. Phys. Chem. A* **1997**, *101*, 9710–9719. [CrossRef]

51. Jabłoński, M. Binding of X-H to the lone-pair vacancy: Charge-inverted hydrogen bond. *Chem. Phys. Lett.* **2009**, *477*, 374–376. [CrossRef]

52. Jabłoński, M. Ten years of charge-inverted hydrogen bonds. *Struct. Chem.* **2020**, *31*, 61–80. [CrossRef]

53. Wrackmeyer, B.; Tok, O.L.; Milius, W.; Khan, A.; Badshah, A. 1-Silacyclopent-2-enes and 1-silacyclohex-2-enes bearing functionally substituted silyl groups in 2-positions. Novel electron-deficient Si–H–B bridges. *Appl. Organomet. Chem.* **2006**, *20*, 99–105. [CrossRef]

54. Pinto, S.S.; Diogo, H.P.; Guedes, R.C.; Costa Cabral, B.J.; Minas da Piedade, M.E.; Martinho Simões, J.A. Energetics of Hydroxybenzoic Acids and of the Corresponding Carboxyphenoxyl Radicals. Intramolecular Hydrogen Bonding in 2-Hydroxybenzoic Acid. *J. Phys. Chem. A* **2005**, *109*, 9700–9708. [CrossRef]

55. Roy, D.; Sunoj, R.B. Quantification of Intramolecular Nonbonding Interactions in Organochalcogens. *J. Phys. Chem. A* **2006**, *110*, 5942–5947. [CrossRef]

56. Hammett, L.P. The Effect of Structure upon the Reactions of Organic Compounds. Benzene Derivatives. *J. Am. Chem. Soc.* **1937**, *59*, 96–103. [CrossRef]

57. Hammett, L.P. *Physical Organic Chemistry*; McGraw-Hill: New York, NY, USA, 1940.

58. Hansch, C.; Leo, A.; Taft, R.W. A Survey of Hammett Substituent Constants and Resonance and Field Parameters. *Chem. Rev.* **1991**, *91*, 165–195. [CrossRef]

59. Krygowski, T.M.; Stępień, B.T. Sigma- and Pi-Electron Delocalization: Focus on Substituent Effects. *Chem. Rev.* **2005**, *105*, 3482–3512. [CrossRef]

60. Lipkowski, P.; Koll, A.; Karpfen, A.; Wolschann, P. An approach to estimate the energy of the intramolecular hydrogen bond. *Chem. Phys. Lett.* **2002**, *360*, 256–263. [CrossRef]

61. Nowroozi, A.; Raissi, H.; Farzad, F. The presentation of an approach for estimating the intramolecular hydrogen bond strength in conformational study of β-Aminoacrolein. *J. Mol. Struct.* **2005**, *730*, 161–169. [CrossRef]

62. Nowroozi, A.; Raissi, H.; Hajiabadi, H.; Mohammadzadeh Jahani, P. Reinvestigation of Intramolecular Hydrogen Bond in Malonaldehyde Derivatives: An Ab initio, AIM and NBO Study. *Int. J. Quant. Chem.* **2011**, *111*, 3040–3047. [CrossRef]

63. Nowroozi, A.; Masumian, E. Comparative study of NH⋯O and NH⋯S intramolecular hydrogen bonds in β-aminoacrolein, β-thioaminoacrolein and their halogenated derivatives by some usual methods. *Struct. Chem.* **2017**, *28*, 587–596. [CrossRef]

64. Tayyari, S.F.; Fazli, M.; Milani-nejad, F. Molecular conformation and intramolecular hydrogen bonding in 4-amino-3-penten-2-one. *J. Mol. Struct.* **2001**, *541*, 11–15. [CrossRef]

65. Buemi, G.; Zuccarello, F.; Venuvanalingam, P.; Ramalingam, M. Ab initio study of tautomerism and hydrogen bonding of β-carbonylamine in the gas phase and in water solution. *Theor. Chem. Acc.* **2000**, *104*, 226–234. [CrossRef]

66. Buemi, G.; Zuccarello, F. Molecular conformations and intramolecular hydrogen bonding of 3-formylmalondialdehyde and 3-formylacetylacetone. An ab initio study. *Electr. J. Theor. Chem.* **1997**, *2*, 118–129. [CrossRef]

67. Buemi, G.; Zuccarello, F.; Venuvanalingam, P.; Ramalingam, M.; Ammal, S.S.C. *Ab initio* study of formazan and 3-nitroformazan. *J. Chem. Soc. Faraday Trans.* **1998**, *94*, 3313–3319. [CrossRef]

68. Buemi, G. Ab initio DFT study of the hydrogen bridges in hexafluoro-acetylacetone, trifluoro-acetylacetone and some 3-substituted derivatives. *J. Mol. Struct.* **2000**, *499*, 21–34. [CrossRef]

69. Buemi, G. Ab initio study of 2,4-dihalosubstituted malonaldehyde and 2-halo-phenols in gas phase and solution. *Chem. Phys.* **2002**, *277*, 241–256. [CrossRef]

70. Buemi, G. Ab initio DFT study of the rotation barriers and competitive hydrogen bond energies (in gas phase and water solution) of 2-nitroresorcinol, 4,6-dinitroresorcinol and 2-nitrophenol in their neutral and deprotonated conformations. *Chem. Phys.* **2002**, *282*, 181–195. [CrossRef]

71. Buemi, G.; Zuccarello, F. DFT study of the intramolecular hydrogen bonds in the amino and nitro-derivatives of malonaldehyde. *Chem. Phys.* **2004**, *306*, 115–129. [CrossRef]

72. Ramalingam, M.; Venuvanalingam, P.; Swaminathan, J.; Buemi, G. Ab initio and DFT studies on conformations, hydrogen bonding and electronic structures of glyoxalmonoxime and its methyl derivatives. *J. Mol. Struct.* **2004**, *712*, 175–185. [CrossRef]

73. Palusiak, M.; Krygowski, T.M. Unusual electron density topology and intramolecular steric π–π interaction in 1,3,5,7-cyclooctatetraene. *Chem. Phys. Lett.* **2009**, *481*, 34–38. [CrossRef]

74. Jabłoński, M.; Palusiak, M. The halogen⋯oxygen interaction in 3-halogenopropenal revisited—The *dimer model* vs. QTAIM indications. *Chem. Phys.* **2013**, *415*, 207–213. [CrossRef]

75. Palusiak, M.; Grabowski, S.J. Do intramolecular halogen bonds exist? Ab initio calculations and crystal structures' evidences. *Struct. Chem.* **2007**, *18*, 859–865. [CrossRef]

76. Hehre, W.J.; Ditchfield, R.; Radom, L.; Pople, J.A. Molecular Orbital Theory of the Electronic Structure of Organic Compounds. V. Molecular Theory of Bond Separation. *J. Am. Chem. Soc.* **1970**, *92*, 4796–4801. [CrossRef]

77. Radom, L.; Hehre, W.J.; Pople, J.A. Molecular Orbital Theory of the Electronic Structure of Organic Compounds. VII. A Systematic Study of Energies, Conformations, and Bond Interactions. *J. Am. Chem. Soc.* **1971**, *93*, 289–300. [CrossRef]

78. Radom, L. *Ab Initio* Molecular Orbital Calculations on Anions. Determination of Gas Phase Acidities. *J. Chem. Soc. Chem. Commun.* **1974**, 403–404. [CrossRef]

79. George, P.; Trachtman, M.; Bock, C.W.; Brett, A.M. An Alternative Approach to the Problem of Assessing Stabilization Energies in Cyclic Conjugated Hydrocarbons. *Theor. Chim. Acta* **1975**, *38*, 121–129. [CrossRef]

80. George, P.; Trachtman, M.; Bock, C.W.; Brett, A.M. Homodesmotic Reactions for the Assessment of Stabilization Energies in Benzenoid and Other Conjugated Cyclic Hydrocarbons. *J. Chem. Soc. Perkin Trans.* **1976**, *2*, 1222–1227. [CrossRef]

81. George, P.; Trachtman, M.; Bock, C.W.; Brett, A.M. An Alternative Approach to the Problem of Assessing Destabilization Energies (Strain Energies) in Cyclic Hydrocarbons. *Tetrahedron* **1976**, *32*, 317–323. [CrossRef]

82. George, P.; Trachtman, M.; Brett, A.M.; Bock, C.W. Comparison of Various Isodesmic and Homodesmotic Reaction Heats with Values derived from Published *Ab initio* Molecular Orbital Calculations. *J. Chem. Soc. Perkin Trans.* **1977**, *2*, 1036–1047. [CrossRef]

83. Hehre, W.J.; Radom, L.; von Ragué Schleyer, P.; Pople, J.A. *Ab Initio Molecular Orbital Theory*; Wiley: New York, NY, USA, 1986.

84. Varnali, T.; Hargittai, I. Geometrical consequences of resonance-assisted hydrogen bonding in 2-nitrovinyl alcohol and indication of a slight attractive $O\cdots H$ interaction in 2-nitroethanol. An ab initio molecular orbital investigation. *J. Mol. Struct.* **1996**, *388*, 315–319.

85. Wiberg, K.B.; Ochterski, J.W. Comparison of Different *Ab Initio* Theoretical Models for Calculating Isodesmic Reaction Energies for Small Ring and Related Compounds. *J. Comput. Chem.* **1997**, *18*, 108–114. [CrossRef]

86. Howard, S.T. Relationship between Basicity, Strain, and Intramolecular Hydrogen-Bond Energy in Proton Sponges. *J. Am. Chem. Soc.* **2000**, *122*, 8238–8244. [CrossRef]

87. Sanz, P.; Mó, O.; Yáñez, M. Characterization of intramolecular hydrogen bonds and competitive chalcogen–chalcogen interactions on the basis of the topology of the charge density. *Phys. Chem. Chem. Phys.* **2003**, *5*, 2942–2947. [CrossRef]

88. Cyrański, M.K. Energetic Aspects of Cyclic Pi-Electron Delocalization: Evaluation of the Methods of Estimating Aromatic Stabilization Energies. *Chem. Rev.* **2005**, *105*, 3773–3811. [CrossRef]

89. Raab, V.; Gauchenova, E.; Merkoulov, A.; Harms, K.; Sundermeyer, J.; Kovačević, B.; Maksić, Z.B. 1,8-Bis(hexamethyltriaminophosphazenyl)naphthalene, HMPN: A Superbasic Bisphosphazene "Proton Sponge". *J. Am. Chem. Soc.* **2005**, *127*, 15738–15743. [CrossRef]

90. Emel'yanenko, V.N.; Strutynska, A.; Verevkin, S.P. Enthalpies of Formation and Strain of Chlorobenzoic Acids from Thermochemical Measurements and from ab Initio Calculations. *J. Phys. Chem. A* **2005**, *109*, 4375–4380. [CrossRef]

91. Deshmukh, M.M.; Suresh, C.H.; Gadre, S.R. Intramolecular Hydrogen Bond Energy in Polyhydroxy Systems: A Critical Comparison of Molecular Tailoring and Isodesmic Approaches. *J. Phys. Chem. A* **2007**, *111*, 6472–6480. [CrossRef]

92. Wheeler, S.E.; Houk, K.N.; Schleyer, P.; Allen, W.D. A Hierarchy of Homodesmotic Reactions for Thermochemistry. *J. Am. Chem. Soc.* **2009**, *131*, 2547–2560. [CrossRef] [PubMed]

93. Łukomska, M.; Rybarczyk-Pirek, A.J.; Jabłoński, M.; Palusiak, M. The nature of NO-bonding in *N*-oxide group. *Phys. Chem. Chem. Phys.* **2015**, *17*, 16375–16387. [CrossRef] [PubMed]

94. Bader, R.F.W. *Atoms in Molecules: A Quantum Theory*; Oxford University Press: New York, NY, USA, 1990.

95. Popelier, P.L.A. *Atoms in Molecules. An Introduction*; Longman: Singapore, 2000.

96. Matta, C.F.; Boyd, R.J. *The Quantum Theory of Atoms in Molecules*; Wiley-VCH: Weinheim, Germany, 2007.

97. Bader, R.F.W. A Quantum Theory of Molecular Structure and Its Appllcatlons. *Chem. Rev.* **1991**, *91*, 893–928. [CrossRef]

98. Bader, R.F.W. A Bond Path: A Universal Indicator of Bonded Interactions. *J. Phys. Chem. A* **1998**, *102*, 7314–7323. [CrossRef]

99. Cioslowski, J.; Mixon, S.T.; Edwards, W.D. Weak Bonds in the Topological Theory of Atoms in Molecules. *J. Am. Chem. Soc.* **1991**, *113*, 1083–1085. [CrossRef]

100. Cioslowski, J.; Mixon, S.T. Topological Properties of Electron Density in Search of Steric Interactions in Molecules: Electronic Structure Calculations on Ortho-Substituted Biphenyls. *J. Am. Chem. Soc.* **1992**, *114*, 4382–4387. [CrossRef]

101. Haaland, A.; Shorokhov, D.J.; Tverdova, N.V. Topological Analysis of Electron Densities: Is the Presence of an Atomic Interaction Line in an Equilibrium Geometry a Sufficient Condition for the Existence of a Chemical Bond? *Chem. Eur. J.* **2004**, *10*, 4416–4421. [CrossRef]

102. Cerpa, E.; Krapp, A.; Vela, A.; Merino, G. The Implications of Symmetry of the External Potential on Bond Paths. *Chem. Eur. J.* **2008**, *14*, 10232–10234. [CrossRef]

103. Poater, J.; Solà, M.; Bickelhaupt, F.M. Hydrogen–Hydrogen Bonding in Planar Biphenyl, Predicted by Atoms-in-Molecules Theory, Does Not Exist. *Chem. Eur. J.* **2006**, *12*, 2889–2895. [CrossRef]

104. Poater, J.; Solà, M.; Bickelhaupt, F.M. A Model of the Chemical Bond Must Be Rooted in Quantum Mechanics, Provide Insight, and Possess Predictive Power. *Chem. Eur. J.* **2006**, *12*, 2902–2905. [CrossRef]

105. Dem'yanov, P.; Polestshuk, P. A Bond Path and an Attractive Ehrenfest Force Do Not Necessarily Indicate Bonding Interactions: Case Study on M_2X_2 (M = Li, Na, K; X = H, OH, F, Cl). *Chem. Eur. J.* **2012**, *18*, 4982–4993. [CrossRef] [PubMed]

106. Demy'anov, P.I.; Polestshuk, P.M. Forced Bonding and QTAIM Deficiencies: A Case Study of the Nature of Interactions in He@Adamantane and the Origin of the High Metastability. *Chem. Eur. J.* **2013**, *19*, 10945–10957. [CrossRef] [PubMed]

107. Tognetti, V.; Joubert, L. On the physical role of exchange in the formation of an intramolecular bond path between two electronegative atoms. *J. Chem. Phys.* **2013**, *138*, 024102. [CrossRef] [PubMed]

108. Tognetti, V.; Joubert, L. On critical points and exchange-related properties of intramolecular bonds between two electronegative atoms. *Chem. Phys. Lett.* **2013**, *579*, 122–126. [CrossRef]

109. Shahbazian, S. Why Bond Critical Points Are Not "Bond" Critical Points. *Chem. Eur. J.* **2018**, *24*, 5401–5405. [CrossRef]

110. Wick, C.R.; Clark, T. On bond-critical points in QTAIM and weak interactions. *J. Mol. Model.* **2018**, *24*, 142. [CrossRef]

111. Jabłoński, M. Hydride-Triel Bonds. *J. Comput. Chem.* **2018**, *39*, 1177–1191. [CrossRef]

112. Jabłoński, M. Bond Paths Between Distant Atoms Do Not Necessarily Indicate Dominant Interactions. *J. Comput. Chem.* **2018**, *39*, 2183–2195. [CrossRef]

113. Jabłoński, M. On the Uselessness of Bond Paths Linking Distant Atoms and on the Violation of the Concept of Privileged Exchange Channels. *ChemistryOpen* **2019**, *8*, 497–507. [CrossRef]

114. Jabłoński, M. Counterintuitive bond paths: An intriguing case of the $C(NO_2)_3^-$ ion. *Chem. Phys. Lett.* **2020**, *759*, 137946. [CrossRef]

115. Espinosa, E.; Molins, E.; Lecomte, C. Hydrogen bond strength revealed by topological analyses of experimentally observed electron densities. *Chem. Phys. Lett.* **1998**, *285*, 170–173. [CrossRef]

116. Blanco, M.A.; Pendás, A.M.; Francisco, E. Interacting Quantum Atoms: A Correlated Energy Decomposition Scheme Based on the Quantum Theory of Atoms in Molecules. *J. Chem. Theory Comput.* **2005**, *1*, 1096–1109. [CrossRef]

117. Guevara-Vela, J.M.; Francisco, F.; Rocha-Rinza, T.; Pendás, A.M. Interacting Quantum Atoms—Review. *Molecules* **2020**, *25*, 4028. [CrossRef] [PubMed]

118. Gatti, C.; May, E.; Destro, R.; Cargnoni, F. Fundamental Properties and Nature of CH··O Interactions in Crystals on the Basis of Experimental and Theoretical Charge Densities. The Case of 3,4-Bis(dimethylamino)-3-cyclobutene-1,2-dione (DMACB) Crystal. *J. Phys. Chem. A* **2002**, *106*, 2707–2720. [CrossRef]

119. Nikolaienko, T.Y.; Bulavin, L.A.; Hovorun, D.M. Bridging QTAIM with vibrational spectroscopy: the energy of intramolecular hydrogen bonds in DNA-related biomolecules. *Phys. Chem. Chem. Phys.* **2012**, *14*, 7441–7447. [CrossRef]

120. Mata, I.; Alkorta, I.; Espinosa, E.; Molins, E. Relationships between interaction energy, intermolecular distance and electron density properties in hydrogen bonded complexes under external electric fields. *Chem. Phys. Lett.* **2011**, *507*, 185–189. [CrossRef]

121. Afonin, A.V.; Vashchenko, A.V.; Sigalov, M.V. Estimating the energy of intramolecular hydrogen bonds from ^{1}H NMR and QTAIM calculations. *Org. Biomol. Chem.* **2016**, *14*, 11199–11211. [CrossRef] [PubMed]

122. Pendás, A.M.; Blanco, M.A.; Francisco, E. The nature of the hydrogen bond: A synthesis from the interacting quantum atoms picture. *J. Chem. Phys.* **2006**, *125*, 184112. [CrossRef]

123. García-Revilla, M.; Francisco, E.; Popelier, P.L.A.; Pendás, A.M. Domain-Averaged Exchange-Correlation Energies as a Physical Underpinning for Chemical Graphs. *ChemPhysChem* **2013**, *14*, 1211–1218. [CrossRef]

124. Guevara-Vela, J.M.; Chávez-Calvillo, R.; García-Revilla, M.; Hernández-Trujillo, J.; Christiansen, O.; Francisco, E.; Pendás, A.M.; Rocha-Rinza, T. Hydrogen-Bond Cooperative Effects in Small Cyclic Water Clusters as Revealed by the Interacting Quantum Atoms Approach. *Chem. Eur. J.* **2013**, *19*, 14304–14315. [CrossRef]

125. Guevara-Vela, J.M.; Romero-Montalvo, E.; Gómez, V.A.M.; Chávez-Calvillo, R.; García-Revilla, M.; Francisco, E.; Pendás, A.M.; Rocha-Rinza, T. Hydrogen bond cooperativity and anticooperativity within the water hexamer. *Phys. Chem. Chem. Phys.* **2016**, *18*, 19557–19566. [CrossRef] [PubMed]

126. Alkorta, I.; Mata, I.; Molins, E.; Espinosa, E. Charged versus Neutral Hydrogen-Bonded Complexes: Is There a Difference in the Nature of the Hydrogen Bonds? *Chem. Eur. J.* **2016**, *22*, 9226–9234. [CrossRef] [PubMed]

127. Guevara-Vela, J.M.; Romero-Montalvo, E.; Costales, A.; Pendás, A.M.; Rocha-Rinza, T. The nature of resonance-assisted hydrogen bonds: A quantum chemical topology perspective. *Phys. Chem. Chem. Phys.* **2016**, *18*, 26383–26390. [CrossRef] [PubMed]

128. Guevara-Vela, J.M.; Romero-Montalvo, E.; del Río Lima, A.; Pendás, A.M.; Hernández-Rodríguez, M.; Rocha Rinza, T. Hydrogen-Bond Weakening through π Systems: Resonance-Impaired Hydrogen Bonds (RIHB). *Chem. Eur. J.* **2017**, *23*, 16605–16611. [CrossRef] [PubMed]

129. Romero-Montalvo, E.; Guevara-Vela, J.M.; Costales, A.; Pendás, A.M.; Rocha-Rinza, T. Cooperative and anticooperative effects in resonance assisted hydrogen bonds in merged structures of malondialdehyde. *Phys. Chem. Chem. Phys.* **2017**, *19*, 97–107. [CrossRef] [PubMed]

130. Ebrahimi, S.; Dabbagh, H.A.; Eskandari, K. Nature of intramolecular interactions of vitamin C in view of interacting quantum atoms: The role of hydrogen bond cooperativity on geometry. *Phys. Chem. Chem. Phys.* **2016**, *18*, 18278–18288. [CrossRef]

131. Pendás, A.M.; Francisco, E.; Blanco, M.A.; Gatti, C. Bond Paths as Privileged Exchange Channels. *Chem. Eur. J.* **2007**, *13*, 9362–9371. [CrossRef]

132. Badger, R.M.; Bauer, S.H. Spectroscopic Studies of the Hydrogen Bond. II. The Shift of the O–H Vibrational Frequency in the Formation of the Hydrogen Bond. *J. Chem. Soc.* **1937**, *5*, 839–851. [CrossRef]

133. Joesten, M.D.; Drago, R.S. The Validity of Frequency Shift-Enthalpy Correlations. I. Adducts of Phenol with Nitrogen and Oxygen Donors. *J. Am. Chem. Soc.* **1962**, *84*, 3817–3821. [CrossRef]

134. Epley, T.D.; Drago, R.S. Calorimetric Studies on Some Hydrogen-Bonded Adducts. *J. Am. Chem. Soc.* **1967**, *89*, 5770–5773. [CrossRef]

135. Odinokov, S.E.; Glazunov, V.P.; Nabiullin, A.A. Infrared Spectroscopic Studies of Hydrogen Bonding in Triethylammonium Salts. Part 3.—Strong Hydrogen Bonding. *J. Chem. Soc. Faraday Trans.* **1984**, *80*, 899–908. [CrossRef]

136. Ratajczak, H.; Orville-Thomas, W.J.; Rao, C.N.R. Charge transfer theory of hydrogen bonds: relations between vibrational spectra and energy of hydrogen bonds. *Chem. Phys.* **1976**, *17*, 197–216. [CrossRef]

137. Jabłoński, M.; Sadlej, A.J. Infrared and Raman Intensities in Proper and Improper Hydrogen-Bonded Systems. *Pol. J. Chem.* **2007**, *81*, 767–782.

138. Pecul, M.; Leszczynski, J.; Sadlej, J. Comprehensive ab initio studies of nuclear magnetic resonance shielding and coupling constants in XH$\cdots$O hydrogen-bonded complexes of simple organic molecules. *J. Chem. Phys.* **2000**, *112*, 7930–7938. [CrossRef]

139. McDowell, S.A.C.; Buckingham, A.D. Shift of proton magnetic shielding in ClH$\cdots$Y dimers (Y = N_2, CO, BF). *Mol. Phys.* **2006**, *104*, 2527–2531. [CrossRef]

140. Scheiner, S. Interpretation of Spectroscopic Markers of Hydrogen Bonds. *ChemPhysChem* **2016**, *17*, 2263–2271. [CrossRef] [PubMed]

141. Scheiner, S. Assessment of the Presence and Strength of H-Bonds by Means of Corrected NMR. *Molecules* **2016**, *21*, 1426. [CrossRef]

142. Fliegl, H.; Lehtonen, O.; Sundholm, D. Kailaz, V.R.I. Hydrogen-bond strengths by magnetically induced currents. *Phys. Chem. Chem. Phys.* **2011**, *13*, 434–437. [CrossRef]

143. Monaco, G.; Della Porta P.; Jabłoński, M.; Zanasi, R. Topology of the magnetically induced current density and proton magnetic shielding in hydrogen bonded systems. *Phys. Chem. Chem. Phys.* **2015**, *17*, 5966–5972. [CrossRef]

144. Busfield, W.K.; Ennis, M.P.; McEwen, I.J. An infrared study of intramolecular hydrogen bonding in α, ω diols. *Spectrochim. Acta A* **1973**, *29*, 1259–1264. [CrossRef]

145. Schaefer, T. A Relationship between Hydroxyl Proton Chemical Shifts and Torsional Frequencies in Some Ortho-Substituted Phenol Derivatives. *J. Phys. Chem.* **1975**, *79*, 1888–1890. [CrossRef]

146. Takasuka, M.; Matsui, Y. Experimental Observations and CNDO/2 Calculations for Hydroxy Stretching Frequency Shifts, Intensities, and Hydrogen Bond Energies of Intramolecular Hydrogen Bonds in *ortho*-Substituted Phenols. *J. Chem. Soc. Perkin Trans. 2* **1979**, 1743–1750. [CrossRef]

147. Shagidullin, R.R.; Chernova, A.V.; Shagidullin, R.R. Intramolecular hydrogen bonds and conformations of the 1,4-butanediol molecule. *Russ. Chem. Bull.* **1993**, *42*, 1505–1510. [CrossRef]

148. Dziembowska, T.; Szczodrowska, B.; Krygowski, T.M.; Grabowski, S.J. Estimation of the OH$\cdots$O interaction energy in intramolecular hydrogen bonds: A comparative study. *J. Phys. Org. Chem.* **1994**, *7*, 142–146. [CrossRef]

149. Vashchenko, A.V.; Afonin, A.V. Comparative estimation of the energies of intramolecular C-H$\cdots$O, N-H$\cdots$O, and O-H$\cdots$O hydrogen bonds according to the QTAIM analysis and NMR spectroscopy data. *J. Struct. Chem.* **2014**, *55*, 636–643. [CrossRef]

150. Iogansen, A.V.; Rassadin, B.V. Dependence of Intensity and Shift in the Infrared Bands of the Hydroxyl Group on Hydrogen Bonding Energy. *J. Appl. Spectrosc.* **1969**, *11*, 1318–1325. [CrossRef]
151. Iogansen, A.V. Direct proportionality of the hydrogen bonding energy and the intensification of the stretching ν(XH) vibration in infrared spectra. *Spectrochim. Acta Part A* **1999**, *55*, 1585–1612. [CrossRef]
152. Yurenko, Y.P.; Zhurakivs.ky, R.O.; Ghomi, M.; Samijlenko, S.P.; Hovorun, D.M. Comprehensive Conformational Analysis of the Nucleoside Analogue 2′-β-Deoxy-6-azacytidine by DFT and MP2 Calculations. *J. Phys. Chem. B* **2007**, *111*, 6263–6271. [CrossRef]
153. Yurenko, Y.P.; Zhurakivs.ky, R.O.; Ghomi, M.; Samijlenko, S.P.; Hovorun, D.M. How Many Conformers Determine the Thymidine Low-Temperature Matrix Infrared Spectrum? DFT and MP2 Quantum Chemical Study. *J. Phys. Chem. B* **2007**, *111*, 9655–9663. [CrossRef]
154. Yurenko, Y.P.; Zhurakivs.ky, R.O.; Ghomi, M.; Samijlenko, S.P.; Hovorun, D.M. Ab Initio Comprehensive Conformational Analysis of 2′-Deoxyuridine, the Biologically Significant DNA Minor Nucleoside, and Reconstruction of Its Low-Temperature Matrix Infrared Spectrum. *J. Phys. Chem. B* **2008**, *112*, 1240–1250. [CrossRef]
155. Gränacher, I. Einfluss der Wasserstoffbrückenbindung auf das Kernresonanzspektrum von Phenolen. *Helv. Phys. Acta* **1961**, *34*, 272–304.
156. Hobza, P.; Halvas, Z. Blue-Shifting Hydrogen Bonds. *Chem. Rev.* **2000**, *100*, 4253–4264. [CrossRef] [PubMed]
157. Joseph, J.; Jemmis, E.D. Red-, Blue-, or No-Shift in Hydrogen Bonds: A Unified Explanation. *J. Am. Chem. Soc.* **2007**, *129*, 4620–4632. [CrossRef] [PubMed]
158. Lippincott, E.R.; Schroeder, R. One-Dimensional Model of the Hydrogen Bond. *J. Chem. Phys.* **1955**, *23*, 1099–1106. [CrossRef]
159. Schroeder, R.; Lippincott, E.R. Potential Function Model of Hydrogen Bonds. II. *J. Phys. Chem.* **1957**, *61*, 921–928. [CrossRef]
160. Musin, R.N.; Mariam, Y.H. An integrated approach to the study of intramolecular hydrogen bonds in malonaldehyde enol derivatives and naphthazarin: Trend in energetic versus geometrical consequences. *J. Phys. Org. Chem.* **2006**, *19*, 425–444. [CrossRef]

Publisher's Note: MDPI stays neutral with regard to jurisdictional claims in published maps and institutional affiliations.

Review

Molecular Tailoring Approach for the Estimation of Intramolecular Hydrogen Bond Energy

Milind M. Deshmukh [1],* and Shridhar R. Gadre [2],*

[1] Department of Chemistry, Dr. Harisingh Gour Vishwavidyalaya (A Central University), Sagar 470003, India

[2] Department of Scientific Computing, Modelling and Simulation, Savitribai Phule Pune University, Pune 411007, India

* Correspondence: mdeshmukh@dhsgsu.edu.in or milind.deshmukh@gmail.com (M.M.D.); gadre@unipune.ac.in (S.R.G.)

Abstract: Hydrogen bonds (HBs) play a crucial role in many physicochemical and biological processes. Theoretical methods can reliably estimate the intermolecular HB energies. However, the methods for the quantification of intramolecular HB (IHB) energy available in the literature are mostly empirical or indirect and limited only to evaluating the energy of a single HB. During the past decade, the authors have developed a direct procedure for the IHB energy estimation based on the molecular tailoring approach (MTA), a fragmentation method. This MTA-based method can yield a reliable estimate of individual IHB energy in a system containing multiple H-bonds. After explaining and illustrating the methodology of MTA, we present its use for the IHB energy estimation in molecules and clusters. We also discuss the use of this method by other researchers as a standard, state-of-the-art method for estimating IHB energy as well as those of other noncovalent interactions.

Keywords: hydrogen bond (HB); intramolecular hydrogen bond (IHB); molecular tailoring approach (MTA); fragmentation methods; bond energy estimation; noncovalent interactions

Citation: Deshmukh, M.M.; Gadre, S.R. Molecular Tailoring Approach for the Estimation of Intramolecular Hydrogen Bond Energy. *Molecules* **2021**, *26*, 2928. https://doi.org/10.3390/molecules26102928

Academic Editor: Mirosław Jabłoński

Received: 17 April 2021
Accepted: 12 May 2021
Published: 14 May 2021

Publisher's Note: MDPI stays neutral with regard to jurisdictional claims in published maps and institutional affiliations.

1. Introduction

The hydrogen bond (HB) is a dominant noncovalent interaction found in chemical and biological systems [1–4]. The term "hydrogen bond" seems to have emerged around 1930, from the works of Pauling [1] and Huggins [5,6]. However, the mention of weak, yet specific interactions involving the hydrogen atom is much older. The dimeric association of molecules with hydroxyl groups was suggested by Nernst in 1892 [7]. The term "Nebenvalenz" by Werner [8] and "weak union" by Moore and Winmill [9] are other early stipulations of this noncovalent interaction. In 1920, Latimer and Rodebush [10] suggested that the hydrogen nucleus in an aqueous solution of amines is held jointly by two octets, constituting a weak bond. Barnes, while studying the structure of ice [11], suggested that the hydrogen atoms were midway between the two oxygen atoms, though he did not explicitly mention the hydrogen bond. Huggins [12] claimed that he was the first to propose the term "H-bond" in 1919. His later usage of the term "hydrogen-bridge" may have led to the German word "Wasserstoffbrücke." The concept of HB gained popularity after Pauling published his classic book, *The Nature of the Chemical Bond* in 1939 [1]. Pimentel and McClellan [13] suggested that an HB exists when (i) there is evidence of a bond and (ii) there is evidence that this bond involves a hydrogen atom already bonded to another atom. The recent definition of HB by IUPAC [14] is similar in spirit to that in Ref. [13]. The former [14] states that "the hydrogen bond is an attractive interaction between a hydrogen atom from a molecule or a molecular fragment X–H in which X is more electronegative than H, and an atom or a group of atoms in the same or a different molecule, in which there is evidence of bond formation."

An HB may be generally represented as X–H⋯Y, where X–H is the proton donor and Y is a proton acceptor. The X–H⋯Y interactions such as O–H⋯O, N–H⋯O, N–H⋯N,

43

S–H$\cdots$O, etc., in neutral molecular systems, exhibit interaction energies lying between ~1 to 20 kcal/mol. Typical H$\cdots$Y distances suggested in the literature fall in the range of ~1.2–3.0 Å and X–H$\cdots$Y angles lie between 100 and 180° [2–4]. The HB's in liquid water are central to water's life-providing properties [1,15]. It is stipulated in the literature [16] that if HBs in water were 7% stronger or 29% weaker, water would not be a liquid at room temperature. HBs provide a significant driving force for the native structures and functions of biomolecules [17,18]. Hence, it is of great importance to reliably estimate these X$-$H$\cdots$Y HB strengths for shedding light on several physicochemical phenomena and life processes.

The theoretical estimation of intermolecular X–H$\cdots$Y HB strength in a complex A$\cdots$B is routinely performed using a supermolecular approach, in which the HB energy (E_{HB}) is estimated as $E_{HB} = E_{A\cdots B} - (E_A + E_B)$. Several methods for estimating the intermolecular interaction energies are reported in the literature (see, e.g., Refs. [19–22]). On the other hand, quantifying an intramolecular hydrogen bond (IHB) strength is not as straightforward as the intermolecular one. The main difficulty lies in isolating the X–H$\cdots$Y interaction present within a molecule than in a dimer or a complex. Many studies for gauging the strength of the IHB in the literature are based on spectroscopic- [23–26] and electron density topological approaches [27–30]. Nevertheless, some empirical, semiempirical, and ab initio procedures [31–39] have also been reported in the literature for estimating the IHB energy. These have been nicely summarized by Jablonski [40] in his article in this Special Issue, and we shall discuss only the aspects of these methods (e.g., their merits and demerits) that are not explicitly covered in Ref. [40].

One of the early approaches is the conformational analysis (CA), in which two different conformers of the reference molecule are considered. These conformers are chosen such that the HB is kept intact in one of the conformers and is broken in another. The energy difference between these two conformers is then taken as the measure of IHB energy [32,40,41]. A significant disadvantage of this method is that the estimated IHB energy is erroneous due to the incorporation of attractive (*syn-anti*) or repulsive (*anti-anti*) additional interaction in one of the conformers [42]. In another similar procedure, viz., the *ortho-para* method [43], the IHB energy of the X-H$\cdots$Y bond formed by two substituents, which are *ortho* to each other, is taken as the energy difference between the *ortho* and *para* forms of the reference molecule. However, this method applies only to aromatic systems in which an HB is present in two substituents, which are *ortho* to each other. The main drawback of this method is that it assumes that the electronic effects caused by the substituent at different positions are similar between the *ortho-* and *para*-conformers [44–46].

Yet, another indirect approach is the isodesmic/homodesmic reactions. In the former, the IHB making/breaking reaction is written so that the number and type of bonds on either side of the reaction are equal, except the HB, which is retained in one of the reactants [33,47,48]. A further assumption is that the atomic hybridization is conserved on both sides of the reaction. In that case, the method is called homodesmic reaction, which is supposed to give more reliable energy estimates than the isodesmic reaction [49–51]. The main drawback of the isodesmic/homodesmic reaction approach is that it does not give HB energy but includes strain energy due to the formation of a ring structure [52]. Another major disadvantage of these indirect methods is that they are applicable only to the evaluation of energy of a single HB present in the system and cannot be employed in a system containing multiple HBs.

Another popular but indirect method is based on the quantum theory of atoms in molecules (QTAIM) [53]. In this method, the presence of a (3, $-$1) bond critical point (BCP) of the molecular electron density (MED) between H$\cdots$Y, is considered as the signature of an HB. The large/small value of the MED at the BCP is seen to correlate with strong/weak X–H$\cdots$Y interaction [54–58]. Espinosa et al. [59] proposed an empirical relation, $E_{HB} = 0.5$ $V(r_{cp})$, where $V(r_{cp})$ is the potential energy density at BCP. Interacting quantum atoms (IQA) [60] framework, leading to QTAIM-compatible energy partition, is another indirect approach wherein the HB energy is calculated as the sum of the classical Coulombic interaction between groups involved in the HB and the exchange-correlation energy. It has been pointed out that there is no check on the reliability of HB energy provided by both of these methods [61,62]. Further, these empirical equations are applicable only when a $(3, -1)$ BCP is present. For instance, a $(3, -1)$ BCP at O–H$\cdots$O bond is conspicuous by its absence in all the polyols having an O–H$\cdots$O interactions between the vicinal -OH groups. However, the weak O–H$\cdots$O HB was confirmed in ethylene glycol-based on the vapor phase OH-stretching overtone spectroscopy [63–65]. Some other variants and indirect empirical equations proposed in the literature for estimating the strength of IHBs are summarized in Ref. [40]. Hence, we skip the discussion of these methods.

A brief review of the above literature suggests that these empirical and indirect methods for estimating IHB energy are generally limited to singly H-bonded systems. These cannot be readily extended to a system containing multiple HBs and hence also to the estimation of HB cooperativity due to an interconnected network of HBs. Most importantly, there is no check on the reliability of the estimated HB energies. Thus, it was felt necessary to come up with a direct theoretical method for a reliable estimation of IHB energy. Deshmukh and Gadre [66,67] proposed such a procedure, based on the in-house developed molecular tailoring approach (MTA) for the ab initio treatment of large molecular systems [68–82]. The MTA currently enables the calculation of one-electron properties, geometry optimization, and the calculation of vibrational infrared and Raman spectra of large molecules/clusters using DFT or correlated methods. Before discussing the application of MTA for the IHB energy estimation, we explain the working principle of MTA, with an illustrative example, in Section 2.

2. Molecular Tailoring Approach

The molecular tailoring approach (MTA) is a fragmentation-based technique developed by Gadre et al. [68–82]. Within MTA, a spatially extended molecular system under consideration is partitioned into a set of overlapping fragments (called the "main fragments") on which ab initio calculations for one-electron property or the energy are carried out. The fragmentation may be carried out automatically or manually. The quality of the fragmentation scheme is gauged by a parameter called R-Goodness (Rg), which may be estimated as follows: Put a sphere of radius R centered on a reference atom i so that all the atoms of the parent system lying within this sphere belong to at least one of the main fragments. The maximum value of R obeying this condition is called the Rg value of atom i in the given fragmentation scheme. The minimum of such atomic Rg values is called the Rg value of the scheme. In general, the larger the Rg value of a fragmentation scheme, the better is the chemical environment of each atom mimicked [69]. After choosing an appropriate fragmentation scheme, molecular energy E, of the spatially extensive parent system is estimated approximately by patching those of the individual fragment energies [69] using Equation (1).

$$E = \sum_i{}' E^{F_i} - \sum_{i<j}{}' E^{F_i \cap F_j} + \cdots + (-1)^k \sum_{i<j<\cdots<n} E^{F_i \cap F_j \cap \cdots \cap F_n} \tag{1}$$

where the energy E of the parent molecule is estimated as the sum of energies of primary fragments $\{F_i\}$ minus the sum of energies of binary overlap fragments $\{F_i \cap F_j\}$ plus the sum of energies of ternary overlap fragments, etc. Here, k stands for the degree of overlap., e.g., for binary overlap, k = 1, for ternary overlap, k = 2, etc. Equation (1) is generalized for estimating an electronic property of the molecule, such as the energy gradients, the Hessian matrix elements, etc.

We now illustrate the fragmentation procedure with a test example, viz., the α-tocopherol molecule (shown as M) in Scheme 1, fragmented into three primary fragments F_1, F_2, and F_3 (shown by appropriate circles). The fragments F_4 and F_5 are the binary overlaps of fragments F_1, F_2, and F_2, F_3, respectively. Here, the term binary overlap means the common structural part of two primary fragments. In fragmentation Scheme 1, the ternary fragment (overlap of three fragments F_1, F_2, and F_3) is absent. The valencies of the cut regions are satisfied by placing the H-atoms at the appropriate C–H distance of

1 Å along the cut C–C bond. The calculations for the single point energy (or the property) of these fragments F_1 to F_5 are performed. For this test case, we report the MTA energy calculation of M at HF/6-31+G(d,p) level theory using Equation (1). It should be noted that the HF method employed here is only for illustrative purposes. MTA method works at any correlated level of theory. All the calculations are performed with the Gaussian 16 package [83].

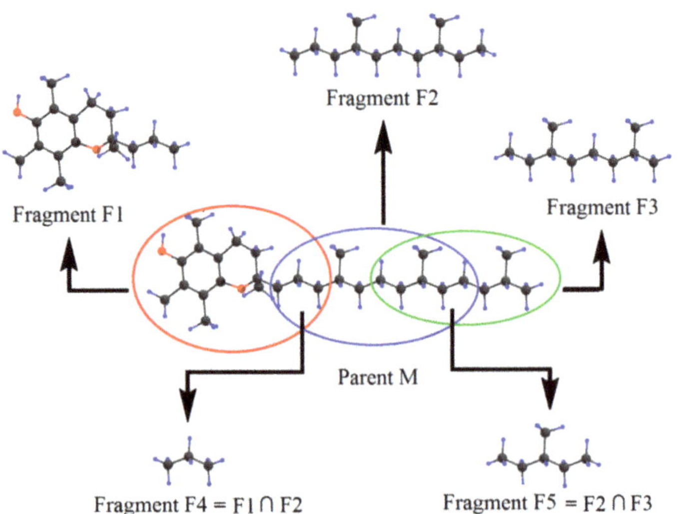

Scheme 1. Illustration of the molecular tailoring approach (MTA)-based fragmentation procedure for α-tocopherol, shown as the parent molecule, M. See text for details.

In the present case, the energy (E_{MTA}) of the parent molecule (M) is obtained by Equation (2) as

$$E_{MTA} = \{E_{F1} + E_{F2} + E_{F3}\} - \{E_{F4} + E_{F5}\} \tag{2}$$

In the present case of α-tocopherol this energy is calculated utilizing the energies of the fragments as $E_{MTA} = \{(-769.56869) + (-469.57421) + (-391.51579)\} - \{(-118.26152) + (-235.36444)\} = -1277.03273$ hartrees (a.u.). The actual energy (E_{FC}) at the HF/6-31+G(d,p) level of theory is $E_{FC} = -1277.03288$ a.u. Thus, the error (E_{Error}) in the estimation of molecular energy is given $E_{Error} = E_{FC} - E_M = -0.00016$ a.u. This error can be further reduced using the so-called grafting procedure embedded in the current version of MTA [77,78].

Gadre et al. first proposed the MTA methodology for estimating the electrostatic properties of large, closed-shell molecules [68]. However, in the last 24 years, the scope of MTA was extended for the estimation of molecular energy [69], geometry optimization [70,71], the estimation of the Hessian matrix [72], the computation of vibrational infrared [73] and Raman spectra [74], and binding energies of large molecular clusters and complexes [75–82]. Since the fragment computations are independent of each other, MTA has the advantage that the energy computation of the parent molecule is intrinsically parallel. Further, MTA can currently work with Gaussian, GAMESS, and NWChem at the back-end, thereby becoming a powerful tool for ab initio treatment of large molecules and clusters, when used in conjunction with a high level of theory, such as MP2 or CCSD(T) employing a large basis set. One important application of MTA is in estimating the IHB energy [52,66,67,84–92]. This will be discussed in the next section.

3. Intramolecular Hydrogen Bond Energy Estimation by Molecular Tailoring Approach

With the above brief introduction to MTA, we now discuss its application for estimating the IHB energy. As discussed in the introduction section, intermolecular X–H$\cdots$Y HB energy in a complex A$\cdots$B is estimated as $E_{HB} = E_{A\cdots B} - (E_A + E_B)$. This estimation is possible because the energies of the two monomers A and B can be separately calculated. In the case of intramolecular X–H$\cdots$Y interaction, such a separation is, in general, difficult. However, the MTA procedure allows the generation of fragments so that the atoms/functional groups involved in the HB formation are parts of two different fragments.

The fragmentation procedure is illustrated in Scheme 2 for the test molecule of 1,2,4-butanetriol. In Scheme 2, the parent test molecule, denoted as M, is shown at the center. The geometry of 1,2,4-butanetriol was optimized at the MP2/6-31+G(d,p) (default option: frozen core) level of theory using the Gaussian package [83]. The energy of the optimized structure is −383.01926 a.u. at MP2/6-31+G(d,p) level. The three oxygen atoms are shown as O1, O2, and O3 (see Scheme 2), with the two HB's, viz., **HB1** (O2-H$\cdots$O1) and **HB2** (O3-H$\cdots$O2) whose energy is to be estimated. For this purpose, the parent molecule is cut into three primary fragments F1, F2, and F3, obtained by replacing −O1H, −O2H, and −O3H groups, respectively, with an H atom each. Dotted circles show these cut regions on the original molecule. The H-atoms are added along the respective C–O bonds (which are cut to form these primary fragments) so that the C–H distance is 1 Å. Hydrogen is the simplest monovalent atom that can be used for satisfying the valencies of cut regions. It is emphasized here that H-atoms placed at slightly different distances (say at 0.9 or 1.1 Å) from the C-atom do not change the results appreciably. This is because of the cancellation of errors in estimating the molecular energy using these fragments. Fragments F4, F5, and F6 are obtained by taking the binary intersection of these primary fragments, i.e., (F1∩F2), (F2∩F3), and (F1∩F3), respectively. Here, intersection means the common structural parts between two primary fragments apart from added H-atoms. For instance, in fragment F4, C1(H_2)–C2(H_2)–C3(H_2)–C4(H_2)O3(H) is the common structural part that is also present in fragments F1 and F2. Similarly, fragment F7 is the common intersection of three primary fragments F1, F2, and F3, i.e., (F1∩F2∩F3). A single point energy evaluation, at MP2/6-31+G(d,p) level of theory, is carried out on all seven fragments obtained by the above fragmentation procedure. The fragment geometries are not optimized to avoid the conformational changes in them so that they lead to reliable estimates of IHB energies. It is necessary first to provide a check on MTA application to the parent molecule, M. As discussed above, the actual energy of the original molecule (M) is E_M = −383.01926 a.u. at the MP2 level of theory. Using the MP2 single point energies of these fragments, the estimated molecular energy of M is: $E_M = \{E_{F1} + E_{F2} + E_{F3}\} - \{E_{F4} + E_{F5} + E_{F6}\} + E_{F7} = \{-307.97304 + (-307.9642) + (-307.97765)\} - \{-232.92410 + (-232.92514) + (-232.93273)\} + (-157.88552) = -383.01844$ a.u. The error, ΔE = | MTA energy - actual energy | in molecular energy indeed turns out to be very small, viz., 0.00082 a.u. This excellent agreement between the MTA-estimated and actual energy suggests that the present fragmentation scheme is reliable for evaluating HB energies.

Now we estimate the energy of two hydrogen bonds **HB1** and **HB2,** in the parent 1,2,4-butanetriol. Recall that the hydroxyl groups involved in the formation of hydrogen bond **HB1** are O1–H and O2–H. These hydroxyl groups were replaced in fragments F1 and F2, respectively, by H-atoms. Putting the geometry of fragment F1 over F2, we regenerate the parent molecule except following two things: (i) the O–H$\cdots$O H-bond, i.e., the **HB1** interaction between O1–H and O2–H present in the parent molecule is missed out and (ii) there is double counting of common structural part between F1 and F2 (viz., the secondary fragment, F4). Upon addition of single-point energies of fragments F1 and F2, followed by subtraction of the energy of fragment F4 would give the energy of the parent molecule except that the energy of the HB, viz., **HB1** is missed out. If the energy of the parent 1,2,4-butanetriol E_M is subtracted from ($E_{F1} + E_{F2} - E_{F4}$), the HB energy E_{HB1} is obtained as $E_{HB1} = (E_{F1} + E_{F2} - E_{F4}) - E_M = 3.84$ kcal/mol. In a similar fashion, E_{HB2} is obtained as $E_{HB2} = (E_{F2} + E_{F3} - E_{F5}) - E_M = 1.60$ kcal/mol. It should be noted here that these estimated HB energies are in the gas phase. However, the MTA-based method in principle can provide HB energies in the solvent phase, wherein the energies of the fragments in solvent (using continuum solvation model) could be employed.

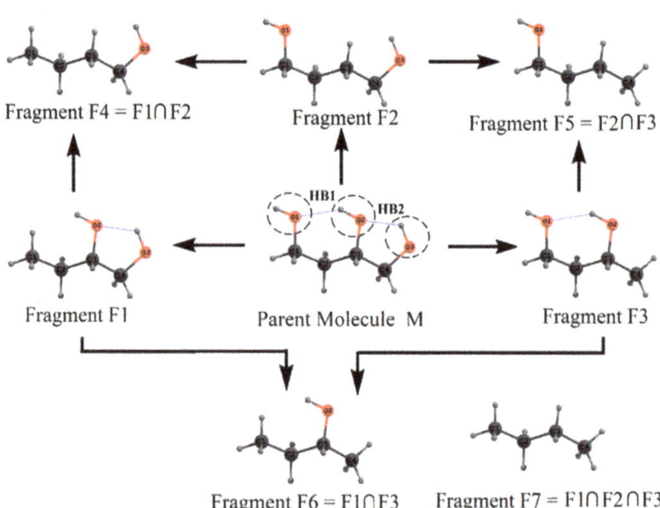

Scheme 2. Fragmentation procedure for estimating the energies of the H-bonds, **HB1**, and **HB2** in 1,2,4-butanetriol (Parent M) molecule. See text for details.

We note that the two HBs, **HB1** and **HB2**, are interconnected, forming an H-bond network. Such networking of H-bonds leads to a phenomenon called cooperativity [67]. In general, it is anticipated that the strengths of **HB1** and **HB2** are enhanced because of this networking effect. To estimate the contribution of cooperativity toward each of these two H-bonds, we reestimated the HB energy of these two HBs by isolating them from each other. The difference between the HB energy estimated earlier (in the presence of network) and the one when they are isolated (in the absence of a network) is the cooperativity contribution toward this HB. For example, consider fragment F3 in which only **HB1** is present and fragment F1 in which **HB2** is present. To estimate the energy of **HB1** in the absence of the networking effect of **HB2**, we consider fragment F3 as our parent molecule. In the present case, fragments F5 and F6 are the two primary fragments that, when placed over each other, would give us the parent fragment F3 except **HB1**, and fragment F7 is the binary overlap of F5 and F6. Therefore, utilizing these fragments' energies, the energy of **HB1** is obtained as $E_{HB1} = (E_{F5} + E_{F6} - E_{F7}) - E_{F3} = 3.32$ kcal/mol. Similarly, the energy of **HB2** in the absence of the networking effect of **HB1** is obtained as $E_{HB2} = (E_{F4} + E_{F6} - E_{F7}) - E_{F1} = 1.09$ kcal/mol. These reestimated HB energies are indeed smaller than those estimated in the presence of the networking effect. The difference in the energy is cooperativity contribution. The cooperativity contribution to **HB1** is $E_{HB1}^{coop} = 3.84 - 3.32 = 0.52$ kcal/mol and that for **HB2** is $E_{HB2}^{coop} = 1.60 - 1.09 = 0.51$ kcal/mol. In the present test case, the estimated cooperativity contributions are not large because only two HBs are present. The later sections will show that cooperativity values in some molecules can indeed be as large as a typical HB energy.

The HB energies obtained by applying the above procedure to some alkanetriol molecules are shown in Table 1 [66]. The estimated HB energies fall in a range between 1.50 and 4.97 kcal/mol (see Table 1). This is the expected energy range from chemical intuition. Further, these HB energies are in a qualitative agreement with those expected from the corresponding HB distances. For instance, the strongest HB in 1,2,5-pentanetriol has an energy of 4.97 kcal/mol, with the corresponding HB distance being the shortest

(1.80 Å) among all the alkanetriols reported in Table 1. One of the noteworthy results in Table 1 is that the error in estimating molecular energies of all the alkanetriols is quite small, viz., between 0.40 to 0.65 kcal/mol. By considering this accuracy, we estimate the maximum error associated with our calculated HB energies to be 0.3 kcal/mol. The present method is thus capable of calculating accurately the IHB energies and cooperativity values of multiply H-bonded systems. This is a significant advantage of the current method over the other indirect approaches reported in the literature.

Table 1. The H-bond (HB) distances (in Å), HB energies (in kcal/mol), and the error in the molecular energy estimation for alkanetriols using similar fragments, $\Delta E = |E_M - E_e|$. The corresponding O–H stretching frequencies (cm^{-1}) and the molecular electron density (MED) value at the $(3, -1)$ bond critical point (BCP) (a.u.) are also shown. The calculations are performed at MP2 (full)/6-311++G(2d,2p) level theory.

Molecule	HB Label	HB Distances [a] (in Å)	HB Energy (kcal/mol)	ΔE (kcal/mol)	O–H Stretch Frequency (cm^{-1})	MED at $(3, -1)$ BCP (a.u.)
1,2,3-propanetriol [b]	HB1	2.16	1.90		3784	
	HB2	2.08	2.47	0.50	3765	0.0201
	HB3	2.58	1.63		3845	
1,2,3-butanetriol [b]	HB1	2.13	2.13		3768	
	HB2	2.05	2.72	0.50	3745	0.0211
	HB3	2.58	1.60		3844	
1,2,4-butanetriol	HB1	1.98	2.90		3789	
	HB2	2.22	1.75	0.40	3828	0.0219
					3875	
1,2,5-pentanetriol	HB1	1.80	4.97		3669	
	HB2	2.25	1.78	0.55	3825	0.0334
					3865	
1,3,5-pentanetriol	HB1	1.94	2.91		3763	0.0225
	HB2	1.96	2.90	0.58	3792	0.0239
					3875	
2,3,4-pentanetriol [b]	HB1	2.12	2.18		3759	
	HB2	2.02	2.94	0.52	3731	0.0223
	HB3	2.56	1.50		3820	
2,4,6-heptanetriol	HB1	1.92	3.02		3753	0.0250
	HB2	1.93	2.94	0.65	3773	0.0242
					3857	

[a] The MP2 (FC)/6-311++G(2d,2p) optimized geometries were employed. Table 1 is partially reproduced from our earlier study reported in Ref. [66]; *Copyright (2006) The American Chemical Society.* [b] The triols wherein three OH groups are present on the successive C-atoms show three H-bonds. See text for details.

4. Critical Comparison of MTA with Other Methods

We now compare the estimated HB strengths in these molecules with those qualitatively estimated by other indirect measures. These measures include the O–H stretching frequency, molecular electron density (MED) value at $(3, -1)$ BCP, and the HB energy estimated using the isodesmic reaction approach (IDRA) and that by using Espinosa's equation. Both the MED value at $(3, -1)$ BCP and the shift in the stretching frequency of the O–H involved in the HB show a good qualitative correlation with the estimated HB energies (see Table 1). For instance, the strongest HB found in 1,2,5-pentanetriol (4.97 kcal/mol) corresponds to the highest MED value (0.0334 au) at the $(3, -1)$ BCP. The stretching frequency of the O–H bond involved in the HB is 3669 cm^{-1}, showing a redshift. The weakest HB is seen in 2,3,4-pentanetriol (1.50 kcal/mol), with the $(3, -1)$ BCP being absent and a large O–H stretch frequency of 3820 cm^{-1}. In general, the calculated HB energies show a good qualitative rank–order relationship with the corresponding O–H stretching frequencies and the MED value at the respective $(3, -1)$ BCP.

We now critically compare our MTA-based results with those obtained by yet another indirect method, viz., the isodesmic reaction approach (IDRA) [33,48]. In IRDA, the IHB-making/breaking reaction is written such that, except for the HB under consideration, the number and type of other bonds are conserved on both sides of the reaction. Within IDRA, all the reactant and product geometries are optimized at the appropriate level of theory. The energy change for such a reaction is taken as the HB energy. The main disadvantage of IDRA is that there is no unique way of writing an isodesmic reaction for estimating the HB energy. For example, in Scheme 3, four possible reactions are shown for evaluating the HB energy E_{HB2} in 1,2,5-pentanetriol, which retain **HB1** on both sides of these isodesmic reactions. See Ref. [52] for details of several other isodesmic reactions for estimating the HB energies of **HB1** and **HB2**.

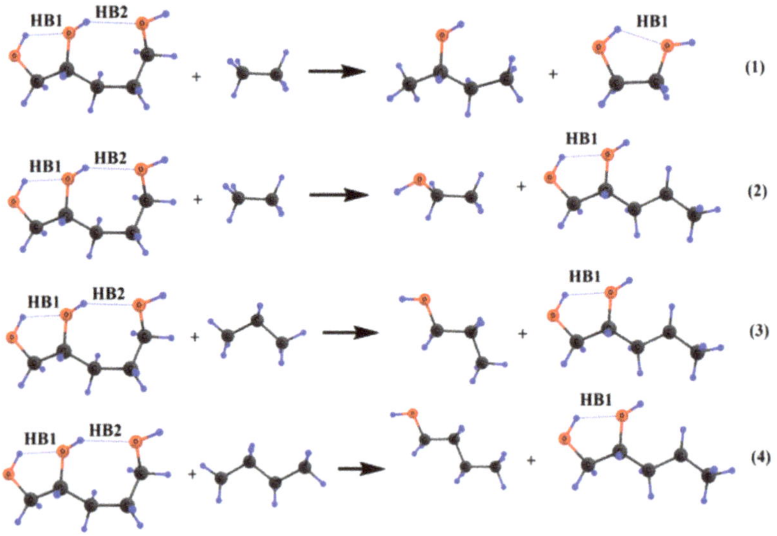

Scheme 3. Some possible isodesmic reactions for the estimation of H-bond energy, E_{HB2} in 1,2,5-Pentanetriol. See text and Ref. [52] for details.

Figure 1 shows a histogram of the estimated HB energy E_{HB2} by these four reactions and also by the MTA-based method. The HB energies estimated by the isodesmic reactions vary significantly across the levels of theory and from each other. These estimated energy values are much smaller (30 to 40%) than the respective MTA-based ones. The reasons for these smaller HB energy values by IDRA can be understood as follows: In IDRA reactions presented in Scheme 3, on the reactant side, the **HB2** bond formation between the -OH groups at C2 and C5 positions leads to a seven-membered ring-like structure involving one O–H⋯O, two C–O, and three C–C bonds. This ring formation has the ring strain effect in the parent 1,2,5-pentanetriol molecule. Since the reactant and product geometries are optimized, this ring strain effect is not preserved on the product side of these reactions. Therefore, the molecules on the product side are more relaxed, losing most of their ring strain due to the loss of the **HB2** bond. It may be noted here that the ring strain may be small or canceled out when one estimates the energy of **HB1** using IDRA. This is because the formation of **HB1** leads to the formation of a five-membered ring-like structure involving one O-H⋯O, two C–O, and only one C–C bonds. This ring (five-membered) backbone is expected to be maintained (to some extent if not fully) on both reactant and product sides as it involves only one C-C bond. For more details about **HB1** energy by IDRA, see Ref. [52].

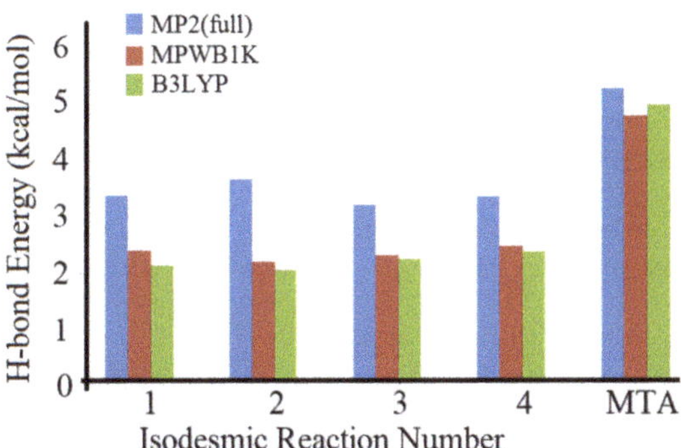

Figure 1. The H-bond energy, E_{HB2}, in 1,2,5-pentanetriol calculated at different levels of theory using isodesmic reactions (see Scheme 3) and also by molecular tailoring approach (MTA). See text for details.

Thus, the finer bonding effects are not mimicked evenly on both sides of the reaction, resulting in poor estimates of HB energy by IDRA. On the contrary, in the MTA method, the geometry of the fragments is not optimized, and the fragment backbone structure is preserved, leading to accurate estimates of the HB energies. For instance, the evaluation of HB energy E_{HB1} as $(E_{F1} + E_{F2} - E_{F4}) - E_M$ in 1,2,4-butanetiols involves the energies of F1, F2, and F4 (see Scheme 2). Thus, the MTA-based method leads to unique isodesmic/homodesmic reaction, viz., M + F4 → F1 + F2. As seen in Scheme 2, the carbon backbone structure is retained on either side of this unique reaction. In other words, as advocated in the literature [49–52], IDRA does not yield the true HB energies. Another drawback of the IDRA is for estimating the HB strengths in multiply hydrogen-bonded systems. Here, one has to write different reactions for different HBs. In summary, the MTA-based method for the estimation of IHB energy is accurate and can be applied to the evaluation of multiple IHB energies in a given molecule. In the following sections, we present and discuss the results of IHB energies obtained by applying the MTA-based method to a variety of systems.

We now present a comparison of the HB energy estimated by the MTA-based method with those by Espinosa's empirical relation [59]. For this purpose, we consider two molecular systems having O–H···O=C HB, viz., 2'-Hydroxyacetophenone, and methyl 2-hydroxybenzoate. The geometries and HB energies by Espinosa's method in these two molecules were taken from the Ref. [93]. The reported HB energies at B3LYP/6-311++G(d,p) level, by Espinosa's method, in 2'-Hydroxyacetophenone and methyl 2-hydroxybenzoate are 15.3 and 11.4 kcal/mol, respectively. The HB energies for these molecules were calculated by us at the given geometries, at the same level of theory, by MTA-based method are 9.7 and 7.8 kcal/mol, respectively. As can be observed, the HB energies estimated by Espinosa's method are significantly overestimated, compared to the HB energies by the MTA method. These results are in agreement with an earlier report that the use of Espinosa's method significantly overestimated the energy of HBs [62,93]. Although our MTA-based method requires extensive calculations and more computational time, reliable direct estimates of the HB energies are thereby obtained. Further, the internal benchmarking of the total energy is possible for the MTA-based method. Such benchmarks are not available in Espinosa's method.

5. Application to Large Molecules and Clusters with Multiple Hydrogen Bonds

We now discuss the application of the MTA-based method for IHB energy estimation in several large systems. One such interesting system is a class of carbohydrates, viz., sugar molecules. These molecules play an important role in biological processes, which are mainly governed by weak interactions such as hydrogen bonding, hydrophobic effects, etc.

Hence, understanding the interactions such as H-bonding is of utmost importance. Further, the O–H groups in carbohydrates form a network of interconnected O–H $\cdots$O HBs. It is suggested that the strengths of these individual O–H$\cdots$O HBs are enhanced due to such networking of HBs.

In our earlier work [67], the IHB energies and the contribution to cooperativity were investigated in eight aldopyranose monosaccharides, which vary in the position of hydroxyl groups [axial (*ax*) or equatorial (*eq*)], as shown in Figure 2. The estimated HB energies are in the range of 1.2 to 4.1 kcal/mol at the MP2(full)/6-311++G(2d,2p) level of theory. It is found that the OH$\cdots$O *eq-eq* interaction energies are between 1.8 and 2.5 kcal/mol, with axial-equatorial ones being stronger (2.0 to 3.5 kcal/mol). The strongest bonds involve nonvicinal *ax-ax* O–H groups (3.0 to 4.1 kcal/mol). The cooperativity contribution to the HBs is seen to fall between 0.1 and 0.6 kcal/mol for *eq-eq* HBs and is seen to be higher (0.5 to 1.1 kcal/mol) for *ax–ax* HBs. This work [67] was one of the first attempts for estimating the IHB energies and the respective cooperativity contributions in sugar molecules.

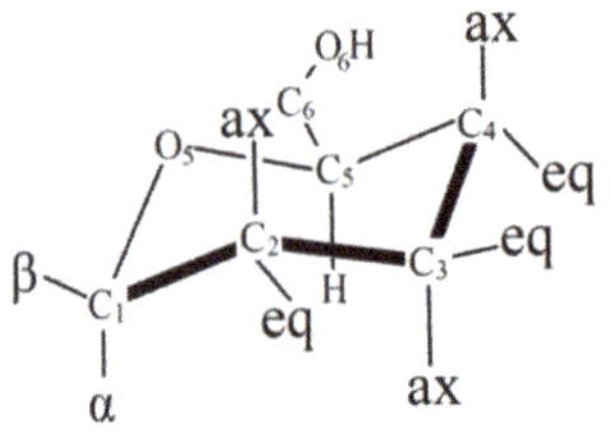

Aldopyranose	Position of hydroxyl group		
	C_2	C_3	C_4
D-glucose	*eq*	*eq*	*eq*
D-mannose	*ax*	*eq*	*eq*
D-allose	*eq*	*ax*	*eq*
D-galactose	*eq*	*eq*	*ax*
D-altrose	*ax*	*ax*	*eq*
D-talose	*ax*	*eq*	*ax*
D-gulose	*eq*	*ax*	*ax*
D-idose	*ax*	*ax*	*ax*

Figure 2. General structure of the aldopyranose sugar. In Table, *ax* represents axial, and *eq* represents equatorial orientations of the hydroxyl group at carbons C_2-C_4. Figure partially reproduced from Ref. [67] with the permission from American Chemical Society (ACS). *Copyright (2008) The American Chemical Society.*

A similar class of compounds having no ring oxygen atom is hexahydroxy–cyclohexane, called inositol. Inositol derivatives function as intracellular signal transduction molecules, playing an important role in biological processes. It is hence important to understand the structure and stability of various inositol conformers [84]. On applying the MTA-based method, the estimated HB energies in the presence of a cooperative HB network are seen to be in the range of 2.2 to 3.8 kcal/mol at the MP2/6-311+G(d,p) level of theory. The sum of all the HB energies in these conformers of inositol falls between 7.2 to 18.1 kcal/mol. The total cooperativity contribution in these conformers is rather large, between 2 to 5 kcal/mol. It increases on going from isomers with more *eq* O-H groups to those with more *ax* ones. Importantly, the highest stability of the *scyllo* isomer in the solvent was attributed [84] to weaker intramolecular OH$\cdots$O HBs between *eq* hydroxyl groups. It is suggested that these weak OH$\cdots$O HBs in *scyllo* isomer may facilitate favorable intermolecular interactions with solvent molecules. In contrast, the inositol isomers with *ax* O-H groups are involved in the formation of relatively strong HBs. Therefore, they are less stable due to large steric factors in the gas phase and unfavorable intermolecular interaction in the solvent phase.

The MTA-based method was also employed for understanding the conformational stability of fructose [85]. The experimental rotational spectroscopic studies suggest that the molecules of fructose and ribose preferentially adopt the β-pyranose structure in the gas phase. It was noted that [85] the relative stability of different conformers of fructose in the gas phase could be explained in terms of three collective effects: (i) the sum of HB energies in a given conformer, (ii) the strain energy of a bare fructose ring, and (iii) the sum of anomeric stabilization (*endo* + *exo*) energies. It was concluded [85] that the small ring strain, sufficiently large sum of the IHB energies, and the higher stabilization due to anomeric interactions in β-fructo-pyranose makes it a conformationally locked predominant structure in the gas phase.

Another molecular system wherein the D-glucose units are joined to each other by 1–4 linkage is cyclodextrins (CDs), which are macrocyclic oligosaccharides. They possess a unique ability to entrap guest molecules in their cavities owing to their bucket/bowl shape. Such inclusion complexes are used in the pharmaceutical industry for a variety of formulations. The most commonly known CDs are α-, β- and γ-CDs which have six, seven, and eight glucose units, respectively (see Figure 3). The IHB's between the secondary O–H groups of glucose units result in the formation of a smaller rim of the CD bowls, whereas the primary O–H groups form the larger ones. The strength of the IHBs is suggested to be an important factor governing the stability of the CDs [94–96]. Moreover, one would expect a larger cooperativity contribution due to the formation of a more extended network of HBs between different types of O–H groups. The estimated IHB energies obtained by using the MTA-based method [86] belonged to a wide range of 1.1 to 8.3 kcal/mol at the B3LYP/6-311++G(d,p) level, suggesting that strong HBs are seen in CDs. For the $O_6H \cdots O_6'$ HBs, HB energies fall between 6.7 to 8.3 kcal/mol, and for $O_3H \cdots O_2'$ HBs, they are between 3.3 to 5.5 kcal/mol and 1.9 to 2.8 kcal/mol for $O_2H \cdots O_3$ HBs. The cooperativity contribution by the $O_6H \cdots O_6'$ HB is larger (1.3 to 2.7 kcal/mol) than that for $O_3H \cdots O_2$ (0.3 to 1.0 kcal/mol) and $O_2H \cdots O_3$ (0.25 to 1.10 kcal/mol) HBs. Note that the cooperativity contribution in glucopyranose (0.1 to 1.1 kcal/mol) is much smaller than that in CDs. The higher HB strength in CDs could be one of the possible reasons for the much lower aqueous solubility of the natural CDs than that of comparable acyclic polysaccharides [97–99].

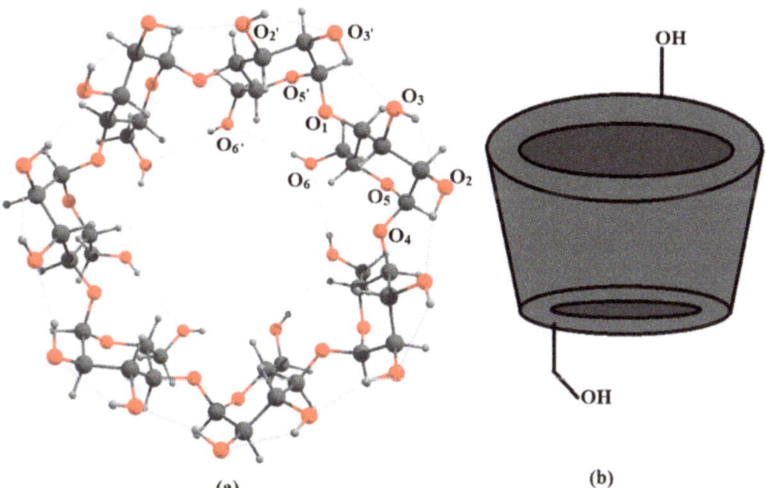

Figure 3. (**a**) A schematic structure of β-cyclodextrin molecule (containing seven glucose units) indicating different types of hydroxyl groups and (**b**) a CD bowl, secondary hydroxyl groups at smaller and primary ones at the larger rim, respectively. See text for details.

We now discuss the HB energies and cooperativity contributions calculated by using MTA in yet another macrocyclic system, viz., *p*-substituted Calix[n]arenes CX[n] (*n* = 4, 5), for exploring substitution effect on the strength of the HBs [87]. The estimated HB energies were between 4.6 to 8.2 kcal/mol, with cooperativity contribution being 0.2 to 2.6 kcal/mol, both calculated at B3LYP/6-311G(d,p) level of theory. The estimated HB energies showed [87] the following order: CX[n]-t-Bu ≈ CX[n]-NH₂ > CX[n]-CH₂Cl >

CX[n] > CX[n]-SO$_3$H > CX[n]-NO$_2$ > S-CX[n]-t-Bu > S-CX[n]. It was also observed that the HB energies in CX[5] derivatives are larger than those in CX[4]. Additionally, as expected, large HB energy values and the respective cooperativity contributions were found in CX[n] hosts with electron-donating substituents.

Recently, we have widened the application of the MTA-based method for the estimation of individual O–H···O HB energy and their cooperativity contribution in water clusters [88]. Many works reported in the literature focus on the global minimum structure of water clusters of variable sizes [73,75,76,100–103]. In spite of a large number of studies, the nature of water clusters at the molecular level is not fully understood. For instance, the strengths of individual O-H···O HB interactions reported in the literature are mostly limited to the water–dimer. Larger clusters are expected to have a cooperativity contribution due to the networking of HBs therein. We recently applied the MTA-based method for the reliable estimation of individual O–H···O HB energies and their cooperativity contribution in small water clusters W$_n$, n = 3 to 8 (see Figure 4).

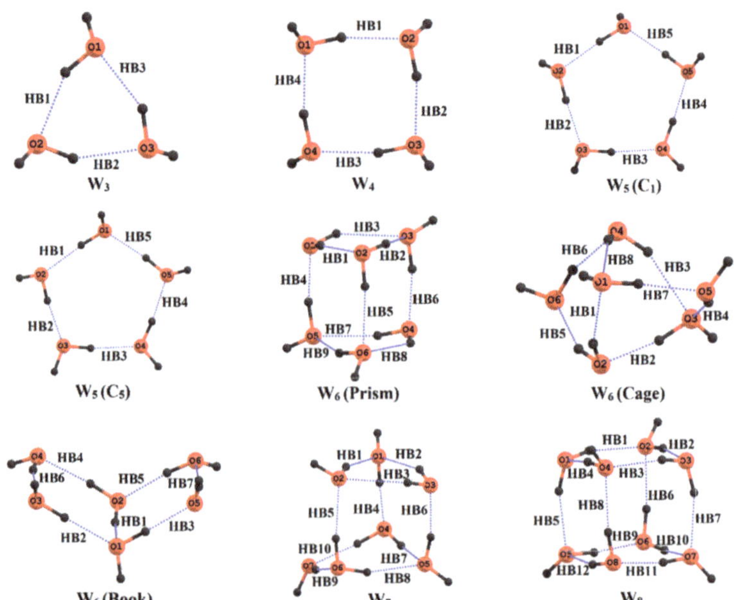

Figure 4. The MP2-optimized geometries of various water clusters (W$_n$). Figure reproduced from Ref. [88] with the permission from American Chemical Society (ACS). *Copyright (2020) The American Chemical Society.*

The fragmentation procedure for the estimation of O–H···O HB energy in W$_n$ is similar to the one for molecules, discussed in the previous section. The difference is that no dummy atoms are needed here because no covalent bond is cut. In the present case, two water molecules involved in the formation of an O–H···O HB were removed for generating the two primary fragments keeping the other water molecules within the cluster intact (see Ref. [88]). The calculated HB energies in W$_n$, for n = 3 to 8, are in the wide range of 0.3 to 10.7 kcal/mol at the CCSD(T)/aug-cc-pVDZ level, with the respective cooperativity contribution being 1.2 to 7.0 kcal/mol. To check the reliability of the results, the sum of all the HB energies for a given cluster was added to the sum of monomers energies. The molecular energy of a water cluster thus estimated agreed well with the actual energy (typical error less than 8.3 mH), suggesting HB energies obtained by our MTA-based method [88] are reliable.

6. Applications to Biomolecules

We now discuss the application of the MTA-based method for the energy estimation of the N–H···O=C, N–H···N, N–H···S, N–H···O, C–H···N, and resonance-assisted O–H···O=C HBs in biomolecules. Quantitative estimates of the strengths of such individual HB interactions are scarcely available, although they play a vital role in the determination of the structure of biomolecules such as polypeptides/proteins. In order to gauge the strengths of the N–H···O=C HBs, a model tetrapeptide was taken as a test case [89]. It has two different N–H···O=C HBs, the fragmentation scheme being similar to that discussed earlier. In the present case, we remove a –(H)N–(O=C)- functional group of amino acids, involved in the formation of N–H···O=C HB to generate the primary fragments (see Ref. [89]). The MTA-based methodology is applied for the estimation of these IHB energies in five different substituted tetrapeptides of polyglycine abbreviated as GGGGG apart from the capped acetyl and NH_2 groups. The five substituted tetrapeptides (viz., GAGGG, GVGGG, GLGGG, and GIGGG) in which the second amino acid residue is replaced by alanine (A), valine (V), leucine (L), and isoleucine (I), respectively, were considered (see Figure 5). The corresponding completely substituted tetrapeptides, AAAAA, VVVVV, LLLLL, and IIIII, were also employed with a view to addressing the effect of substituents on the strengths of the different types of N–H···O=C HBs (Chain 1 and Chain 2 in Figure 5). The estimated HB energies at B3LYP/D95(d,p) level were in the range of 4 to 6 kcal/mol in the partially and fully substituted polypeptides. These values were in concurrence with the geometric parameters and reflected the subtle substituents effects for the substituted polypeptides. The MTA-based procedure was thus considered to be applicable for the IHB energy estimation in polypeptides and it would be fascinating to apply it to an actual protein.

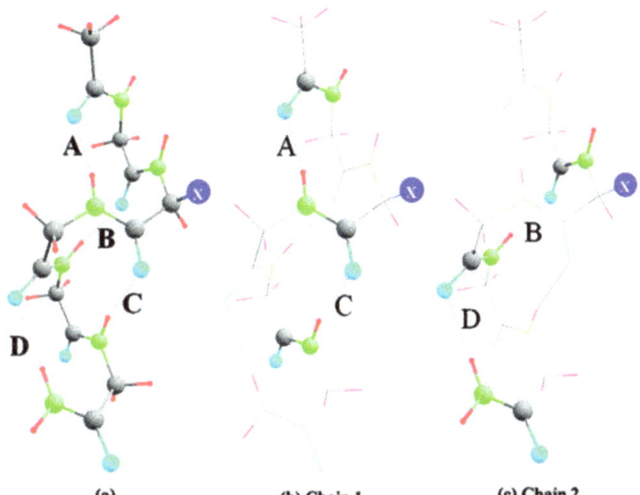

Figure 5. Hydrogen-bonded chains in the helical peptide structures. "X" represents the substituent at the second position. A, B, C and D are four hydrogen bonds. See text for details. Reprinted from Ref. [89] with permission from American Chemical Society. *Copyright (2009) The American Chemical Society.*

Recently, the MTA-based procedure was applied to another biologically important class of molecules, viz., porphyrin analogs called meta-benziporphodimethenes (**1**) and N-confused isomers containing γ-lactam ring (**2** and **3**) [90,91]. In γ-lactam-containing isomers **3** (not shown in Figure 6) the ring O-atom of lactam ring is down, whereas, in isomer **2**, O-atom is up (see Figure 6). The substitution at meso sp^3 (R′) and on the benzene at sp^2 carbons (R_1 to R_5) may affect the strengths of N–H···N, N–H···S, N–H···O, C–H···N HBs, which, in turn, decide the stability of these conformers. Hence, the IHB strengths in substituted **1**, **2**, and **3** were determined by the MTA-based methodology. The estimated N–H···N HB energies were between 11.0 to 15.6 kcal/mol at B3LYP/6-311+G(d,p) level, the individual bifurcated N–H···N interactions being ~6.0 kcal/mol. The N-H···O (~10.0 kcal/mol) and N–H···S (11.2 kcal/mol) HBs were found to be weaker than the N–H···N HBs, with the C-H···N HBs being the weakest (0.1 to 4.4 kcal/mol) [90]. The substituent effect on IHB strength was also investigated [90,91].

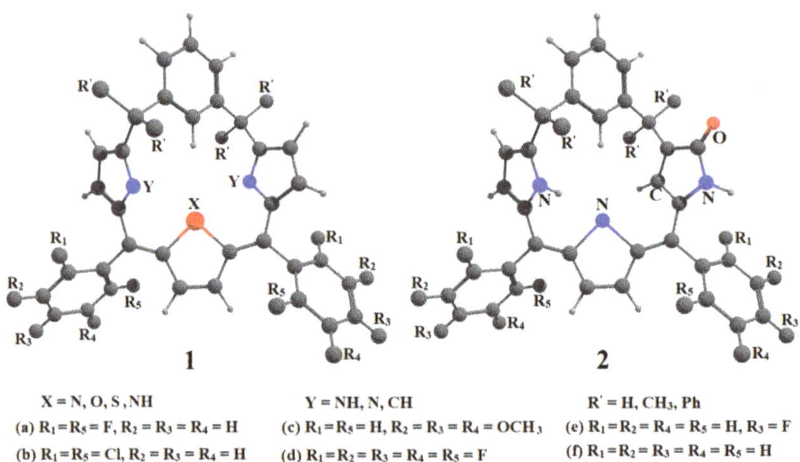

Figure 6. Structures of meta-benziporphodimethene **1** and N-confused meta-benziporphodimethene containing γ-lactam ring isomer (O-up, **2**). See text for details.

It has been suggested that intramolecular O–H···O and resonance-assisted (RA) O–H···O=C HBs may have a pronounced effect on the antioxidant activity of the natural products [104–107]. Hence, an attempt was made to obtain reliable estimates of IHB energies in these molecules [92]. The presence of an RA IHB was seen in all the antioxidant molecules studied. In some molecules, this type of HB was found to be in the nature of two bifurcated HBs along with IHBs such as O–H···OCH₃, O–H···OH, and O–H···O(ring) also being present. The energies of O–H···OCH₃ HBs were found to lie between 2.2 to 2.4 kcal/mol, with the O–H···O (ring) HB energies being much smaller (~0.5 to 0.6 kcal/mol) at the MP2/6-311G(d,p) level. The energy of RA O–H···O=C IHB being largest (10.0 to 17.9 kcal/mol.) Recently, Restrepo et al. [108] suggested that the weak HBs are directly related to the antioxidant properties of these molecules. The reliable estimates of the strengths of these weak HBs by the MTA-based method could be useful for quantitative structure–activity relationship (QSAR) studies on a larger set of antioxidant molecules.

7. Use of the MTA-Based Method by Other Researchers

We summarize here the use of our MTA-based method for the estimation of IHB energy by other researchers [109–117]. In an early use of the method, Lopes Jesus et al. [109] studied the 65 local minima conformers of 1,4-butanediol molecule. The H···O IHB energies in the two lowest energy conformers (c1, c2) were estimated using the MTA method to be 5.5 and 4.1 kcal/mol, respectively. The IHB energies were also obtained by two other methods, viz., conformational analysis (CA) [33,41,42] and use of the empirical Iogansen equation (IE) to spectroscopic data [110]. The IHB energy values turned out to be CA: (4.5 and 3.1 kcal/mol) and IE: (4.1 and 2.2 kcal/mol), in good qualitative agreement with their MTA counterparts.

Rusinska-Roszak et al. extensively used the MTA-based method [111–113]. In an early work [111], they estimated the O–H···O=C IHB energies in several aliphatic systems. However, their fragmentation scheme consisted of two appropriately made primary fragments and an overlapping fragment. The IHB energy was estimated, by using the energies of these three fragments and the actual energy of the parent system, ignoring the ternary and quaternary overlap fragments. The estimated IHB energies at MP2/6-311++(2d,2p) level in the saturated hydroxyl compounds were between 1.4 to 7.0 kcal/mol and 13.7 kcal/mol in the unsaturated ones [111].

In another work [112], they estimated the O−H···O=C IHB energies in 299 structures of hydroxycarbonyl aliphatic compounds involving resonance-assisted HB. The estimated IHB energies showed a wide range, from 8.2 to 26.3 kcal/mol. The HB energies showed a good qualitative correlation with other indirect measures of HB strength, viz., geometrical parameters, the O–H stretching frequencies, and the MED values at the BCP [112].

These authors [113] also applied the MTA-based method for the estimation of Ar–O–H···O=C HBs in mono-, di-, and triphenols substituted (by electron-donating/withdrawing groups) at the ortho- position by carbonyl-containing functional groups. The estimated HB energies for phenolic O–H···O=C (six-membered ring) fall in the range of 5.4 to 15.4 kcal/mol at MP2/6-311++G(2d,2p) level. These HBs energies are smaller (4.6 to 9.6 kcal/mol) when the O=C group is a part of seven or eight-membered rings. These HBs' energy range is similar to corresponding HBs involving saturated carbonyl substituted alcohols.

Very recently, Afonin et al. [114] applied our method for estimating the energy of push–pull effect in β-diketones. In this push–pull system, the intramolecular charge transfer (ICT) occurs as a result of interaction between the π-donor and -acceptor parts, joined by a π linker. Further, the IHB is also present in the Z-conformation of β-diketones. Here, basic idea is to estimate the π component of conjugation energy in these systems. For this purpose, an E conformation of the parent β-diketones was considered wherein this HB is not present. The molecule in this configuration was fragmented using the MTA-based method. The donor and acceptor groups involved in ICT interaction were separated into two primary fragments, F1 and F2, and the overlapping fragment, F3, is the conjugation unit connecting these groups. The energy expression for the π conjugation energy is similar to that discussed in Section 3, viz., $E_\pi(MTA) = [(E_{F1} + E_{F2} - E_{F3}) - E_M]$, E_M being the energy of the parent molecule in E configuration. The effect of electron-donating and -withdrawing groups on the π component of conjugation energy was also estimated. The authors stated [114] that "although the choice of the fragmentation scheme and the computational protocol used did not play a decisive role in estimating the energies of the same IMHB, this issue also needs to be investigated when estimating the conjugation energy."

Afonin et al. [115] applied the methodology for the quantitative decomposition of RAHB energy in β-diketones into resonance and hydrogen bonding components. They also compared the estimated HB energy by the MTA method with that obtained by the functional-based approach (FBA). In FBA, the HB energy is written as an empirical function of HB descriptors, $E_{HB} = f(D)$, the parameter D is one of the H-bond descriptors (i.e., geometrical, topological, and/or spectral characteristics of the H-bond). For details of these empirical FBA equations, see Ref. [115]. It has been shown by these authors that the FBA method evaluates "the component of RAHB interaction corresponding to the energy of the pure H-bond without resonance component." However, the MTA method implicitly takes into account the π component in the RAHB interaction and hence "the difference in the energy of the IMHB as evaluated by means of the MTA and FBA yields a quantitative estimate of the resonance component in the case of the resonance-assisted hydrogen bonds" [115]. The resonance component energy was reported to be 6 to 7 kcal/mol for the weak to strong RAHB, respectively. The HB component varied in the wide range from 2 to 20 kcal/mol in a series of the β-diketone molecules [115]. In summary, the energy of IHB by the MTA-based method provides reliable values for the RAHB, including both the resonance and HB components.

Recently the IHB energies estimated by the MTA-based method were compared with those obtained by other indirect HB descriptors for the wide range of malonaldehyde derivatives by Nowroozi et al. [116]. The latter included structural, spectroscopic, topological, and molecular orbital parameters in the intramolecular RAHBs. The substituent effect of electron-donating and -accepting groups on the HB energy was also estimated by the MTA-based method [116]. Further, the significance of π-electron delocalization (π-ED) of RAHB rings was evaluated by the geometrical factor and the harmonic oscillator model of aromaticity (HOMA). The authors [116] stated that "the excellent linear correlations with MTA energies, which may be implied on the validity of RAHB theory". Further, it was emphasized by authors that "On the basis of these results, one can claim that the MTA method is reliable for estimation of IMHB energies of RAHB systems."

A significant application of our MTA-method was in the determination of the push–pull π^+/π^- (PP$\pi\pi$) effect in the Henry reaction [117], an organocatalytic reaction catalyzed by squaramide [118]. In this reaction, several noncovalent interactions such as HBs, $\pi\cdots$H, $\pi\cdots$O, etc. were observed between the squaramide and benzaldehyde. It was suggested that the reaction proceeds via an unprecedented mode of activation, modulated by $\pi\cdots$H ($\pi\cdots\delta^+$) and $\pi\cdots$O ($\pi\cdots\delta^-$) interactions formed with the two rings of a naphthyl group and benzaldehyde. These authors employed the MTA-based approach for determining the energies of these two π interactions [118]. Here, the naphthyl group involved in the PP$\pi\pi$ interactions, observed in the intermediates (INTs) and transition state (TS) structures, along the most favorable pathway (P1), was replaced with an H atom. Thus, generated INTs and TS structures along the modeled pathway (P1–H) have the same noncovalent interactions as P1, except the two π interactions whose energy is to be determined. The interaction energy of these two π interactions in the INTs and TS structures were estimated as the energy difference in the total interaction energies between the catalyst and the substrates in P1 and that in modeled P1$-$H pathways. The estimated sum of two π interactions in INTs and TS were found to be between 2.7 to 4.7 kcal/mol at the ωB97X-D level of theory. In summary, the MTA-based method proposed by us has been successfully employed by several other active research groups for exploring the strengths of IHBs and other intramolecular noncovalent interactions in a variety of systems.

8. Summary and Concluding Remarks

This review article has summarized a direct and simple procedure for the estimation of X–H$\cdots$Y IHB energy employing a fragmentation method, viz., the molecular tailoring approach (MTA). It has been applied to a variety of systems having multiple HBs over the last one and half decades. A plus point of the method is that it provides the reliable energy of every individual HB and can also estimate the corresponding cooperativity contribution as a result of interconnected networks of HBs. In this present review article, we have discussed the application of the MTA-based method for the estimation of X–H$\cdots$Y (X–H = O–H, N–H, C–H, etc. and Y = N, O, S, OH, OCH$_3$, O=C, etc.) HB energies in a variety of systems. The systems covered ranged from small alkanediols to large systems such as cyclodextrins and biomolecules such as polypeptides, meta-benzoporphodimethenes, and antioxidant molecules. Further, being of general nature, our method has been utilized as a standard method for a reliable estimation of HB energies by other research groups. It may be emphasized here that the MTA-based method is applicable for the estimation of other noncovalent interactions as well. In recent years, this approach has been applied for the estimation of O–H$\cdots$O HB energies in water clusters, the π component of the conjugation energy in resonance-assisted, hydrogen-bonded push–pull systems, etc. In a recent impressive application, the method is employed for determining the favorable organocatalytic reaction pathways [118].

It may be noted that the present method can, in principle, be applied to the estimation of IHB energy in larger molecular systems. However, with the increase in the size of a molecule, the size of the fragments would also increase, making the evaluation of HB energies computationally rather demanding. This difficulty can, in principle, be overcome by the use of the MTA methodology discussed in Section 2. For instance, the energy of the parent and fragment molecules can be reliably calculated using MTA at a correlated level of theory which may be further used for estimating the HB energies in larger molecular systems.

Being general in nature, the MTA-based method can be employed for exploring other intramolecular interactions such as $\pi \cdots \pi$ [119], C-H$\cdots\pi$ [120], the so-called halogen bonds [121], dihydrogen bonds [122], sulfur bonds [123], metal-H$\cdots$S and metal-H$\cdots$Se bonds [124], etc. Although identified in the recent literature by these labels, they are cut from the same cloth called the *non-covalent interactions*. The prowess of the MTA-based method is that it can be applied to all such inter- and intramolecular interactions.

Author Contributions: M.M.D. and S.R.G. contribute equally to this article, conceptualization, methodology, writing and review. All authors have read and agreed to the published version of the manuscript.

Funding: The publication of this research received no external funding.

Institutional Review Board Statement: Not applicable.

Informed Consent Statement: Not applicable.

Data Availability Statement: The data presented in this study are available on request from the corresponding authors.

Acknowledgments: M.M.D. thanks the University Grant Commission (UGC), New Delhi, for the Start-up grant (F.30-56/2014/BSR). S.R.G. is thankful to the National Supercomputing Mission (NSM) for support under the project [CORP: DG:3187], which enabled the use of computational resources at the PARAM Shivay facility at IIT, BHU.

Conflicts of Interest: The authors declare no conflict of interest.

References

1. Pauling, L. *The Nature of the Chemical Bond*; Oxford University Press: London, UK, 1939.
2. Jeffery, G.A. *An Introduction to Hydrogen Bonding*; Oxford University Press: New York, NY, USA, 1997.
3. Desiraju, G.R.; Steiner, T. The Weak Hydrogen Bond. In *Structural Chemistry and Biology*; Oxford University Press: New York, NY, USA, 1999.
4. Scheiner, S. *Hydrogen Bond: A Theoretical Perspective*; Oxford University Press: New York, NY, USA, 1997.
5. Huggins, M.L. Hydrogen Bridges in Ice and Liquid Water. *J. Phys. Chem.* **1936**, *40*, 723–731. [CrossRef]
6. Huggins, M.L. Hydrogen Bridges in Organic Compounds. *J. Org. Chem.* **1936**, *1*, 407–456. [CrossRef]
7. Nernst, W. Über die Löslichkeit von Mischkrystallen. *Z. Phys. Chem.* **1892**, *9*, 137–142. [CrossRef]
8. Werner, A. Ueber Haupt und Nebenvalenzen und die Constitution der Ammoniumverbindungen. *Liebigs Ann. Chem.* **1902**, *322*, 261–296. [CrossRef]
9. Moore, T.S.; Winmill, T.F. The State of Amines in Aqueous Solution. *J. Chem. Soc.* **1912**, *101*, 1635–1676. [CrossRef]
10. Latimer, W.M.; Rodebush, W.H. Polarity and Ionization from the Standpoint of the Lewis Theory of Valence. *J. Am. Chem. Soc.* **1920**, *42*, 1419–1443. [CrossRef]
11. Barnes, W.H. The Crystal Structure of Ice between 0 °C and −183 °C. *Proc. R. Soc. Lond. A* **1929**, *125*, 670–693.
12. Huggins, M.L. 50 Years of Hydrogen Bond Theory. *Angew. Chem. Int. Ed. Engl.* **1971**, *10*, 147–152. [CrossRef]
13. Pimentel, G.C.; McClellan, A.L. *The Hydrogen Bond*; Freeman: San Francisco, CA, USA, 1960.
14. Arunan, E.; Desiraju, G.R.; Klein, R.A.; Sadlej, J.; Scheiner, S.; Alkorta, I.; Clary, D.C.; Crabtree, R.H.; Dannenberg, J.J.; Hobza, P.; et al. Defining the Hydrogen Bond: An Account (IUPAC Technical Report). *Pure Appl. Chem.* **2011**, *83*, 1619–1636. [CrossRef]
15. Pauling, L. The Nature of the Chemical Bond. Applications of Results Obtained from the Quantum Mechanics and from a Theory of Paramagnetic Susceptibility of the Structure of Molecules. *J. Am. Chem. Soc.* **1931**, *53*, 1367–1400. [CrossRef]
16. Chaplin, M.F. *Water and Life: The Unique Properties of H_2O*; Lynden-Bell, R.M., Morris, S.M., Barrow, J.D., Finney, J.L., Harper, C., Eds.; CRC Press: Boca Raton, FL, USA, 2010; pp. 69–86.
17. Laage, D.; Elsaesser, T.; Hynes, J.T. Water Dynamics in the Hydration Shells of Biomolecules. *Chem. Rev.* **2017**, *117*, 10694–10725. [CrossRef] [PubMed]
18. Bellissent-Funel, M.-C.; Hassanali, A.; Havenith, M.; Henchman, R.; Pohl, P.; Sterpone, F.; van der Spoel, D.; Xu, Y.; Garcia, A.E. Water Determines the Structure and Dynamics of Proteins. *Chem. Rev.* **2016**, *116*, 7673–7697. [CrossRef] [PubMed]
19. Keutsch, F.N.; Cruzan, J.D.; Saykally, R.J. The Water Trimer. *Chem. Rev.* **2003**, *103*, 2533–2578. [CrossRef]

20. Sánchez-García, E.; George, L.; Montero, L.A.; Sander, W. 1:2 Formic Acid/Acetylene Complexes: Ab Initio and Matrix Isolation Studies of Weakly Interacting Systems. *J. Phys. Chem. A* **2004**, *108*, 11846–11854. [CrossRef]
21. Munshi, P.; Row, T.N.G. Exploring the Lower Limit in Hydrogen Bonds: Analysis of Weak C−H···O and C−H···π Interactions in Substituted Coumarins from Charge Density Analysis. *J. Phys. Chem. A* **2005**, *109*, 659–672. [CrossRef]
22. Cockroft, S.L.; Hunter, C.A.; Lawso, K.R.; Perkins, J.; Urch, C.J. Electrostatic Control of Aromatic Stacking Interactions. *J. Am. Chem. Soc.* **2005**, *127*, 8594–8595. [CrossRef] [PubMed]
23. Grabowski, S.J. Hydrogen Bonding Strength-Measures Based on Geometric and Topological Parameters. *J. Phys. Org. Chem.* **2004**, *17*, 18–31. [CrossRef]
24. Bellamy. *Advances in Infrared Group Frequencies*; Methuen: London, UK, 1968; p. 241.
25. Glasel, J.A. *Water: A Comprehensive Treatise*; Frank, F., Ed.; Plenum Press: New York, NY, USA; London, UK, 1982; Volume 1, p. 223, Chapter 6.
26. Hobza, P.; Havlas, Z. Blue-Shifting Hydrogen Bonds. *Chem. Rev.* **2000**, *100*, 4253–4264. [CrossRef]
27. Fuster, F.; Silvi, B. Does the Topological Approach Characterize the Hydrogen Bond? *Theor. Chem. Acc.* **2000**, *104*, 13–21. [CrossRef]
28. Espinosa, E.; Alkorta, I.; Elguero, J.; Molins, E. From Weak to Strong Interactions: A Comprehensive Analysis of the Topological and Energetic Properties of the Electron Density Distribution Involving X–H···F–Y Systems. *J. Chem. Phys.* **2002**, *117*, 5529–5542. [CrossRef]
29. Klein, R.A. Hydrogen Bonding in Diols and Binary Diol–Water Systems Investigated using DFT Methods. II. Calculated Infrared OH-Stretch Frequencies, Force Constants, and NMR Chemical Shifts Correlate with Hydrogen Bond Geometry and Electron Density Topology. A Reevaluation of Geometrical Criteria for Hydrogen Bonding. *J. Comput. Chem.* **2003**, *24*, 1120–1131.
30. Deshmukh, M.M.; Sastry, N.V.; Gadre, S.R. Molecular Interpretation of Water Structuring and Destructuring Effects: Hydration of Alkanediols. *J. Chem. Phys.* **2004**, *121*, 12402–12410. [CrossRef]
31. Kovács, A.; Szabo, A.; Hargittai, I. Structural Characteristics of Intramolecular Hydrogen Bonding in Benzene Derivatives. *Acc. Chem. Res.* **2002**, *35*, 887–894. [CrossRef]
32. Chung, G.; Kwon, O.; Kwon, Y. Theoretical Study on 1,2-Dihydroxybenzene and 2-Hydroxythiophenol: Intramolecular Hydrogen Bonding. *J. Phys. Chem. A* **1997**, *101*, 9415–9420. [CrossRef]
33. Rozas, I.; Alkorta, I.; Elguero, J. Intramolecular Hydrogen Bonds in ortho-Substituted Hydroxybenzenes and in 8-Susbtituted 1-Hydroxynaphthalenes: Can a Methyl Group Be an Acceptor of Hydrogen Bonds? *J. Phys. Chem. A* **2001**, *105*, 10462–10467. [CrossRef]
34. Lipkowski, P.; Koll, A.; Karpfen, A.; Wolschann, P. An Approach to Estimate the Energy of the Intramolecular Hydrogen Bond. *Chem. Phys. Lett.* **2002**, *360*, 256–263. [CrossRef]
35. Buemi, G.; Zuccarello, F. Is the Intramolecular Hydrogen Bond Energy Valuable from Internal Rotation Barriers? *J. Mol. Struct. Theochem.* **2002**, *581*, 71–85. [CrossRef]
36. Korth, H.-G.; de Heer, M.I.; Mulder, P. A DFT Study on Intramolecular Hydrogen Bonding in 2-Substituted Phenols: Conformations, Enthalpies, and Correlation with Solute Parameters. *J. Phys. Chem. A* **2002**, *106*, 8779–8789. [CrossRef]
37. Kjaergaard, H.G.; Howard, D.L.; Schofield, D.P.; Robinson, T.W.; Ishiuchi, S.; Fujii, M. OH- and CH-Stretching Overtone Spectra of Catechol. *J. Phys. Chem. A* **2002**, *106*, 258–266. [CrossRef]
38. Zhang, H.-Y.; Sun, Y.-M.; Wang, X.-L. Substituent Effects on O-H Bond Dissociation Enthalpies and Ionization Potentials of Catechols: A DFT Study and Its Implications in the Rational Design of Phenolic Antioxidants and Elucidation of Structure–Activity Relationships for Flavonoid Antioxidants. *Chem. Eur. J.* **2003**, *9*, 502–508. [CrossRef]
39. Bakalbassis, E.G.; Lithoxoidou, A.T.; Vafiadis, A.P. Theoretical Calculation of Accurate Absolute and Relative Gas- and Liquid-Phase O-H Bond Dissociation Enthalpies of 2-Mono- and 2,6-Disubstituted Phenols, Using DFT/B3LYP. *J. Phys. Chem. A* **2003**, *107*, 8594–8606. [CrossRef]
40. Jabłoński, M.A. Critical Overview of Current Theoretical Methods of Estimating the Energy of Intramolecular Interactions. *Molecules* **2020**, *25*, 5512. [CrossRef]
41. Jabłoński, M.; Kaczmarek, A.; Sadlej, A.J. Estimates of the Energy of Intramolecular Hydrogen Bonds. *J. Phys. Chem. A* **2006**, *110*, 10890–10898. [CrossRef] [PubMed]
42. Jabłoński, M. Full vs. Constrained Geometry Optimization in the Open-Closed Method in Estimating the Energy of Intramolecular Charge-Inverted Hydrogen Bonds. *Chem. Phys.* **2010**, *376*, 76–83. [CrossRef]
43. Estacio, S.G.; Cabral do Counta, P.; Costa Cabral, B.J.; Minas Da Piedade, M.E.; Martinho Simoes, J.A. Energetics of Intramolecular Hydrogen Bonding in Di-substituted Benzenes by the ortho-para Method. *J. Phys. Chem. A* **2004**, *108*, 10834–10843. [CrossRef]
44. Hammett, L.P. The Effect of Structure upon the Reactions of Organic Compounds. Benzene Derivatives. *J. Am. Chem. Soc.* **1937**, *59*, 96–103. [CrossRef]
45. Hammett, L.P. *Physical Organic Chemistry*; McGraw-Hill: New York, NY, USA, 1940.
46. Suresh, C.H.; Gadre, S.R. A Novel Electrostatic Approach to Substituent Constants: Doubly Substituted Benzenes. *J. Am. Chem. Soc.* **1998**, *120*, 7049–7055. [CrossRef]
47. Hehre, W.J.; Ditchfield, R.; Radom, L.; Pople, J.A. Molecular Orbital Theory of the Electronic Structure of Organic Compounds. V. Molecular Theory of Bond Separation. *J. Am. Chem. Soc.* **1970**, *92*, 4796–4801. [CrossRef]

48. Suresh, C.H.; Koga, N. An Isodesmic Reaction Based Approach to Aromaticity of a Large Spectrum of Molecules. *Chem. Phys. Lett.* **2006**, *419*, 550–556. [CrossRef]

49. Wheeler, S.E.; Houk, K.N.; Schleyer, P.; Allen, W.D. A Hierarchy of Homodesmotic Reactions for Thermochemistry. *J. Am. Chem. Soc.* **2009**, *131*, 2547–2560. [CrossRef]

50. George, P.; Trachtman, M.; Brett, A.M.; Bock, C.W. Comparison of Various Isodesmic and Homodesmotic Reaction Heats with Values derived from Published *Ab initio* Molecular Orbital Calculations. *J. Chem. Soc. Perkin Trans.* **1977**, *2*, 1036–1047. [CrossRef]

51. Howard, S.T. Relationship between Basicity, Strain, and Intramolecular Hydrogen-Bond Energy in Proton Sponges. *J. Am. Chem. Soc.* **2000**, *122*, 8238–8244. [CrossRef]

52. Deshmukh, M.M.; Suresh, C.H.; Gadre, S.R. Intramolecular Hydrogen Bond Energy in Polyhydroxy Systems: A Critical Comparison of Molecular Tailoring and Isodesmic Approaches. *J. Phys. Chem. A* **2007**, *111*, 6472–6480. [CrossRef] [PubMed]

53. Bader, R.F.W. *Atoms in Molecules: A Quantum Theory*; Oxford University Press: New York, NY, USA, 1990.

54. Bader, R.F.W. A Bond Path: A Universal Indicator of Bonded Interactions. *J. Phys. Chem. A* **1998**, *102*, 7314–7323. [CrossRef]

55. Haaland, A.; Shorokhov, D.J.; Tverdova, N.V. Topological Analysis of Electron Densities: Is the Presence of an Atomic Interaction Line in an Equilibrium Geometry a Sufficient Condition for the Existence of a Chemical Bond? *Chem. Eur. J.* **2004**, *10*, 4416–4421. [CrossRef]

56. Poater, J.; Solà, M.; Bickelhaupt, F.M. A Model of the Chemical Bond Must Be Rooted in Quantum Mechanics, Provide Insight, and Possess Predictive Power. *Chem. Eur. J.* **2006**, *12*, 2902–2905. [CrossRef]

57. Dem'yanov, P.; Polestshuk, P. A Bond Path and an Attractive Ehrenfest Force Do Not Necessarily Indicate Bonding Interactions: Case Study on M_2X_2 (M = Li, Na, K; X = H, OH, F, Cl). *Chem. Eur. J.* **2012**, *18*, 4982–4993. [CrossRef]

58. Jabłoński, M. Bond Paths Between Distant Atoms Do Not Necessarily Indicate Dominant Interactions. *J. Comput. Chem.* **2018**, *39*, 2183–2195. [CrossRef]

59. Espinosa, E.; Molins, E.; Lecomte, C. Hydrogen Bond Strength Revealed by Topological Analyses of Experimentally Observed Electron Densities. *Chem. Phys. Lett.* **1998**, *285*, 170–173. [CrossRef]

60. Guevara-Vela, J.M.; Francisco, F.; Rocha-Rinza, T.; Pendás, A.M. Interacting Quantum Atoms-Review. *Molecules* **2020**, *25*, 4028. [CrossRef]

61. Gatti, C.; May, E.; Destro, R.; Cargnoni, F. Fundamental Properties and Nature of CH···O Interactions in Crystals on the Basis of Experimental and Theoretical Charge Densities. The Case of 3,4-Bis(dimethylamino)-3-cyclobutene-1,2-dione (DMACB) Crystal. *J. Phys. Chem. A* **2002**, *106*, 2707–2720. [CrossRef]

62. Nikolaienko, T.Y.; Bulavin, L.A.; Hovorun, D.M. Bridging QTAIM with Vibrational Spectroscopy: The Energy of Intramolecular Hydrogen Bonds in DNA-related Biomolecules. *Phys. Chem. Chem. Phys.* **2012**, *14*, 7441–7447. [CrossRef] [PubMed]

63. Klein, R.A. Electron Density Topological Analysis of Hydrogen Bonding in Glucopyranose and Hydrated Glucopyranose. *J. Am. Chem. Soc.* **2002**, *124*, 13931–13937. [CrossRef]

64. Klein, R.A. Ab initio conformational studies on diols and binary diol-water systems using DFT methods. Intramolecular hydrogen bonding and 1:1 complex formation with water. *J. Comput. Chem.* **2002**, *23*, 585–599.

65. Howard, D.L.; Jørgensen, P.; Kjaergaard, H.G. Weak Intramolecular Interactions in Ethylene Glycol Identified by Vapor Phase OH-Stretching Overtone Spectroscopy. *J. Am. Chem. Soc.* **2005**, *127*, 17096–17103. [CrossRef]

66. Deshmukh, M.M.; Gadre, S.R.; Bartolotti, L.J. Estimation of Intramolecular Hydrogen Bond Energy via Molecular Tailoring Approach. *J. Phys. Chem. A* **2006**, *110*, 12519–12523.

67. Deshmukh, M.M.; Bartolotti, L.J.; Gadre, S.R. Intramolecular Hydrogen Bonding and Cooperative Interactions in Carbohydrates via the Molecular Tailoring Approach. *J. Phys. Chem. A* **2008**, *112*, 312–321. [CrossRef]

68. Gadre, S.R.; Shirsat, R.N.; Limaye, A.C. Molecular Tailoring Approach for Simulation of Electrostatic Properties. *J. Phys. Chem.* **1994**, *98*, 9165–9169. [CrossRef]

69. Ganesh, V.; Dongare, R.K.; Balanarayan, P.; Gadre, S.R. Molecular Tailoring Approach for Geometry Optimization of Large Molecules: Energy Evaluation and Parallelization Strategies. *J. Chem. Phys.* **2006**, *125*, 104109–104110. [CrossRef]

70. Elango, M.; Subramanian, V.; Rahalkar, A.P.; Gadre, S.R.; Sathyamurthy, N. Structure, Energetics, and Reactivity of Boric Acid Nanotubes: A Molecular Tailoring Approach. *J. Phys. Chem. A* **2008**, *112*, 7699–7704. [CrossRef]

71. Gadre, S.R.; Ganesh, V. Molecular Tailoring Approach: Towards PC-based ab Initio Treatment of Large Molecules. *J. Theor. Comput. Chem.* **2006**, *5*, 835–855. [CrossRef]

72. Rahalkar, A.P.; Ganesh, V.; Gadre, S.R. Enabling Ab-initio Hessian and Frequency Calculations of Large Molecules. *J. Chem. Phys.* **2008**, *129*, 234101. [CrossRef]

73. Sahu, N.; Gadre, S.R. Accurate Vibrational Spectra via Molecular Tailoring Approach: A Case Study of Water Clusters at MP2 Level. *J. Chem. Phys.* **2015**, *142*, 014107. [CrossRef]

74. Sahu, N.; Gadre, S.R. Vibrational Infrared and Raman Spectra of Polypeptides: Fragments-in-fragments Within Molecular Tailoring Approach. *J. Chem. Phys.* **2016**, *144*, 114113. [CrossRef]

75. Furtado, J.P.; Rahalkar, A.P.; Shankar, S.; Bandyopadhyay, P.; Gadre, S.R. Facilitating Minima Search for Large Water Clusters at the MP2 Level via Molecular Tailoring. *J. Phys. Chem. Lett.* **2012**, *3*, 2253–2258. [CrossRef] [PubMed]

76. Sahu, N.; Singh, G.; Nandi, A.; Gadre, S.R. Toward an Accurate and Inexpensive Estimation of CCSD (T)/CBS Binding Energies of Large Water Clusters. *J. Phys. Chem. A* **2016**, *120*, 5706–5714. [CrossRef]

77. Gadre, S.R.; Yeole, S.D.; Sahu, N. Quantum Chemical Investigations on Molecular Clusters. *Chem. Rev.* **2014**, *114*, 12132–12173. [CrossRef]

78. Singh, G.; Nandi, A.; Gadre, S.R. Breaking the Bottleneck: Use of Molecular Tailoring Approach for the Estimation of Binding Energies at MP2/CBS Limit for Large Water Clusters. *J. Chem. Phys.* **2016**, *144*, 104102. [CrossRef] [PubMed]

79. Yeole, S.D.; Gadre, S.R. On the Applicability of Fragmentation Methods to Conjugated π-Systems within Density Functional Framework. *J. Chem. Phys.* **2010**, *132*, 094102. [CrossRef]

80. Yeole, S.D.; Gadre, S.R. Molecular Cluster Building Algorithm: Electrostatic Guidelines and Molecular Tailoring Approach. *J. Chem. Phys.* **2011**, *134*, 084111. [CrossRef]

81. Khire, S.S.; Gadre, S.R. Pragmatic Many-body Approach for Economic MP2 Energy Estimation of Molecular Clusters. *J. Phys. Chem. A* **2019**, *123*, 5005–5011. [CrossRef] [PubMed]

82. Khire, S.S.; Bartolotti, L.J.; Gadre, S.R. Harmonizing Accuracy and Efficiency: A Pragmatic Approach to Fragmentation of Large Molecules. *J. Chem. Phys.* **2018**, *149*, 064112. [CrossRef] [PubMed]

83. Frisch, M.J.; Trucks, G.W.; Schlegel, H.B.; Scuseria, G.E.; Robb, M.A.; Cheeseman, J.R.; Scalmani, G.; Barone, V.; Petersson, G.A.; Nakatsuji, H.; et al. *Gaussian 16, Revision A.03*; Gaussian, Inc.: Wallingford, CT, USA, 2016.

84. Siddiqui, N.; Singh, V.; Deshmukh, M.M.; Gurunath, R. Structures, Stability and Hydrogen Bonding in Inositol Conformers. *Phys. Chem. Chem. Phys.* **2015**, *17*, 18514–18523. [CrossRef] [PubMed]

85. Deshmukh, M.M.; Gadre, S.R.; Cocinero, E.M. Stability of Conformationally Locked Free Fructose: Theoretical and Computational Insights. *New J. Chem.* **2015**, *39*, 9006–9018. [CrossRef]

86. Deshmukh, M.M.; Bartolotti, L.J.; Gadre, S.R. Intramolecular Hydrogen Bond Energy and Cooperative Interactions in α-, β-, and γ-cyclodextrin Conformers. *J. Comput. Chem.* **2011**, *32*, 2996–3004. [CrossRef]

87. Khedkar, J.K.; Deshmukh, M.M.; Gejji, S.P.; Gadre, S.R. Intramolecular Hydrogen Bonding and Cooperative Interactions in Calix[n]arenes (n = 4,5). *J. Phys. Chem. A* **2012**, *116*, 3739–3744. [CrossRef]

88. Ahirwar, M.B.; Gadre, S.R.; Deshmukh, M.M. Direct and Reliable Method for Estimating the Hydrogen Bond Energies and Cooperativity in Water Clusters Wn, n = 3 to 8. *J. Phys. Chem. A* **2020**, *124*, 6699–6706. [CrossRef]

89. Deshmukh, M.M.; Gadre, S.R. Estimation of N-H···O=C Intramolecular Hydrogen Bond Energy in Polypeptides. *J. Phys. Chem. A* **2009**, *113*, 7927–7932. [CrossRef] [PubMed]

90. Kumar, A.; Hung, C.-H.; Rana, S.; Deshmukh, M.M. Study on the Structure, Stability, and Tautomerism of meta-Benziporphodimethenes and N-Confused Isomers Containing γ-Lactam Ring. *J. Mol. Struct.* **2019**, *1187*, 138–150. [CrossRef]

91. Ahluwalia, D.; Kumar, A.; Warkar, S.G.; Deshmukh, M.M. Effect of Substitutions on the Geometry and Intramolecular Hydrogen Bond Strength in meta-benziporphodimethenes: A New Porphyrin Analogue. *J. Mol. Struct.* **2020**, *1220*, 128773. [CrossRef]

92. Singh, V.; Ibnusaud, I.; Gadre, S.R.; Deshmukh, M.M. Fragmentation Method Reveals a Wide Spectrum of Intramolecular Hydrogen Bond Energies in Antioxidant Natural Products. *New J. Chem.* **2020**, *44*, 5841–5849. [CrossRef]

93. Afonin, A.V.; Vashchenkoa, A.V.; Sigalov, M.V. Estimating the energy of intramolecular hydrogen bonds from ^{1}H NMR and QTAIM calculations. *Org. Biomol. Chem.* **2016**, *14*, 11199–11211. [CrossRef]

94. Lipkowitz, K.B. Applications of Computational Chemistry to the Study of Cyclodextrins. *Chem. Rev.* **1998**, *98*, 1829–1874. [CrossRef]

95. Crini, G. Review: A History of Cyclodextrins. *Chem. Rev.* **2014**, *114*, 10940–10975. [CrossRef] [PubMed]

96. Pinjari, R.V.; Joshi, K.A.; Gejji, S.P. Theoretical Studies on Hydrogen Bonding, NMR Chemical Shifts and Electron Density Topography in α, β and γ-Cyclodextrin Conformers. *J. Phys. Chem. A* **2007**, *111*, 13583–13589. [CrossRef] [PubMed]

97. Higuchi, T.; Connors, K.A. *Advances in Analytical Chemistry and Instrumentation*; Reilly, C.N., Ed.; Wiley-Interscience: New York, NY, USA, 1965; Volume 4, pp. 117–212.

98. Loftsson, T.; Brewster, M.E. Pharmaceutical Applications of Cyclodextrins. 1. Drug Solubilization and Stabilization. *J. Pharm. Sci.* **1996**, *85*, 1017–1025. [CrossRef] [PubMed]

99. Loftsson, T.; Jarho, P.; Másson, M.; Järvine, T. Cyclodextrins in Drug Delivery. *Expert Opin. Drug Deliv.* **2005**, *2*, 335–351. [CrossRef] [PubMed]

100. Yoo, S.; Aprà, E.; Zeng, X.C.; Xantheas, S.S. High-Level Ab Initio Electronic Structure Calculations of Water Clusters $(H_2O)_{16}$ and $(H_2O)_{17}$: A New Global Minimum for $(H_2O)_{16}$. *J. Phys. Chem. Lett.* **2010**, *1*, 3122–3127. [CrossRef]

101. Maheshwary, S.; Patel, N.; Sathyamurthy, N.; Kulkarni, A.D.; Gadre, S.R. Structure and Stability of Water Clusters $(H_2O)_n$, n =8−20: An Ab Initio Investigation. *J. Phys. Chem. A* **2001**, *105*, 10525–10537. [CrossRef]

102. Kazachenko, S.; Thakkar, A.J. Improved Minima-Hopping. TIP4P Water Clusters, $(H_2O)_n$ with n $\leq$ 37. *Chem. Phys. Lett.* **2009**, *476*, 120–124. [CrossRef]

103. Sahu, N.; Gadre, S.R.; Rakshit, A.; Bandyopadhyay, P.; Miliordos, E.; Xantheas, S.S. Low Energy Isomers of $(H_2O)_{25}$ from a Hierarchical Method Based on Monte Carlo Temperature Basin Paving and Molecular Tailoring Approaches Benchmarked by MP2 Calculations. *J. Chem. Phys.* **2014**, *141*, 164304. [CrossRef] [PubMed]

104. Foti, M.C.; Johnson, E.R.; Vinqvist, M.R.; Wright, J.S.; Barclay, L.R.C.; Ingold, K.U. Naphthalene Diols: A New Class of Antioxidants Intramolecular Hydrogen Bonding in Catechols, Naphthalene Diols, and Their Aryloxyl Radicals. *J. Org. Chem.* **2002**, *67*, 5190–5196. [CrossRef] [PubMed]

105. Foti, M.C.; Barclay, L.R.C.; Ingold, K.U. The Role of Hydrogen Bonding on the H-Atom-Donating Abilities of Catechols and Naphthalene Diols and on a Previously Overlooked Aspect of Their Infrared Spectra. *J. Am. Chem. Soc.* **2002**, *124*, 12881–12888. [CrossRef] [PubMed]

106. Amorati, R.; Lucarini, M.; Mugnaini, V.; Pedulli, G.F. Antioxidant Activity of o-Bisphenols: The Role of Intramolecular Hydrogen Bonding. *J. Org. Chem.* **2003**, *68*, 5198–5204. [CrossRef] [PubMed]

107. Martínez-Cifuentes, M.; Weiss-López, B.E.; Santos, L.S.; Araya-Maturana, R. Intramolecular Hydrogen Bond in Biologically Active o-Carbonyl Hydroquinones. *Molecules* **2014**, *19*, 9354–9368. [CrossRef]

108. Llano, S.; Gómez, S.; Londoño, J.; Restropo, A. Antioxidant activity of Curcuminoids. *Phys. Chem. Chem. Phys.* **2019**, *21*, 3752–3760. [CrossRef]

109. Lopes Jesus, A.J.; Rosado, M.T.S.; Reva, I.; Fausto, R.; Eusébio, M.E.S.; Redinha, J.S. Structure of Isolated 1,4-Butanediol: Combination of MP2 Calculations, NBO Analysis, and Matrix-Isolation Infrared Spectroscopy. *J. Phys. Chem. A* **2008**, *112*, 4669–4678. [CrossRef]

110. Iogansen, A.V. Direct Proportionality of the Hydrogen Bonding Energy and the Intensification of the Stretching ν(XH) Vibration in Infrared Spectra. *Spectrochim. Acta Part A* **1999**, *55*, 1585–1612. [CrossRef]

111. Rusinska-Roszak, D.; Sowinski, G. Estimation of the Intramolecular O−H···O=C Hydrogen Bond Energy via the Molecular Tailoring Approach. Part I: Aliphatic Structures. *J. Chem. Inf. Model.* **2014**, *54*, 1963–1977. [CrossRef]

112. Rusinska-Roszak, D. Intramolecular O−H···O=C Hydrogen Bond Energy via the Molecular Tailoring Approach to RAHB Structures. *J. Phys. Chem. A* **2015**, *119*, 3674–3687. [CrossRef] [PubMed]

113. Rusinska-Roszak, D. Energy of Intramolecular Hydrogen Bonding in ortho-Hydroxybenzaldehydes, Phenones and Quinones. Transfer of Aromaticity from ipso-Benzene Ring to the Enol System(s). *Molecules* **2017**, *22*, 481. [CrossRef]

114. Afonin, A.V.; Rusinska-Roszak, D. Molecular Tailoring Approach–New Guide to Quantify the Energy of Push-Pull Effect: Case Study on the (E) 3-(1H-pyrrol-2-yl)prop-2-enones. *Phys. Chem. Chem. Phys.* **2020**, *22*, 22190–22194. [CrossRef]

115. Afonin, A.V.; Vashchenko, A.V. Quantitative Decomposition of Resonance-assisted Hydrogen Bond Energy in β-diketones into Resonance and Hydrogen Bonding (π- and σ-) Components using Molecular Tailoring and Function-based Approaches. *J. Comput. Chem.* **2020**, *41*, 1285–1298. [CrossRef] [PubMed]

116. Keykhaei, A.; Nowroozi, A. On the Performance of Molecular Tailoring Approach for Estimation of the Intramolecular Hydrogen Bond Energies of RAHB Systems: A Comparative Study. *Struct. Chem.* **2020**, *31*, 423–433. [CrossRef]

117. Henry, L. Formation synthétique d'alcools nitrés. *Ir. J. Med. Sci.* **1895**, *120*, 1265–1268.

118. Alegre-Requena, J.V.; Marqueés-Loópez, E.; Herrera, R.P. "Push−Pull π+/π−" (PPππ) Systems in Catalysis. *ACS Catal.* **2017**, *7*, 6430–6439. [CrossRef]

119. Tsuzuki, S.; Honda, K.; Uchimaru, T.; Mikami, M.; Tanabe, K. Origin of Attraction and Directionality of the π/π Interaction: Model Chemistry Calculations of Benzene Dimer Interaction. *J. Am. Chem. Soc.* **2002**, *124*, 104–112. [CrossRef]

120. Mishra, B.K.; Deshmukh, M.M.; Ramanathan, V. C-H···π interactions and the nature of the donor carbon atom. *J. Org. Chem.* **2014**, *79*, 8599–8606. [CrossRef]

121. Desiraju, G.; Ho, P.S.; Kloo, L.; Anthony, C.L.; Marquardt, R.; Metrangolo, P.; Politzer, P.; Resnati, G.; Rissanen, K. Definition of the halogen bond (IUPAC Recommendations 2013). *Pure Appl. Chem.* **2013**, *85*, 1711–1713. [CrossRef]

122. Grabowski, S.J. Dihydrogen bond and X-H···σ interaction as sub-classes of hydrogen bond. *J. Phys. Org. Chem.* **2013**, *26*, 452–459. [CrossRef]

123. Biswal, H.S.; Wategaonkar, S. Sulfur, not too far behind O, N, and C: SH···π hydrogen bond. *J. Phys. Chem. A* **2009**, *113*, 12774–12782. [CrossRef] [PubMed]

124. Sahoo, D.K.; Jena, S.; Dutta, J.; Rana, A.; Biswal, H.S. Nature and strength of M−H···S and M−H···Se (M = Mn, Fe, & Co) hydrogen bond. *J. Phys. Chem. A* **2019**, *123*, 2227–2236. [PubMed]

Review

A Spectroscopic Overview of Intramolecular Hydrogen Bonds of NH . . . O,S,N Type

Poul Erik Hansen

Department of Science and Environment, Roskilde University, Universitetsvej 1, DK-4000 Roskilde, Denmark; poulerik@ruc.dk

Abstract: Intramolecular NH . . . O,S,N interactions in non-tautomeric systems are reviewed in a broad range of compounds covering a variety of NH donors and hydrogen bond acceptors. ^{1}H chemical shifts of NH donors are good tools to study intramolecular hydrogen bonding. However in some cases they have to be corrected for ring current effects. Deuterium isotope effects on ^{13}C and ^{15}N chemical shifts and primary isotope effects are usually used to judge the strength of hydrogen bonds. Primary isotope effects are investigated in a new range of magnitudes. Isotope ratios of NH stretching frequencies, νNH/ND, are revisited. Hydrogen bond energies are reviewed and two-bond deuterium isotope effects on ^{13}C chemical shifts are investigated as a possible means of estimating hydrogen bond energies.

Keywords: intramolecular hydrogen bonds; deuterium isotope effects on chemical shifts; isotope ratios; hydrogen bond energies

Citation: Hansen, P.E. A Spectroscopic Overview of Intramolecular Hydrogen Bonds of NH . . . O,S,N Type. *Molecules* **2021**, *26*, 2409. https://doi.org/10.3390/molecules26092409

Academic Editor: Mirosław Jabłoński

Received: 4 January 2021
Accepted: 18 April 2021
Published: 21 April 2021

Publisher's Note: MDPI stays neutral with regard to jurisdictional claims in published maps and institutional affiliations.

1. Introduction

NH . . . O,S,N (in the following called NH . . . X for simplicity) intramolecular hydrogen bonds are very important building blocks in biomolecules, in self-organizing materials, in drugs, in switching molecules and in chemistry as such. Examples are given in this review. As the title indicates, this review is dealing with intramolecular hydrogen bonding. Recent reviews cover this subject [1,2]. Review [1] concentrates on Schiff bases made from salicylaldehyde and TRIS. Rules are set up to predict the predominant tautomer based on linear free energy relationships. Review [2] is focused on aromatic systems such as *o*-hydroxy Schiff bases and Mannich bases and mainly deals with tautomerism. However, in order to make it feasible within the limits of a review of this type, tautomeric systems as such are not dealt with. Nevertheless, NH . . . X systems are of course one of the two forms of a tautomeric system involving NH and may as such provide useful information in the study of tautomeric systems. Older reviews covering hydrogen bonding and tautomeric systems can also be found [3–5]. Energetics are also treated and in that relation the question of hydrogen bond strength will be touched upon. The title may be misleadingly broad. By spectroscopic techniques NMR and infrared spectroscopy are primarily meant, as they are central in these studies. Within NMR, ^{1}H-, ^{13}C- and ^{15}N-chemical shifts, isotope effects on chemical shifts, one-bond NH and long-range coupling constants are included, whereas for IR spectroscopy mainly NH stretching frequencies are explored. Theoretical calculations are included in cases when they supplement experimental results although they are not the focal point for this review. A goal is to give some guidance to which spectroscopic tool to use in a given situation. NH . . . X hydrogen bonds have been investigated less than OH...X intramolecular hydrogen bonds. For an overview of the latter, see [6].

Intramolecular hydrogen bonds can be quite different, as seen in Figures 1 and 2. An important feature is the linker between the NH donor and the hydrogen bond acceptor. If this is a double bond or part of an aromatic system, the system has been termed resonance-assisted hydrogen bonding (RAHB) [6] and this clearly influences the type and the strength of the hydrogen bond. In other cases, e.g., proteins, intramolecular hydrogen bonds are not very different from intermolecular ones, except for the fact that the protein may be keeping the donor and the acceptor close to each other. This type of hydrogen bond is clearly very important in proteins, both in defining α-helices, β-sheets and turns. In proteins many hydrogen bonds may be present. Therefore methods to identify specific pairs and to characterize the individual hydrogen bonded pairs is needed. For DNA and RNA the hydrogen bonds are very similar.

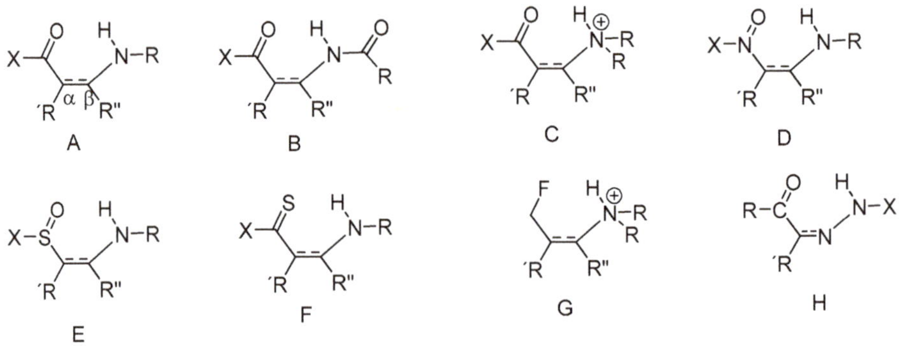

Figure 1. Intramolecular hydrogen bond scheme of RAHB type or charge assisted type. The bond between the α- and the β-carbon can be a double- or an aromatic bond. (**A**) R, R′,R″=H, alkyl or aryl; X=H, C, O, N or S. (**B**) R, R′,R″=H, C,O,N; X=H or alkyl or aryl. (**C**) R=H or C; X=H or C or OR. (**D**) R, R′and R″=H or alkyl or aryl; X=lone pair or O⁻ (nitrogen is positively charged, as it is a nitro group). (**E**) R, R′ and R″=H or alkyl or aryl; X=alkyl or aryl. (**F**) R, R′ and R″=alkyl or aryl; X + R′=benzene ring. (**G**) R, R′and R″=alkyl. (**H**) R and R′=alkyl, X=Ph.

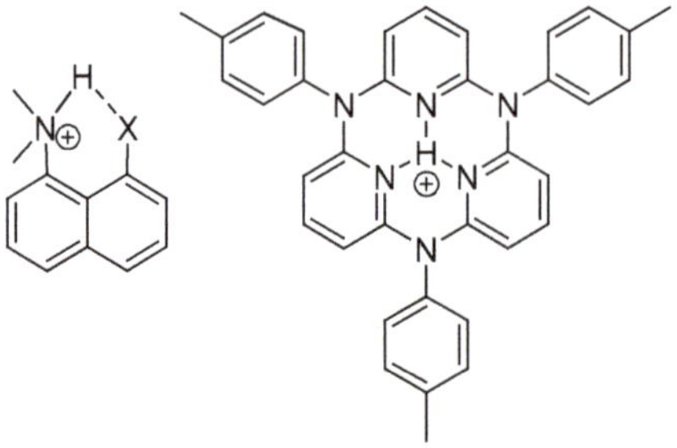

Figure 2. Non-RAHB system. The left hand molecule with X=N(CH$_3$)$_2$ is the well known DMANH⁺ proton sponge. With X=pyrrole in Figure 4 hydrogen bonding to the π-electron system is found, [7] whereas with X=N(CH$_3$)C=OCH$_3$ hydrogen bonding to nitrogen have been tested [8]. The right hand molecule, N,N′,N″-tris(p-tolyl)azacalix [3](2,6)pyridine (TAPH), shows an extremely low field NH proton chemical shift of 22.1 ppm [9,10].

In the following, a number of typical hydrogen bond donors and acceptors and pairs of these are compared. It is of course not possible to mention all compounds with intramolecular NH hydrogen bonds. General trends will be given together with typical examples.

2. NMR

2.1. HN Chemical Shifts

Primary amines may pose a problem in those cases in which rotation around the C-N bond occurs. This leads to averaged NH chemical shifts or as seen later, to averaged one-bond NH coupling constants. In some cases rotation can be stopped at low temperature as seen e.g., in bis(6-amino-1,3-dimethyluracil-5-yl)-methane derivatives (Figure 3). The chemical shifts of the hydrogen bonded NH protons in these compounds are in the 8–9 ppm range [11].

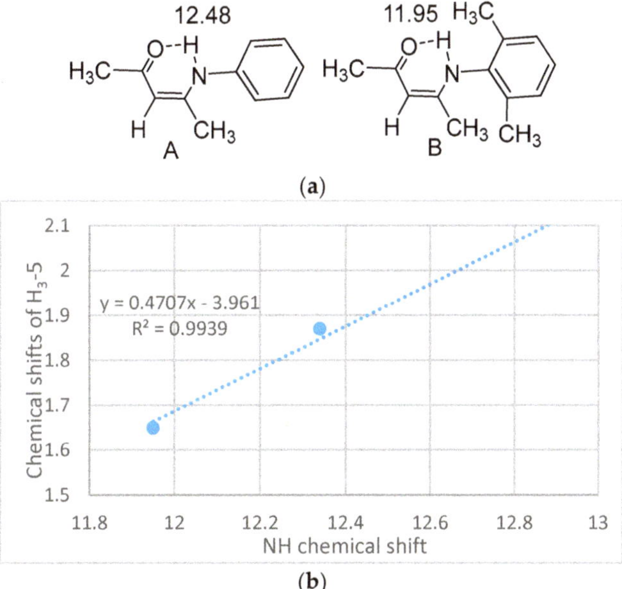

Figure 3. Bis(6-amino-1,3-dimethyluracil-5-yl)-methane derivatives. R being ethyl, pyridine or *p*-dimethylaminopyridine. Taken from [11].

In compounds with aromatic rings close to the NH proton, the NH chemical shift has to be corrected for ring-current effects (for an example of ring current effects see Figure 4) in order to use this to characterize the NH ... X hydrogen bond. In A ring a current is present, whereas it is absent in B.

(a)

(b)

Figure 4. (**a**). Demonstration of possible ring current effects. In A the ring is twisted 34° out to the double bond plane, whereas in B the twist angle is 89°. Data from [12]. (**b**) Plot of ^{1}H chemical shifts of CH_3-5 vs. NH chemical shifts for enaminones with following substituent at nitrogen phenyl substituent, *o*-methyl, *O,O*-dimethyl and 4-isopropyl. Data from [13–15].

The plot of Figure 4b demonstrates the low frequency shift of both the NH and the CH_3-5 chemical shift as the phenyl group in the ortho-substituted phenyl ring is twisted.

In β-enaminones (Figure 1A), a hydroxyl group at the phenyl ring next to the carbonyl group clearly competes with the NH group and the NH chemical shift drops to 11.89 ppm [12], whereas an OH group in the ortho-position of a phenyl group at the nitrogen also leads to drop [16], but in this case it is a combination of a twist of the phenyl ring and a field effect caused by the OH group. In case of the *o*-hydroxyphenyl derivative, Rodríguez et al. [17] also report the finding of the keto-form at low concentration in the ^{1}H-NMR spectrum but not in the ^{13}C-NMR spectrum. *o*-Hydroxyaromatic Schiff bases T are usually either on the OH-form or being tautomeric [18]. However, recently a large number of Schiff bases based on salicylaldehyde and TRIS have been shown to be on the enamine form in the solid state. This is true for the following salicylaldehydes with substituents as follows: 5-nitro, 5-methylcarboxylate, 4-fluor, 4-choro, 4-bromo, 4-methoxy, 4-amino and 5-phenylazo [1]. A few other examples of compounds entirely on the NH form are Schiff bases of 1,3,5-triacyl-2,4,6-trihydroxybenzene [19–21], 1,3,5-triformyl-2,4-6-trihydroxybenzene [19], of gossypol [22,23] or more recently of primarily on the NH form (2-(anilinemethylidenen)cyclohexane-1,3-dione) [24].

A classic comparison is that of hydrogen bonding involving a ketone or an ester is seen in Figure 5A,B. Another comparison can be found in Ref. [25].

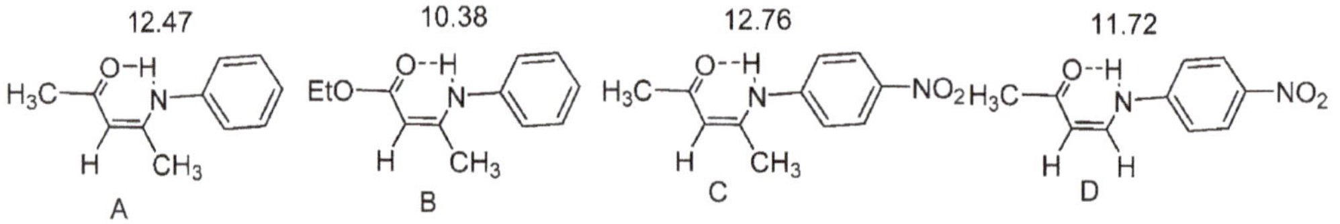

Figure 5. Comparison of different acceptors (**A,B**) and demonstration of steric compression (**C,D**). The numbers are NH chemical shifts in ppm. A and B from [15,26]. For the corresponding N-methyl derivatives, the chemical shifts are 10.90 ppm and 8.55 ppm. (**C**) is from [12] and (**D**) from [27].

It is obvious that substitution at nitrogen plays a role. Furthermore, steric compression is also an important feature as seen by comparing the 3-methyl derivative C with the corresponding non-substituted compound D. The introduction of the nitro groups leads to a slightly more acid NH group and to a stronger hydrogen bond (A vs. C). By comparison of the following compounds Figure 6, it is also clear that ring-size plays a role.

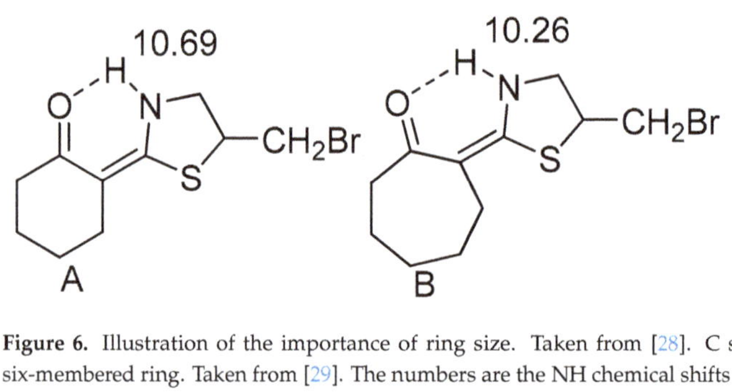

Figure 6. Illustration of the importance of ring size. Taken from [28]. C shows the effect of a six-membered ring. Taken from [29]. The numbers are the NH chemical shifts in ppm.

Another comparison is made in Figure 7, in which the NH is hydrogen bonded to a carbonyl group or to an amide group. The acceptor amide as acceptor clearly leads to the weaker hydrogen bond.

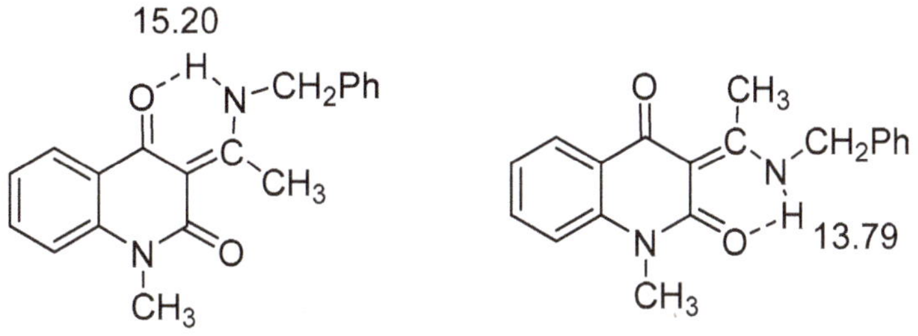

Figure 7. Comparison of ketone and amide groups as acceptors. Data from [30]. The numbers are the NH chemical shifts in ppm.

Amides and thioamides as donors are compared in Figure 8 [31].

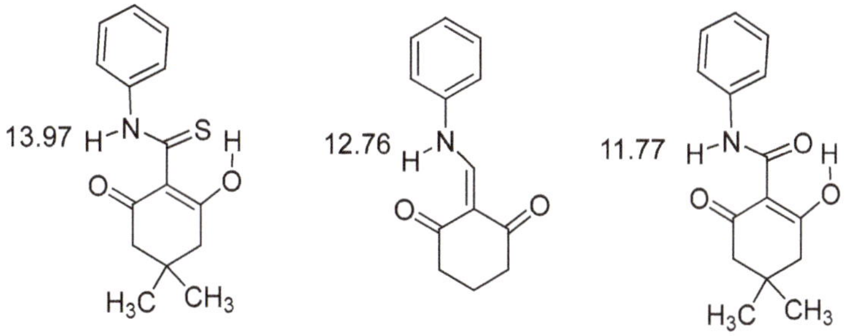

Figure 8. Comparison of amides and thioamides as donors. The numbers are the NH chemical shifts in ppm. For the amide and thioamide the methyl derivatives have chemical shifts of 9.69 and 12.20 ppm. Taken from [31]. The amine is from [24].

It is seen from Figure 6 that the thioamide is the strongest donor followed by the amine and in third place the amide. Hydrogen bonding to a S=O acceptor is demonstrated in Figure 9.

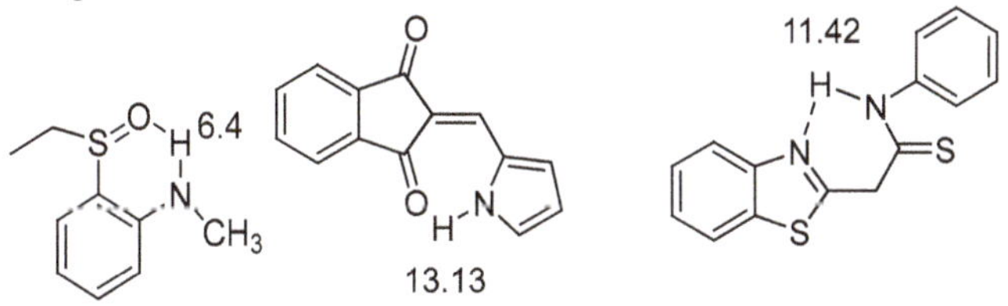

Figure 9. Hydrogen bonding with different motifs. The numbers are NH chemical shifts in ppm. The sulfoxide is from [32]. The indole derivative is from [33]. Other similar motifs is seen in this reference. The benzo[d]thiazol is from [28].

The low chemical shift of the benzo[d]thiazol is due to the lack of conjugation between the donor and the acceptor. In Figure 10 are comparisons again done between different acceptors. NH chemical shifts may for hydrogen bonded hydrazo compounds reach values as high as 15.8 ppm when the NH is hydrogen-bonded to a pyridine nitrogen. In the minor isomer, in which the NH is hydrogen bonded to an ethyl ester group, the chemical shift is 12.8 ppm [34]. In the other pair it is obvious that the stronger hydrogen bond is to a CH₃C=O rather than to a PhC=O. The ability of aromatic nitrogens to form hydrogen bonds is also demonstrated in (Z)-5-((phenylamino)methylene)quinoxaline-6-(5H)-one, 13.15 ppm in

DMSO-d$_6$. This value drops to 11.15 ppm in in (*Z*)-4-((phenylamino)methylene)thiadiazol-5-(4*H*)-one, which has a sulphur instead of a CH$_2$=CH$_2$ unit and hydrogen bonding nitrogen is now part of a five-membered ring [35]. This isomer is with hydrogen bonding to nitrogen is the minor form. The authors discuss hydrogen bonding in terms of quasi-aromaticity.

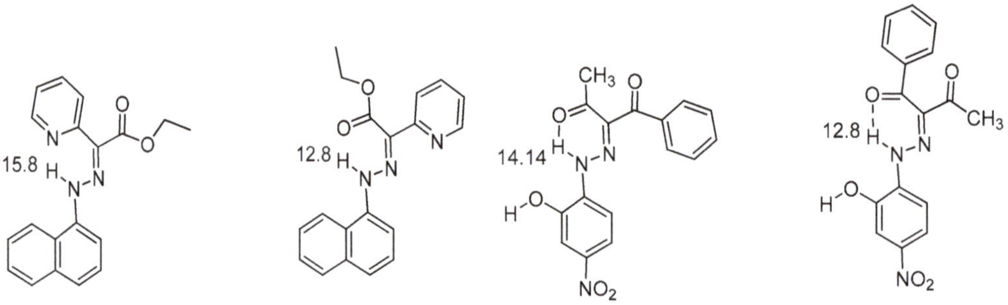

Figure 10. Comparison of rotamers. The numbers are NH chemical shifts in ppm. The chemical shifts are from [36]. The NH chemical shifts for the corresponding benzene derivative (benzene instead of naphthalene) are 14.6 ppm instead of 15.8 ppm. [34] If the ring is a 8-benzoquinoline the chemical shift is 15.36 ppm [37].

Pyrroles can also be hydrogen bond donors as seen in a series of compounds (Figure 11). The NH chemical shifts vary from 10.16 to 13.07 ppm [38]. In a similar case but with an OH group as the acceptor, and two pyrroles present, one hydrogen bonded the other not, the chemical shift drops to 9.39 ppm [39].

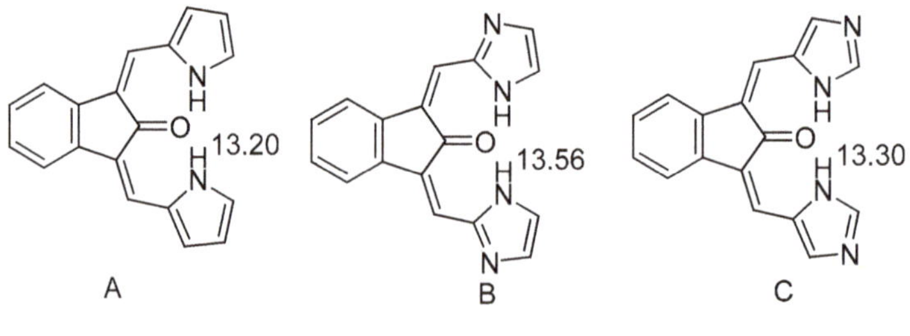

Figure 11. Bifurcated hydrogen bonds. Taken from [40]. The numbers are the NH chemical shifts in ppm.

Bifurcated intramolecular hydrogen bonds were found in azoylmethylidene derivatives of 2-indanone (Figure 11) [40]. Similar kind of molecules have been used to establish a correlation between NH chemical shifts and hydrogen bond energy (see Section 3). It can be seen how the nature of the donor influences the chemical shifts slightly. In case of C the corresponding compound with only one intramolecular hydrogen bond has a chemical shift of 13.73 ppm illustrating the effect of bifurcation. The benzene ring seems to have little effects as the compound corresponding to A simply with the cyclopentanone unit also has a chemical shift of 13.20 ppm. However, by inserting a cyclohexanone ring the chemical shift drops slightly [38]. A different kind of bifurcation can be found in 5-(4-substituted phenylazo)-1-carboxymethyl-3-cyano-6-hydroxy-4-methyl-2-pyridones [41].

Not so common motifs are seen in Figure 12. The question is if the rather high chemical shifts in the N-oxide are caused by a strong hydrogen bond or an electric field effect from the N$^+$-O$^-$ bond. A fact is that the deuterium isotope effects on C=O and CH$_3$ carbon chemical are rather small [42] (for a general discussion of isotope effects see Section 2.4.1). The thio-Schiff base in Figure 12 is drawn as a neutral molecule and as a zwitterionic structure. The latter contributes quite considerably. The NH chemical shifts vary from 18.06 ppm for R and R' = CH$_3$ to 19.26 ppm for R= PhN(CH$_3$)$_2$ and R'=CH$_3$ [43]

or 17.33 ppm for CH$_3$=PhCH$_3$, R′=CH$_3$ or 18.2 ppm for R=PhOCH$_3$ and R′=CH$_3$ [44]. Similar values were also obtained for derivatives in which R′ is H [45].

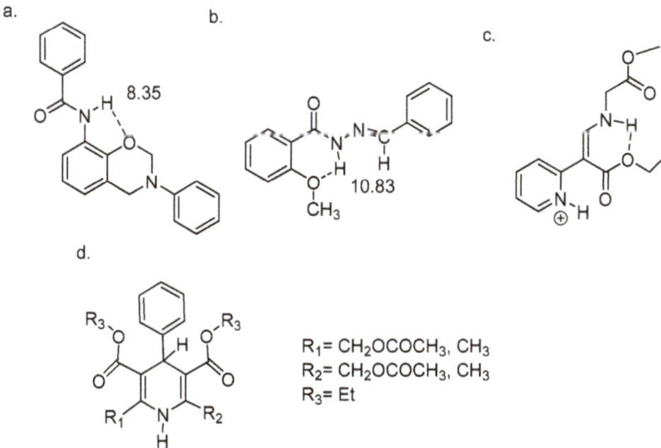

Figure 12. The amino N-oxide demonstrates hydrogen bonding to a charged acceptor. Taken from [42] For the thio-Schiff base resonance forms are demonstrated. Taken from [43]. The number is the NH chemical shift in ppm.

In Figure 13 is seen a comparison of an aldehyde or an imine as acceptor and a donor being either an amide or a sulfamide. The imine is the best acceptor and as the amide is a better donor than the sulfamide.

Figure 13. Comparison of different amides as acceptors. In A Y=O the NH chemical shift is 10.97 ppm, whereas when Y=N-butyl it is 18.83 ppm. In B the NH chemical shift is 10.50 vs. 12.74 ppm. Taken from [46].

Hydrogen bonding is clearly weaker when the hydrogen bonding is to a sp^3 hybridized oxygen and the hydrogen system is a five membered ring as demonstrated in the benzoxazine in Figure 14. Although the hydrogen bond is not so strong it is concentration independent [47].

Figure 14. Hydrogen bonding to single bonded oxygen. (**a**). Benzoxazine with intramolecular hydrogen bonding from [47]. (**b**). N′-benzylidenbenzohydrazide from [48]. (**c**). Protonated enamine molecular switch from [49]. (**d**). 1,4-dihydropyridine derivatives from [50]. The numbers are the NH chemical shifts in ppm.

In the case of N′-benzylidenbenzohydrazide, as seen in Figure 14b, a weak hydrogen bond to the methoxy group is formed. This is clearly stronger than when a classic amide is the hydrogen bond donor. In Figure 14c an amine is hydrogen bonded to an ester oxygen [49].

Charged species show large NH chemical shifts. The protonated DMAN's are tautomeric, but as long as they are symmetric this will not influence the NH chemical shifts. For a review of these see [51]. Very recently an amide type has been investigated. The R substituent has a small effect [8]. Recently motifs A and C have been combined [52].

It is obvious from the chemical shifts seen in Figure 15 and in Figure 2, that the charged systems have high NH chemical shifts. These systems are typically tautomeric and show strong intramolecular hydrogen bonds.

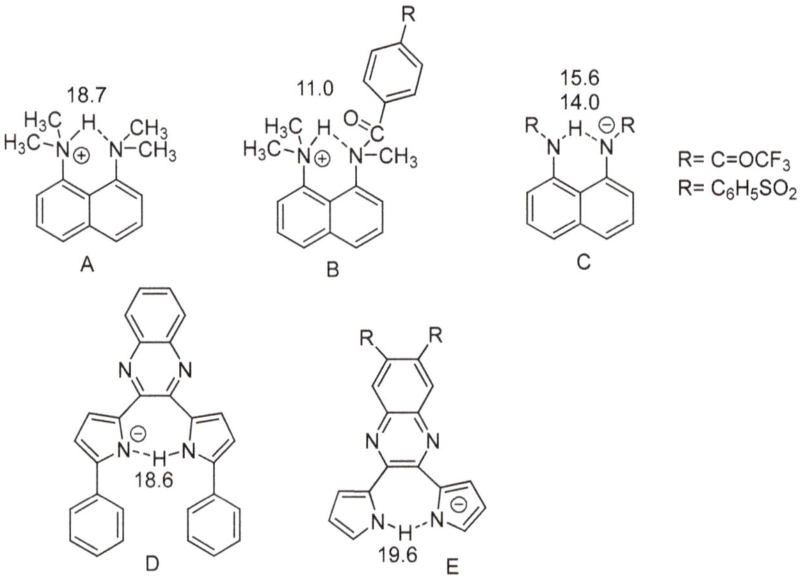

Figure 15. Hydrogen bonding of DMAN types (**A–C**). Counter ions are left out, but the numbers vary slightly with the counter ion. (**C,D**) are hydrogen bonding to a negative acceptor. Numbers are NH chemical shifts. (**A**) from [53], (**B,C**) from [54]. (**D**) from [55]. (**E**) from [56].

Motifs involving urea can be found in a review by Osmialowski [57]. Urea is versatile, as it can act both as an amide type donor and acceptor. The intramolecular nature of the hydrogen bonds were among other things established by measuring the temperature dependence. Temperature dependence was also used to distinguish between NH and OH hydrogen bonds [16]. However, this technique is by no means a reliable tool [58]. Oxamides and thioamides NH temperature coefficients have also been investigated to distinguish intra from inter molecular hydrogen bonding [59].

A number of non-RAHB cases are seen in Figure 14. In addition, proteins often offer many intramolecular hydrogen bonds (for use of coupling constants see Section 2.2). To use NH chemical shifts these should be corrected. This subject has been treated in dipeptides by Scheiner [60].

The results of Figures 4–14 can be summarized in the following way: thioamides seem to be better than hydrazo groups as donors. They are slightly better than aromatic amines, which again are better than aliphatic amines, amides and sulfamides in that order. The pyrroles are not so easy to fit into this scheme. Even when the hydrogen bond is part of a seven membered ring, they are clearly forming strong hydrogen bonds. As acceptors thiones are better than pyridines and other nitrogen containing rings, imines are better than ketones, which are better than amides and esters. Sulfoxides are rather poor. Single bonded oxygens are even poorer. For charged systems the number of cases is limited, but seems to follow the neutral ones. However, the chemical shifts are much higher both in cases with the donor being positive charged or the acceptor being negatively charged as compared to the neutral cases.

2.2. Coupling Constants

Two types of couplings are immediately useful, $^1J(N,H)$ and for derivatives of aldehydes, $^3J(NH,CH)$. $^1J(N,H)$, one-bond hydrogen nitrogen couplings show often a numerical value of around 90 Hz. This coupling is of course negative. Dudek and Dudek showed a small difference between hydrogen bonded and non-hydrogen bonded cases [61]. $^1J(N,H)$ couplings have also been calculated by DFT methods. A recent study optimized for secundary amines the functional and basis set as follows: B3LYP/6-311++G** for structure optimization in chloroform (PCM approach) and APFD/6-311++G**(mixed) for calculation of $^1J(^{15}N,H)$ coupling constants. A very good agreement with experimental values was found. The shorter the bond the larger the coupling constant [62]. A number of useful trends were found to complement the not so many experimental data. Using a simpler basis set, B3LYP-6-31G, it was found that one has to distinguish between primary and secondary amines. For the primary amine cases dissolved in a hydrogen bonding solvent like dimethylsulfoxide, a sulfoxide molecule has to be hydrogen bonded to the "free" NH in order to obtain good results [3].

The $^3J(NH,CH)$ coupling is for a non-tautomeric case close to 12 Hz [61]. The observation of a coupling of this magnitude or $^1J(N,H)$ of around 90 Hz is a clear indication that one is actually dealing with a NH … X hydrogen bond and not with an OH … X one or a tautomeric system. The access to reliable calculations of $^1J(N,H)$ enables one to calculate values for tautomeric systems, but also to estimate the influence of substituents.

2.3. Non-RAHB Cases. Couplings across Hydrogen Bonds

The NH … X bond is central both to proteins, DNA and RNA. A breakthrough was the observation of couplings across hydrogen bonds in RNA [63]. The presence of large J couplings (6–7 Hz) between the hydrogen bond (H-bond) donating and accepting ^{15}N nuclei in Watson–Crick base pairs in double-stranded RNA was found [64]. For proteins, both in α-helices and in β-sheets they are ubiquitous. These hydrogen bonds are generally not very strong. However, in the stronger cases very interesting coupling constants across hydrogen bonds between ^{15}N and $^{13}C=O$ (also referred to as C') have been observed [65,66] enabling pairing of hydrogen bond donors and acceptors. Correlations with bond lengths $^{3h}J_{NC'} = -59000 \exp(-4R_{NO}) \pm 0.09$ Hz, or $R_{NO} = 2.75 - 0.25 \ln(-^{3h}J_{NC'}) \pm 0.06$ Å have been established [67]. Normally such coupling can only be observed in proteins below 10 kD. However, with perdeutration $^{3h}J_{NC'}$ scalar couplings across hydrogen bonds could be observed in the uniformly $^2H/^{13}C/^{15}N$-enriched 30 kDa ribosome inactivating protein MAP30 [68].

A study of lysine interactions in ubiquitine with carbonyl backbone revealed that the NH_3^+ groups of Lys29 and Lys33 exhibit measurable $^{h3}J_{N\zeta C'}$ couplings arising from hydrogen bonds with backbone carbonyl groups of Glu16 and Thr14, respectively. For an example see Figure 16. $^3J_{N\zeta C\gamma}$-coupling constants could also be measured, these together with relaxation studies showed that the NH_3^+ groups are involved in a transient and highly dynamic interaction [69].

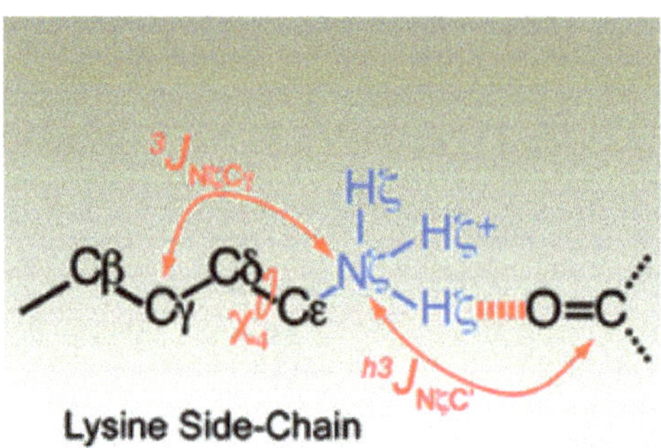

Figure 16. Coupling from at carbonyl carbon to a side-chain lysine N via the hydrogen bond. Taken from [69] with permission from The American Chemical Society.

A plot of ^1J(N,H) vs. dNH for DNA and RNA demonstrated that the N1 ... N3 hydrogen bonds are stronger in dsRNA A:U than in dsDNA A:T bases pairs [70]. Both two-bond ^{1}H-^{31}P and three bond ^{15}N-^{31}P couplings have also been seen from a histidine to the phosphate group of DNA in a zink finger (see Figure 17) [71].

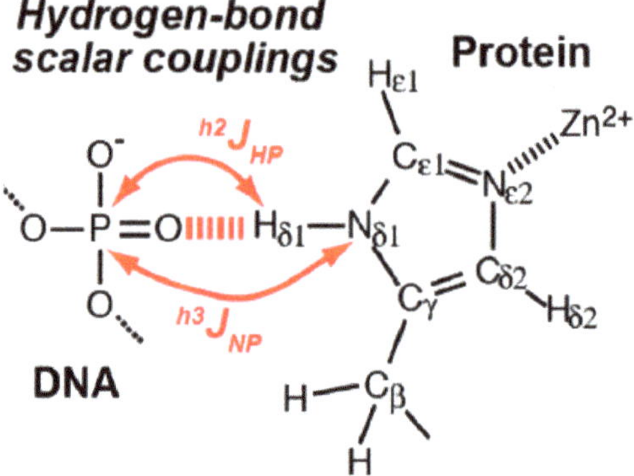

Figure 17. Hydrogen bond scalar coupling involving a phosphate and a histidine. Taken from [71] with permission from the American Chemical Society.

A very large N ... N coupling of 40 Hz through a hydrogen bond is seen in the compound in Figure 15E. The N..N distance is calculated as 2.54 Å. The coupling is the largest of this kind so far reported in a symmetric system [56].

Couplings across hydrogen bonds have also been calculated and summarized by Del Bene [72]. The couplings are dominated by the Fermi contribution and depends on the distance between the heavy atoms. Relationships are noted between hydrogen bond type, X–Y distances, NMR spin–spin coupling constants, and infrared proton-stretching frequencies. This also nicely reflects the experimental findings.

2.4. Isotope Effects on Chemical Shifts

Three different types of deuterium isotope effects on chemical shift are useful, $^n\Delta$C(ND), $^1\Delta$N(D), $^n\Delta$H(ND) and in principle $^n\Delta^{17}$O(ND) in the study of intramolecular hydrogen bonds [73]. Isotope effects is in the present review defined as: $^n\Delta = \delta$X(H) $- \delta$X(D).

2.4.1. RAHB Cases (a "Double Bond" Connecting Cα and Cβ)

$^{n}\Delta C(ND)$ were early on studied in enaminones [74]. Later this study was extended 12,18,32,75,76]. An advantage of studying enaminones is that a number of these may exist both in an *E*- and a *Z*-form. The former without an intramolecular hydrogen bond, the latter with and thus giving a genuine reference compound. This kind of study clearly showed that the two-bond deuterium isotope effect on ^{13}C chemical shifts, $^{2}\Delta C(ND)$, are larger in the intramolecular hydrogen bonded case (Figure 18). This was ascribed to resonance assistance. Having e.g., a substituent at the C-β carbon can introduce steric strain. This will lead to a larger two-bond deuterium isotope effect as seen by comparing number from Figures 14 and 16 (see later) and to a stronger hydrogen bond. An interesting feature in such systems is also the observation of isotope effects at the carbon involved in the intramolecular hydrogen bond and even the carbon attached to the carbonyl group and beyond (see Figure 14) making this a tool for establishing pairs of hydrogen bonds in systems with several possibilities.

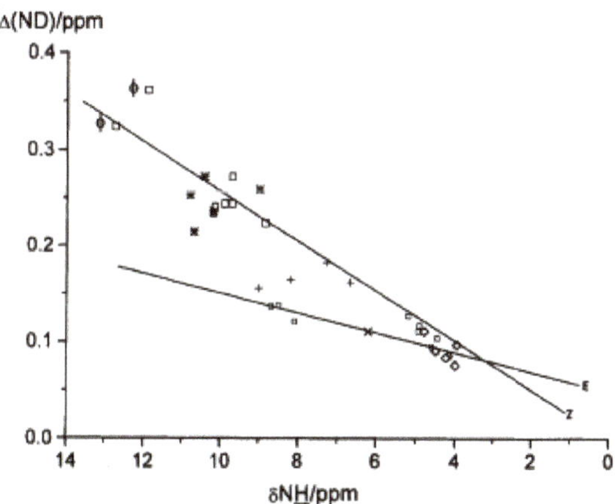

Figure 18. Deuterium isotope effects on ^{13}C chemical shifts in ppm. Taken from Ref. [12].

Isotope effects have often been plotted vs. XH chemical shifts. [75] In the present case, two-bond deuterium isotope effects (TBDIE) are plotted vs. NH chemical shifts (Figure 19). It is clear that *E*- and *Z*-derivatives can be distinguished. The correlation covers enaminones, nitro- and sulphinyl derivatives [32].

Figure 19. Plot of two-bond deuterium isotope effects on ^{13}C chemical shifts vs. NH chemical shifts for enamines. Open squares (*Z*)-enaminones, closed squares (*E*)-enaminones, + enamino esters, * (*Z*)-nitroenamines, crosses (*Z*)-sulphinylenamines and diamonds (*E*)-sulphinylenamines, φ indicates N-phenyl groups. Taken from [32] with permission from Wiley.

One-bond deuterium isotope effects, $^1\Delta N(D)$, depend strongly on hydrogen bonding (the geometry) and related to that RAHB (see Figure 20). [32]

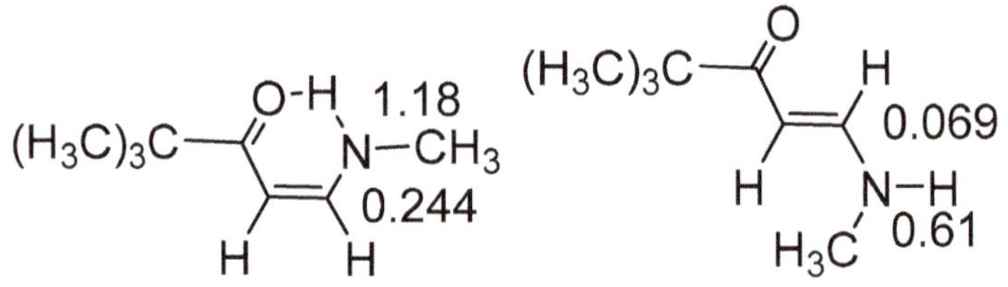

Figure 20. One-bond deuterium isotope effects on ^{15}N chemical shifts and two-bond deuterium isotope effects at carbon in ppm. From [76].

2.4.2. Non-RAHB Cases

Deuterium isotope effects on ^{15}N chemical shifts have been studied in the mono anion of 2,3-dipyrrol-2-ylquinoxaline and its 6,7-dintro derivative (see Figure 15E) [56]. The effects 1.13 ppm and 0.88 ppm (signs have been changed from the original publication) are rather large and indicate a strong hydrogen bond.

Deuterium isotope effects on ^{15}N and ^1H chemical shifts have been used to judge whether certain salt bridges, which are observed in the solid also exist in solution. An example is protein G, B1 domain. The isotope effect demonstrated that two salt bridges found in the X-ray structure did not exist in solution [77]. This approach was also used in Barnase [78]. Both types of isotope effects have also been treated theoretically [79].

In ubiquitin the one-bond deuterium isotope effects of hydrogen bonded NH of the back-bone is correlated to the back bone angles and the angle between the acceptor oxygen and the NH bond (Figure 21) [80].

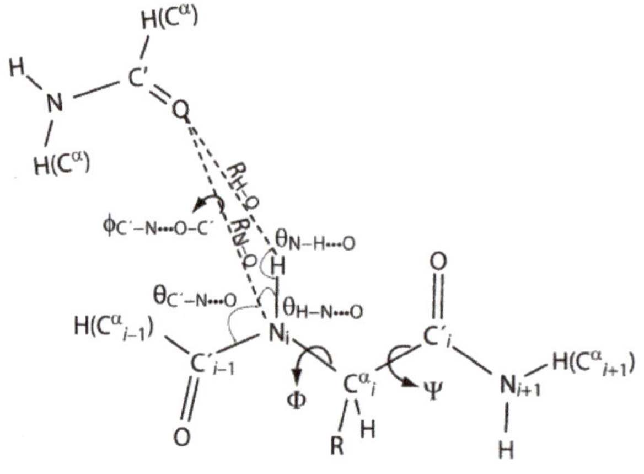

Figure 21. Angles and distances for an inter-chain hydrogen bond. Taken from [80] with permission from Springer.

In DNA and RNA through hydrogen bond isotope effects on chemical shifts can be seen from H-3 to C-2 between adenine and thymine respectively uracil. The isotope effects on chemical shifts are found to be sensitive to the N1-N3 distance suggesting that the isotope effect is sensitive to hydrogen bond strength (see Figure 22) [81,82].

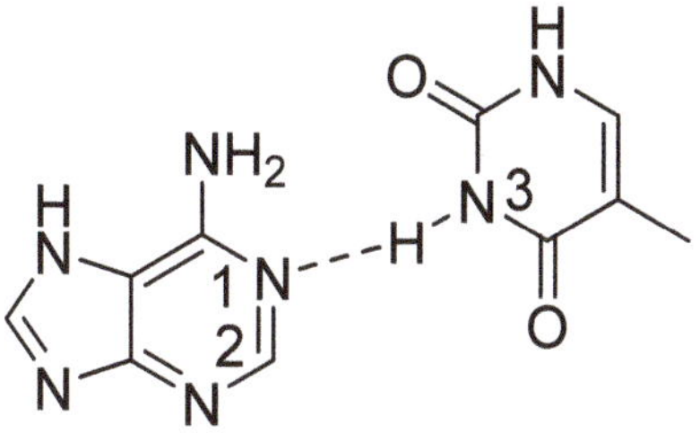

Figure 22. Example of hydrogen bonding in an adenine:thymine base pair.

Very long range isotope effects due to deuteriation at NH have been observed in N-substituted 3-(cycloamino)thioproionamides [83]. These effects were ascribed to electric field effects (Figure 23). The use of nuclei with a large chemical shift range like ^{19}F makes this kind of effect very useful even for weaker hydrogen bonds.

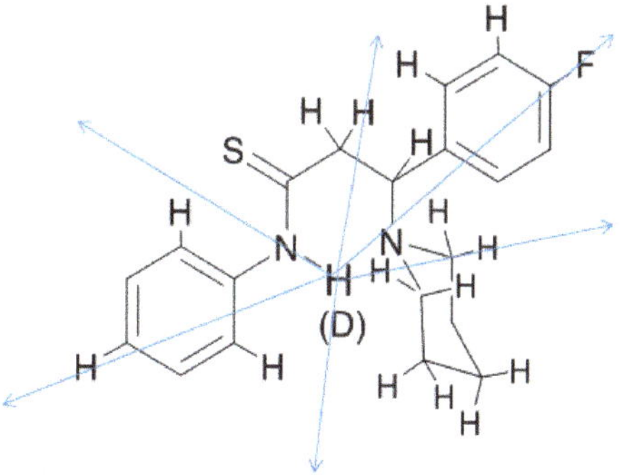

Figure 23. Example of electric field isotope effect through space in N-substituted 3-(cycloamino)-thioproionamides.

2.4.3. Primary Isotope Effects

$^{P}\Delta^{1}$H(ND), and $^{P}\Delta^{1}$H(NT), in which the NH is replaced by either deuterium or tritium may also be used to gauge intramolecular hydrogen bonding. Only a few examples are available [84].

Protonated DMAN's (DMANH^{+}) (Figure 2) show tautomerism. However, for symmetric compounds this will not influence the primary isotope effects [56]. For DMANH^{+} itself the $^{P}\Delta^{1}$H(ND), is 0.69 ppm. Similar values were found for a 4,5-CH$_2$OCH$_2$ derivative as well as the one in which the methyl groups are changed to ethyl groups. However, for 2,7-substituted derivatives the values are much smaller, dichloro, dibromo, bisdimethylamino, dimethoxy and ditrimethylsilyl gave values of 0.30, 0.23, 0.34, 0.31 and 0.14 ppm. [85] The much smaller values in these derivatives were ascribed to a buttressing effect leading to a shorter N ... N distance [86]. These data and more together with data from Ref. [84] are plotted in Figure 24. Interesting points falling in-between at chemical shifts 14.03 and 14.84 ppm are N-phenyl derivatives, so these chemical shifts must be corrected for ring current effects. Others are also N-phenyl derivatives, but substituted in the 2-position so the rings are twisted heavily out of the double bond plane and thereby reducing the ring-current effects. The primary isotope effects were also related to IR isotope ratios [73].

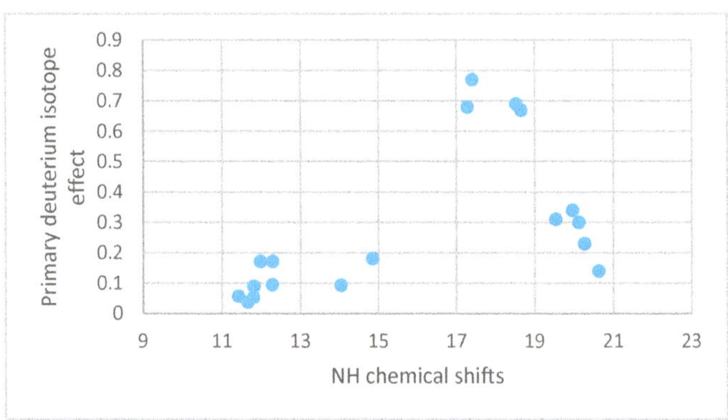

Figure 24. Plot of primary deuterium isotope effects vs. NH chemical shifts. Data from [56,85].

As the review deals with NH hydrogen bonds the obvious isotope to use is deuterium. Secondary deuterium isotope effects over two-bonds tell in a qualitative way about the hydrogen bond strength. The larger the isotope effect, the stronger the hydrogen bond. For a relation to hydrogen bond energy see Section 3. In case of one bond deuterium isotope effects on ^{15}N two different scenarios are found. In RAHB cases the isotope effect increases with increasing hydrogen bond strength, whereas for intramolecular hydrogen bond cases with no direct connection between donor and acceptor (like in DNA) the magnitude of the isotope effect decreases as the distance between heavy atoms decreases.

The primary isotope effects deuterium isotope effects show a more irregular behavior. It would be interesting in the future to compare primary isotope effects for the systems shown above with the one bond secondary isotope effects on ^{15}N.

3. Energy

Hydrogen bond energies for intramolecular hydrogen bonds of the RAHB type are difficult to determine experimentally. Nevertheless, experimental values are necessary in order to have a gauge for theoretical calculations. Spectroscopic data can be useful in this context. The question of calculating the hydrogen bond energy for NH ... X bonds were treated by Reuben [87]. He suggested to calculate the hydrogen bond energy by extending the equation originally suggested by Schaefer [88] (in this case based on OH ... O intramolecular hydrogen bond). It is important to remember that Schaefer said a tentative equation and "rather involved but approximate calculations of electric field effects of the chemical shifts of the hydroxyl proton in intermolecularly hydrogen bonded phenol predict a very nearly linear relationship between the chemical shift and the energy".

The energy was obtained from a correlation with NH chemical shifts:

$$\Delta\delta NH = -1.06 + E_H \tag{1}$$

$\Delta\delta NH$ is in the Reuben case referred to the NH chemical shift of N-methylaniline in CDCl$_3$. Chiara et al. [89] used these equations to obtain hydrogen bond energies of nitro-substituted enaminones of the order of 29 to 34 KJ/mole. A very comprehensive overview is given by Afonin et al. [33] but mostly for weak interactions. Afonin et al. used a slightly different correlation:

$$E_{HB}(\Delta\delta) = \Delta\delta + (0.4 \pm 0.2) \text{ energy in Kcal/mol}$$

Afonin et al. [33] used in their study the NH donor in a pyrrole ring together with OH ... X and CH ... X intramolecular hydrogen bond. In this case the reference compound was pyrrole with a chemical shift of 9.25 ppm. Recently other theoretical approaches have been used. Tupikina et al. used ^{1}H chemical shifts of NH$_2$ groups using the non-hydrogen bonded NH as reference for aniline derivatives and looking mainly at intermolecular hydrogen bonds. Unfortunately, the only RAHB system, an *o*-amino Schiff base, falls off the correlation line obtained [90].

The hydrogen bond and out scheme [91] used for intermolecular hydrogen bonds cannot really be used for NH ... X intramolecular hydrogen bonds. A scheme has been set up by Jablonski et al. [92] for NH_2 groups as donors and later used for APO and 3-methyl APO [93]. A simpler method is to use the "in" and 90 degrees approach in which the energy of the latter is subtracted from the hydrogen bonded one. An example is hydrazone switches. A correlation is found between the hydrogen bond energy and the long range NH coupling across the hydrogen bridge [94]. A different method is to use the electron density at the bond critical point as suggested by Rozas et al. [95]. Using this method for 3-aminopropenal Vakili et al. [93] found 26.6 KJ/mol in good agreement with those of Jablonski et al. using MP2/6-31G** and MP2/6-311++G** [92]. The electron density at the bond critical point is used to estimate the hydrogen bond strength in a number of strategic intramolecular hydrogen bonds of enaminones. Inspired by the Reuben approach [87] energies have been related to two-bond deuterium isotope effects at carbons [96,97]. Recently, two-bond deuterium isotope effects (TBDIE) have been correlated to hydrogen bond energies in *o*-hydroxy aromatic aldehydes in which the hydrogen bond energies were calculated by the hb and out method [98]. The use of TBDIE has the advantage that no reference is needed. The hydrogen bond energies expressed as electron density at the bond critical point are plotted vs. two-bond deuterium isotope effects on ^{13}C chemical shifts in Figure 25 for a small set of enaminones. The ring critical points were calculated using the B3LYP/6-311++G(d,p) functional [99] and the AIM program [100,101]. A reasonable correlation is obtained considering that both ketones, esters and nitro groups are acceptors and compounds are both linear and cyclic and substituents at nitrogen both aliphatic (methyl and *t*-butyl) and aromatic. It is obvious that the cyclic compounds fall on a line of their own.

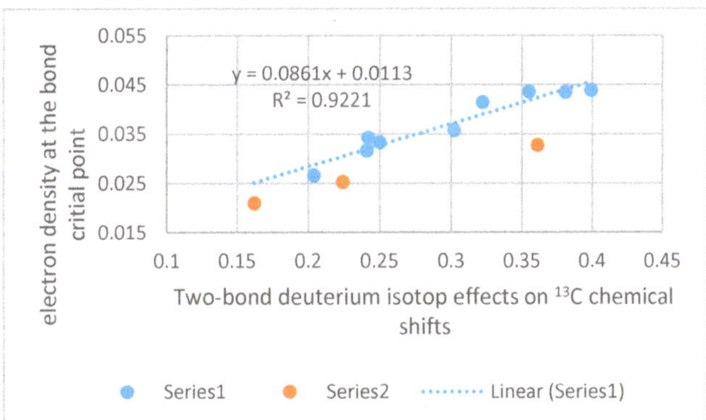

Figure 25. Plot of electron densities at the bond critical point vs. two-bond deuterium isotope effects on ^{13}C chemical shifts in ppm. Series 1 include linear compounds with ketones and nitro groups as acceptors, Series 2 include cyclic compounds both 5- and 6-membered rings, ketones and esters. Isotope effects in ppm from [76,89,102,103]. The ring critical points were calculated using the B3LYP/6-311++G(d,p) functional and the AIM program.

Considering the correlation between the TBDIE on ^{13}C and the electron density at the bond critical point (Figure 25) it is obvious that the large isotope effect is correlated to a stronger hydrogen bond. As TBDIE are also correlated to NH chemical shifts, a series of parameters may be used to predict hydrogen bond strength.

4. Hydrogen Bond Strength

Hydrogen bond strength can be judged from the hydrogen bond energy (see Section 3). The work on predicting hydrogen bond strength using the pK_a slide should be mentioned as a reference method although not based on spectroscopic methods [104]. Gorobets et al. have suggested the difference between ^{1}H chemical shifts of the chemical shifts of primary amides as a simple index of hydrogen bond strength in primary amides as demonstrated in Figure 26 [105]. The hydrogen bond strength can best be described by experimental trends like NH stretching frequencies, the lower the NH stretching frequency the stronger the hydrogen bond, for NH chemical shifts and two-bond deuterium isotope effects on ^{13}C chemical shifts or one-bond deuterium isotope effects on ^{15}N chemical shifts, the larger the stronger the hydrogen bond. A list of criteria also including structural parameters can be found in [106]. Martyniak et al. [107] also find that the asymmetry of the potential curve is a measure of hydrogen bond strength.

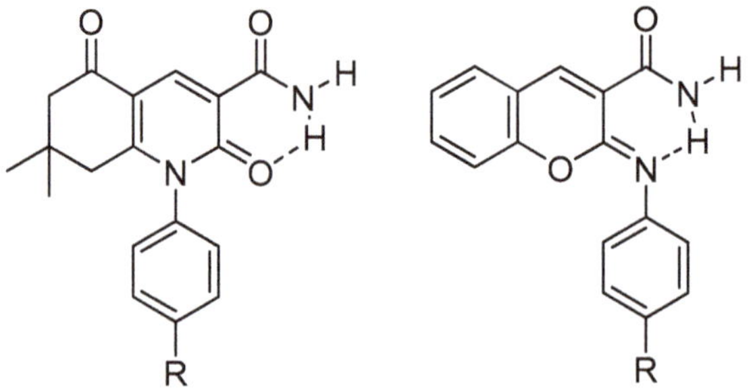

Figure 26. Primary amides used to estimate hydrogen bond strength. Taken from [105].

5. Assignments

5.1. CD Stretching Frequencies

A blue shift of $C_\delta D_2$ stretching frequencies of prolines in the Src homology 3 domain was interpreted as a correlation with a hydrogen bond $N_{i+1}H \ldots N_i$ interaction. The blue shifts were 10 and 17 cm^{-1} for Pro 165 and 21 and 28 cm^{-1} for Pro183 relative to two other prolines not affected. This was further supported by model studies of methyl terminated proline dipeptides similar to those of the SRC homology 3 domain. DFT calculations and NBO analysis supported this type of hydrogen bond [108].

5.2. Assignments

As NH stretching vibrations can be difficult to assign, deuteriation is often a good tool. As most NH stretching frequencies are found around 3000 cm^{-1} the ND stretching frequencies will be around 2100 cm^{-1}, which is in a region with few other resonances. However, the NH/ND ratio is not fixed. This was originally studied by Novak [109] and later supplemented by Sobczyk et al. [110] primarily using OH stretching frequencies. Novak included mostly intermolecular hydrogen bonds whereas Sobczyk et al. added intramolecular hydrogen bonds. The broad trend is a decrease of the isotope ratio as the NH stretching frequency decreases. However, as seen in Figure 27 based on data by Chiara et al. [89,103] for nitro-substituted enaminones and nitro-substituted enamino esters and supplemented with data on enamino esters the picture is not so clear cut at all. This is also seen in the review by Sobczyk et al. [73] but is to some extent masked by the inclusion of a correlation line.

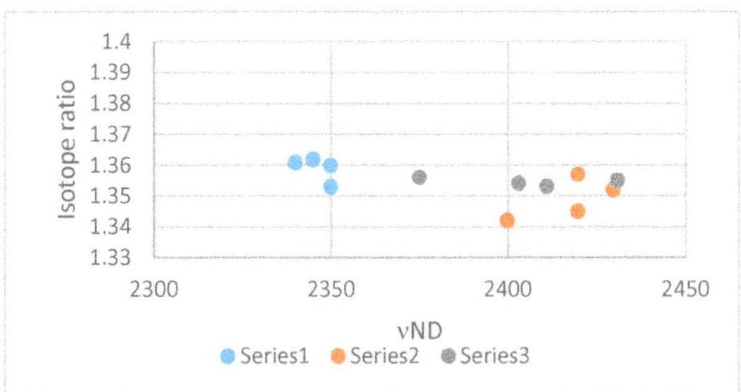

Figure 27. Plot of the νNH/νND ratio vs. νND. Series 1 β-amino-α-nitro-α,β-unsturated ketones, hydrogen bonding to the keto group, Series 2 Hydrogen bonding to the nitro group from [89] and Series 3 enaminoesters from [89,103].

6. Conclusions

In conclusion it is useful to deal with the three different types of intramolecular hydrogen bonds separately. For intramolecular hydrogen bonds of RAHB type hydrogen bond energies are difficult to obtain, resulting in very few or none experimental results are available. A useful method, although not tested to a large extent, on intramolecularly hydrogen NH … X systems, is electron densities at the ring critical point. The latter may also be correlated to deuterium isotope effects on ^{13}C chemical shifts. In view of this, empirical parameters, NH chemical shifts, deuterium isotope effects on ^{13}C or ^{15}N chemical shifts may been used as indicators. NH chemical shifts have to be corrected for ring current effects, if the substituent at the nitrogen is an aromatic rings. Based on these parameters a large range of both donors and acceptors are investigated and rated with their ability to form intramolecular hydrogen bonds. NH chemical shifts and TBDIE have been correlated. However, NH chemical shifts have to be corrected for possible ring current effects, solvent effects and should be measured at as low concentration as possible. The TBDIE have the advantage of being measured as a difference and therefore being dependent on the just mentioned effects. Furthermore, deuterium isotope effects over hydrogen bonds may be used to identify hydrogen bonded pairs in case of multiple possibilities. The finding that ^{1}J(N,H) is rather invariant makes this a good gauge for checking for tautomerism.

Charged systems with connecting bonds between donor and acceptor of the intramolecular hydrogen bonds, but no conjugation, show very large NH chemical shifts.

For intramolecular hydrogen bonds in proteins and nucleic acids coupling through the hydrogen bond can be measured and the magnitude increases with the shorter the heavy atom is. Also ^{1}J(N,H) couplings are useful in the characterization of intramolecular hydrogen bonds.

Theoretical calculations are very useful in calculation of energies, coupling constants, isotope effects on chemical shifts and the finding that NH stretching frequencies can be calculated routinely means that they can be used more easily to identify infra red bands due to NH stretching vibrations. The use of νH/νD ratios to predict NH stretching frequencies based on ND stretching frequencies probably need more investigations.

Funding: This research received no external funding.

Institutional Review Board Statement: Not applicable.

Informed Consent Statement: Not applicable.

Data Availability Statement: Not applicable.

Acknowledgments: The author would warmly like to thank R. Rutu and B.A. Saeed, Basra University, Iraq for their help in calculating electron densities at the bond critical points.

Conflicts of Interest: The author declares no conflict of interest.

References

1. Martinéz, R.F.; Matamoros, E.; Cintas, P.; Palacios, J.C. Imine or Enamine? Insights and Predictive Guidelines from the electronic Effect of Substituents in H-Bonded Salicylimines. *J. Org. Chem.* **2020**, *85*, 5838–5862. [CrossRef]
2. Jezierska, A.; Tolstoy, P.M.; Panek, J.J.; Filarowski, A. Intramolecular Hydrogen Bonds in Selected Aromatic Compounds: Recent Developments. *Catalysts* **2019**, *9*, 909. [CrossRef]
3. Hansen, P.E. Methods to distinguish tautomeric cases from static ones. In *Tautomerism: Ideas, Compounds, Applications*; Antonov, L., Ed.; Wiley-VCH: Weinheim, Germany, 2016.
4. Hansen, P.E.; Spanget-Larsen, J. NMR and IR investigations of strong intramolecular hydrogen bonds. *Molecules* **2017**, *22*, 552 [CrossRef]
5. Sobczyk, L.; Chudoba, D.; Tolstoy, P.M.; Filarowski, A. Some Brief Notes on Theoretical and Experimental Investigations of Intramolecular Hydrogen Bonding. *Molecules* **2016**, *21*, 1657. [CrossRef] [PubMed]
6. Gilli, P.; Bertolasi, V.; Ferretti, V.; Gilli, G. Evidence for intramolecular N-H ... O Resonance–Assisted Hydrogen Bonding in β-Enaminones and Related Heterodienes. A Combined Crystal-Structural, IR and NMR Spectroscopic, and Quantum-Mechanical Investigation. *J. Am. Chem. Soc.* **2000**, *122*, 10405–10417. [CrossRef]
7. Pozharskii, A.F.; Ozeryanskii, V.A.; Filatova, E.A.; Dyablo, O.V.; Pogosa, O.G.; Borodkin, G.S.; Filarowski, A.; Steglenko, D.V. Neutral Pyrrole nitrogen Atom as a π- and Mixed N,π-Donor in Hydrogen Bonding. *J. Org. Chem.* **2019**, *84*, 726–737. [CrossRef]
8. Mikshiev, V.Y.; Pozharskii, A.F.; Filarowski, A.; Novikov, A.S.; Antonov, A.S.; Tolstoy, P.M.; Vovk, M.A.; Khoroshilova, O.V. How Strong is Hydrogen Bonding to Amide Nitrogen? *ChemPhysChem* **2020**, *21*, 1–9. [CrossRef] [PubMed]
9. Zhou, S.; Wang, L. Quantum effects and ^{1}H NMR chemical shifts of a bifurcated short hydrogen bond. *J. Chem. Phys.* **2020**, *153*, 114301. [CrossRef]
10. Kanbara, T.; Suzuki, Y.; Yamamoto, T. New proton-sponge-like macrocyclic compound: Synergistic hydrogen bonds of aminopyridine. *Eur. J. Org. Chem.* **2006**, 3314–3316. [CrossRef]
11. Sigalov, M.V.; Pylaeva, S.A.; Tolstoy, P.M. Hydrogen Bonding in Bis(6-amino-1,3-dimethyluracil-5-yl)-methane Derivatives Dynamic NMR and DFT Evaluation. *J. Phys. Chem. A* **2016**, *120*, 2737–2748. [CrossRef]
12. Zheglova, D.K.; Genov, D.G.; Bolvig, S.; Hansen, P.E. Deuterium Isotope Effects on ^{13}C Chemical Shifts of Enaminones. *Acta Chem. Scand.* **1997**, *51*, 1016–1023. [CrossRef]
13. Chen, X.; She, J.; Shang, Z.; Wu, J.; Wu, H.; Zhang, P. Synthesis of Pyrazoles, Diazepines, Enaminones, and Enamino Esters Using 12-Tungstenphosphoric Acid as a Reusable Catalyst in Water. *Synthesis* **2008**, *21*, 3478–3486.
14. Kidwai, M.; Bardway, S.; Mishra, N.K.; Bansai, V.; Kumar, A.; Mozumdar, S. A novel method for the synthesis of β-enaminones using Cu-nanoparticles as catalyst. *Catal. Commun.* **2009**, *10*, 1514–1519. [CrossRef]
15. Gholap, A.R.; Chakor, N.S.; Daniel, T.; Lahoti, R.J.; Srinivasan, K.V. A remarkable rapid regioselective synthesis of β-enaminones using silica chloride in a heterogenouts as well as an ionic liquid in a homogenenous medium at room temperature. *J. Mol. Catal. A Chem.* **2006**, *245*, 37–46. [CrossRef]
16. Lasić, V.; Jurković, M.; Jednacak, T.; Hrenar, T.; Vuković, J.P.; Novak, P. Intra-and intermolecular hydrogen bonding in acetylacetone and benzoylaceton derived enaminone derivatives. *J. Mol. Struct.* **2015**, *1079*, 243–249. [CrossRef]
17. Rodriguez, M.; Santillan, R.; López, Y.; Farrán, N.; Barba, V.; Nakatani, K.; Garcia-Baez, E.V.; Padilla-Martinéz, J.J. N-H ... C Assisted Structural Changes Induced on Ketoenamine Systems. *Supramol. Chem.* **2007**, *19*, 641–653. [CrossRef]
18. Hansen, P.E.; Rozwadowski, Z.; Dziembowska, T. NMR Studies of Hydroxy Schiff bases. *Curr. Org. Chem.* **2009**, *13*, 194–215 [CrossRef]
19. Sauer, M.; Yeung, C.; Chong, J.H.; Patrick, B.O.; MacLachalan, M.J. N-Salicylideneanilines: Tautomers for Formation of Hydrogen Bonded Capsules, Clefts and Chains. *J. Org. Chem.* **2006**, *71*, 775–788. [CrossRef]
20. Hansen, P.E.; Filarowski, A. Characterisation of the PT-form of o-Hydroxy Acylaromatic Schiff bases by NMR Spectroscopy and DFT Calculations. *J. Mol. Struct.* **2004**, *707*, 69–75. [CrossRef]
21. Kwocz, A.; Panek, J.J.; Jezierska, A.; Hetmańczyk, Ł.; Pawlukojć, A.; Kochel, A.; Lipkowski, P.; Filarowski, A. A molecular roundabout: Triple cyclically arranged hydrogen bonds in light of experiment and theory. *New J. Chem.* **2018**, *42*, 19467–19477. [CrossRef]
22. Prbybylski, P.; Wojciechowski, G.; Schilf, W.; Brezezinski, B.; Bartl, F. Spectroscopic study and PM5 semiempirical calculations of tautomeric forms of gossypol Schiff base with n-butylamine in the solid state and in the solution. *J. Mol. Struct.* **2003**, *646*, 161–168. [CrossRef]
23. Quang, T.T.; Phung, K.P.; Hansen, P.E. Schiff Bases of Gossypol. A NMR and DFT Study. *Magn. Reson. Chem.* **2005**, *43*, 302–308.
24. Dobosz, R.; Mućko, J.; Gawinecki, R. Using Chou´s 5-step rule to Evaluate the Stability of Tautomers: Suseptibility of 2-[Phenylimino)-methyl]-cyclohexane-1,3-dione to Tautomerization Based on the Calculated Gibbs Free Energy. *Energies* **2020**, *13*, 183. [CrossRef]
25. Durgarao, N.; Hilemlan, B.; Cox, J.D.; Scott, H.; Hoang, P.; Robbins, A.; Bowers, K.; Tsebaot, L.; Mioa, K.; Cataneda, M.; et al. Acceleration of the Eschenmoser coupling reaction by sonication: Efficient synthes of enaminones. *RSC Adv.* **2013**, *3*, 181–188.

26. Chen, J.-X.; Zhang, C.-F.; Gao, W.-X.; Jin, H.-L.; Dinga, J.-C.; Wu, H.-Y. B2O3/Al2O3 as a New, Highly Efficient and Reusable Heterogeneous Catalyst for the Selective Synthesis of β-Enamino Ketones and Esters under Solvent-Free Conditions. *J. Braz. Chem. Soc.* **2010**, *21*, S1–S28. [CrossRef]

27. Jebari, M.; Pasturaud, K.; Picard, B.; Maddaluno, J.; Rezgui, F.; Chataignera, I.; Legros, J. "On water" reaction of deactivated anilines with 4-methoxy-3-buten-2-one, an effective butynone surrogate. *Org. Biomol. Chem.* **2016**, *14*, 11085–11087. [CrossRef]

28. Hansen, P.E.; Duus, F.; Neumann, R.; Jagodzinski, T.S. Deuterium Isotope Effects of 0-Hydroxythioamides, Thiazolines and 5-Acyl-2-thiobarbituric Acids. *Pol. J. Chem.* **2000**, *74*, 409–420.

29. Bai, J.; Wang, P.; Cao, W.; Chen, X. Tautomer-selective derivatives of enolate, ketone and enaminone by addition reaction of picolyl-type anions with nitriles. *J. Mol. Struct.* **2017**, *1128*, 645–652. [CrossRef]

30. Hansen, P.E.; Bolvig, S.; Kappe, T. Intramolecular Hydrogen Bonding and Tautomerism of Acyl Pyran-2,4-diones, 2,4,6-Triones and Pyridiones and Benzannelated Derivatives. Deuterium Isotope Effects on 13C NMR Chemical Shifts. *J. Chem. Soc. Perkin Trans.* **1995**, *2*, 1901–1907. [CrossRef]

31. Hansen, P.E.; Duus, F.; Bolvig, S.; Jagodzinski, T.S. Intra-molecular Hydrogen Bonding of the Enol forms of β-Ketoamides and β-Ketothioamides. Deuterium Isotope Effects on 13C Chemical Shifts. *J. Mol. Struct.* **1996**, *378*, 45–59. [CrossRef]

32. Kozerski, L.; Kawecki, R.; Hansen, P.E. Data for Characterization of the Geometrical Isomers of β -Sulfinylenamines. *Magn. Reson. Chem.* **1994**, *32*, 517–524. [CrossRef]

33. Afonin, A.V.; Vashchenko, A.V.; Sigalov, M.V. Estimating the energy of intramolecular hydrogen bonds from ^{1}H NMR and QTAIM calculations. *Org. Biomol. Chem.* **2016**, *14*, 11199–11211. [CrossRef] [PubMed]

34. Landge, S.M.; Thatchouuk, E.; Benitez, D.; Lanfranchi, D.A.; Elhabari, M.; Goodard, W.A., III; Aprahamian, I. Isomerization Mechanism in Hydrazone-Based Rotary Swithces: Lateral Shift, Rotation, or Tautomerization? *J. Am. Chem. Soc.* **2011**, *133*, 9812–9823. [CrossRef] [PubMed]

35. Nguyen, Y.H.; Lampkin, B.J.; Venkateh, A.; Ellern, A.; VanVeller, B. Open-Resonance-assisted Hydrogen Bonds and Competing quasiaromaticity. *J. Org. Chem.* **2018**, *83*, 9850–9857. [CrossRef]

36. Hristova, S.; Kamounah, F.S.; Molla, N.; Hansen, P.E.; Nedeltcheva, D.; Antonov, L. The possible tautomerism of the rotary switch 2-(2-(2-Hydroxy-4-nitrophenyl)hydrazono)-1-phenylbutane-1,3dione. *Dyes Pigments* **2017**, *144*, 249–261. [CrossRef]

37. Su, X.; Aprahamian, I. Switching Around Two Axles: Controlling the Configuration and Conformation of a Hydrazone-Based Switch. *Org. Lett.* **2011**, *13*, 30–33. [CrossRef]

38. Sigalov, M.V.; Shainyan, B.A.; Chipanina, N.N.; Oznobikhina, L.; Strashnikova, N.; Sterkhova, I. Moleclular Structure and Photoinduced Hydrogen Bonding in 2-pyrrolemethylidene Cyclohexanones. *J. Org. Chem.* **2015**, *80*, 10521–10535. [CrossRef] [PubMed]

39. Maeda, H.; Takeda, Y.; Haketa, Y.; Morimoto, Y.; Yasuda, N. Ion-pairing Assemblies of p-Electronic Anions formed by Intramolecular Hydrogen bonding. *Chem. Eur. J.* **2018**, *24*, 8910–8916. [CrossRef]

40. Sigalov, M.V.; Shainyan, B.A.; Chipanina, N.N.; Oznobikhina, L.P. Molecular Structure, Intramolecular Hydrogen Bonding, Solvent-Induced Isomerization, and Tautomerism in Azoylmethylidene Derivatives of 2-Indanone. *Eur. J. Org. Chem.* **2017**, 1353–1364. [CrossRef]

41. Ladarevic, J.; Bozic, B.; Natovic, L.; Nedeljkovic, B.B.; Minj, D. Role of the bifurcated intramolecular hydrogen bond on the physico-chemical profile of the novel azo pyridone dyes. *Dyes Pigments* **2019**, *162*, 562–572. [CrossRef]

42. Diembowska, T.; Rozwadowski, A.; Hansen, P.E. Intramolecular Hydrogen bonding of 8- hydroxyquinoline N-oxides, Quinaldinic Acid N-oxides and Quinaldinamide N-oxide. Deuterium Isotope Effects on 13C Chemical Shifts. *J. Mol. Struct.* **1997**, *436–437*, 189–199. [CrossRef]

43. Saeed, B.A.; Elias, R.S.; Kamounah, F.S.; Hansen, P.E. A NMR, MP2 and DFT Study of Thiophenoxyketenimines (o-ThioSchiff bases). *Magn. Reson. Chem.* **2018**, *56*, 172–182. [CrossRef] [PubMed]

44. Krinsky, J.L.; Arnold, J.; Bergman, R.G. Platinum Group Thiophenoxyimine Complexes, Syntheses and Crystallographic/Computational Studies. *Organometalics* **2007**, *26*, 897–909. [CrossRef]

45. Corrigan, M.F.; Rae, I.D.; West, B.O. The Constitution of N-Substituted 2-(Iminomethyl) benzenethiols (o-mercaptobenzaldimines). *Aust. J. Chem.* **1978**, *31*, 587–594. [CrossRef]

46. Feng, Z.; Jia, S.; Chen, H.; You, L. Modulation of imine chemistry with intramolecular hydrogen bonding. Effects from *ortho*-OH to NH. *Tetrahedron* **2020**, *76*, 131128. [CrossRef]

47. Froimowicz, P.; Zhang, K.; Ishida, H. Intramolecular Hydrogen Bonding in Benzoxazines: When Structural design becomes functional. *Chem. Eur. J.* **2016**, *22*, 2691–2707. [CrossRef]

48. Arya, N.; Mishra, S.K.; Suryaprakash, M. Intramolecular hydrogen bond directed distribution of conformational populations in the derivatives of N´-benzylidenebenzohydrazide. *New. J. Chem.* **2019**, *43*, 13134–13142. [CrossRef]

49. Ren, Y.; Kravchenko, O.; Ramström, O. Configurational and Constitutional Dynamics of Enamine Molecular Swithches. *Chem. Eur. J.* **2020**, *26*, 15654–15663. [CrossRef]

50. Petrova, M.; Mahamadejev, R.; Vigante, B.; Duburs, G.; Liepinsh, E. Intramolecular hydrogen bonds in 1,4-dihydropyridine derivatives. *R. Soc. Open. Sci.* **2018**, *5*, 1800088.

51. Piękoś, P.; Jezierska, A.; Panek, J.J.; Goremychkin, E.A.; Pozharskii, A.F.; Antonov, A.S.; Tolstoy, P.M.; Filarowski, A. Symmetry/Asymmetry of the NHN Hydrogen Bond in Protonated 1,8-Bis(dimethylamino)naphthalene. *Symmetry* **2020**, *12*, 1924. [CrossRef]

52. Vlasenko, M.P.; Ozeryanski, V.A.; Pozharskii, A.F.; Khoroshilova, O.V. Positively and negatively charged NHN hydrogen bonds in one molecule: Synergistic strengthening effect superbasicity and acetonitrile capture. *New J. Chem.* **2019**, *43*, 7557–7561. [CrossRef]

53. Dega-Szafran, Z.; Nowak-Wydra, B.; Szafran, M. Reinvestigtion of ^{1}H and ^{13}C NMR Spectra of 1,8-Bis(dimethylamino)naphthalene Complexes with Mineral Acids. *Magn. Reson. Chem.* **1993**, *31*, 726–730. [CrossRef]

54. Pietrzak, M.; Grech, E.; Nowicka-Scheibe, J.; Hansen, P.E. Deuterium isotope effects on 13C chemical shifts of negatively charged NH . . . N systems. *Magn. Reson. Chem.* **2013**, *51*, 683–688.

55. Yamakoda, R.; Ishibashi, H.; Motoyoshi, Y.; Yasuda, N.; Maeda, H. Ion-pairing assemblies based on p-extended dipyrrolylquinoxalines. *Chem. Commun.* **2019**, *55*, 326–329. [CrossRef]

56. Pietrzak, M.; Try, A.C.; Andriolette, B.; Sessler, J.L.; Anzenbacher, P., Jr.; Limbach, H.-H. The largest ^{15}N-^{15}N Coupling Constant across an NHN Hydrogen bond. *Angew. Chem. Int. Ed.* **2008**, *47*, 1123–1126. [CrossRef]

57. Osmialowski, B. Conformational equilibrium and substituent effects in hydrogen bonded complexes. *Curr. Org. Chem.* **2018**, *22*. [CrossRef]

58. Sosnicki, J.G.; Hansen, P.E. Temperature coefficients of NH chemical Shifts of Thioamides in Relation to Structure. *Mol. Struct.* **2004**, *700*, 91–103. [CrossRef]

59. Desseyn, H.O.; Perlepes, S.P.; Clou, K.; Blaton, N.; van der Veken, B.J.; Dommisse, R.; Hansen, P.E. Theoretical, Structural, Vibribrational, NMR and Thermal evidence of the Inter versus Intramolecular Hydrogen Bonding in Oxamides and Thiooxamides. *J. Phys. Chem. A* **2004**, *108*, 175–182. [CrossRef]

60. Scheiner, S. Assesment of the Presence and Strength of H-Bonds by Means of Corrected NMR. *Molecules* **2016**, *21*, 1426. [CrossRef] [PubMed]

61. Dudek, G.; Dudek, E.P. Spectroscopic Studies of Keto-Enol Equilibria. Part XIII. ^{15}N Substituted Imines. *J. Chem. Soc. B* **1971**, 1356–1360. [CrossRef]

62. Hansen, P.E.; Saeed, B.A.; Rutua, R.S.; Kupka, T. One-bond $^1J(^{15}$N,H) coupling constants at sp^2-hybridized nitrogen of Schiff bases, enaminones and similar compounds: A theoretical study. *Magn. Reson. Chem.* **2020**, *58*, 750–762. [CrossRef] [PubMed]

63. Dingley, A.J.; Grzesiek, S. Direct observation of hydrogen bonds in nucleic acid base pairs by internucleotide (2)J(NN) couplings. S. *J. Am. Chem. Soc.* **1998**, *120*, 8293–8297. [CrossRef]

64. Dingley, A.J.; Masse, J.E.; Peterson, R.D.; Barfield, M.; Feigon, J.; Grzesiek, S. Internucleotide Scalar Couplings across Hydrogen Bonds in Watson-Crick and Hoogsteen Base Pairs of a DNA Triplex. *J. Am. Chem. Soc.* **1999**, *121*, 6019–6027. [CrossRef]

65. Grzesiek, S.; Cordier, F.; Jaravine, V.; Barfield, M. Insights into biomolecular hydrogen bonds from hydrogen bond scalar couplings. *Prog. Nuc. Magn. Reson. Spectrosc.* **2004**, *45*, 275–300. [CrossRef]

66. Cornilescu, G.; Hu, J.S.; Bax, A. Identification of the hydrogen bonding network in a protein by scalar couplings. *J. Am. Chem. Soc.* **1999**, *121*, 2949–2950. [CrossRef]

67. Cornilescu, G.; Ramirez, B.E.; Frank, M.K.; Clore, G.M.; Gronenborn, A.M.; Bax, A. Correlation between 3hJ$_{NC'}$ and Hydrogen Bond Length in Proteins. *J. Am. Chem. Soc.* **1999**, *121*, 6275–6279. [CrossRef]

68. Wang, Y.-X.; Jacob, J.; Cordier, F.; Wingfield, P.; Stahj, S.J.; Lee-Huang, S.; Torchia, D.; Grzesiek, S.; Bax, A. Measurement of 3hJNC' connectivities across hydrogen bonds in a 30 kDa protein. *J. Biomol. NMR* **1999**, *14*, 181–184. [CrossRef] [PubMed]

69. Zandarashvili, L.; Li, D.-W.; Wang, T.; Brüschweiler, R.; Iwahara, J. Signature of Mobile Hydrogen Bonding of Lysine Side Chains from Long-Range 15N–13C Scalar J-Couplings and Computation. *J. Am. Chem. Soc.* **2011**, *133*, 9192–9195. [CrossRef]

70. Manalo, M.N.; Kong, X.; LiWang, A. ^{1}JNH Values show that N1...N3 Hydrogen Bonds are stronger in dsRNA A:U than dsDNA A:T Base pairs. *J. Am. Chem. Soc.* **2005**, *127*, 17974–17975. [CrossRef]

71. Chattopadhyay, A.; Esadze, A.; Roy, S.; Iwahara, J. NMR Scalar Couplings across Intermolecular Hydrogen Bonds between Zinc-Finger Histidine Side Chains and DNA Phosphate Groups. *J. Phys. Chem. B* **2016**, *120*, 10679–10685. [CrossRef]

72. Del Bene, J. *Two-Bond NMR Spin–Spin Coupling Constants across Hydrogen Bonds. Encyclopedia of Theoretical Chemistry*; Wiley: Weinheim, Germany, 2004.

73. Sobczyk, L.; Obzrud, M.; Filarowski, A. H/D Isotope Effects in Hydrogen Bonded Systems. *Molecules* **2013**, *18*, 4467–4476. [CrossRef] [PubMed]

74. Kozerski, L.; von Philipsborn, W. N-15 NMR spectroscopy. 9. Solvent induced Deuterium isotope effects in C-13 NMR and N-15 NMR spectra of enaminones. *Helv. Chim. Acta* **1982**, *65*, 2077–2087. [CrossRef]

75. Hansen, P.E. Isotope effects on chemical shift in the study of intramolecular hydrogen bonds. *Molecules* **2015**, *20*, 2405–2424. [CrossRef] [PubMed]

76. Hansen, P.E.; Kawecki, R.; Krowczynski, A.; Kozerski, L. Deuterium Isotope Effects on ^{13}C and ^{15}N Nuclear Shielding in Intramoleculary Hydrogen-bonded Compounds. II. Investigation of Enamine Derivatives. *Acta Chem. Scand.* **1990**, *44*, 826–832. [CrossRef]

77. Tomlinson, J.H.; Ullah, S.; Hansen, P.E.; Williamson, M.P. Characterisation of salt bridges to lysines in the Protein G B1 domain. *J. Am. Chem. Soc.* **2009**, *131*, 4674–4684. [CrossRef]

78. Williamson, M.P.; Hounslow, A.M.; Ford, J.; Fowler, K.; Hebditch, M.; Hansen, P.E. Detection of salt bridges to lysines in barnase in solution. *Chem. Commun.* **2013**, *49*, 9824–9826. [CrossRef]

79. Ullah, S.; Ishimoto, T.; Williamson, M.P.; Hansen, P.E. Ab initio Calculations of Deuterium Isotope Effects on Chemical Shifts of Salt bridged lysines. *J. Phys. Chem. B* **2011**, *115*, 3208–3215. [CrossRef]

80. Abildgaard, J.; LiWang, A.; Manalo, M.N.; Hansen, P.E. Peptide Deuterium Isotope Effects on Backbone [15]N Chemical Shifts in Proteins. *J. Biomol. NMR* **2009**, *44*, 119–124. [CrossRef]

81. Kim, Y.-I.; Manalo, M.N.; Peréz, L.M.; LiWang, A. Computation and empirical trans-hydrogen bond deuterium isotope shifts suggest that n1-N3 A.U hydrogen bonds of RNA are shorter than those of A:T bonds of DNA. *J. Biomol. NMR* **2006**, *34*, 229–236. [CrossRef]

82. Manalo, M.N.; Péréz, L.M.; LiWang, A. Hydrogen-Bonding and π-π Base-stacking Interactions are Coupled in DNA as Suggested by Calculated and Experimental Trans-Hbond Deuterium Isotope Shifts. *J. Am. Chem. Soc.* **2007**, *129*, 11298–11299. [CrossRef]

83. Sosnicki, J.G.; Langaard, M.; Hansen, P.E. Long-range Deuterium Isotope Effects on ^{13}C chemical Shifts of intramolecularly hydrogen bonded N-substituted-3-(cycloamino) thiopropionamides or amides: A case of electric field effects. *J. Org. Chem.* **2007**, *72*, 4108–4116. [CrossRef]

84. Bolvig, S.; Hansen, P.E.; Morimoto, H.; Wemmer, D.; Williams, P. Primary Tritium and Deuterium Isotope Effects on Chemical Shifts of Compounds having an Intramolecular Hydrogen bond. *Magn. Reson. Chem.* **2000**, *38*, 525–535. [CrossRef]

85. Chmielewski, P.; Ozeryanski, V.A.; Sobczyk, L.; Pozharskii, A.F. primary 1H/2H isotope effect in the NMR chemical shift of HClO4 salts of 1,8-bis(dimethylamino)naphthalene derivatives. *J. Phys. Org. Chem.* **2007**, *20*, 643–648. [CrossRef]

86. Pozharski, A.F.; Ryabtsova, O.V.; Ozeryanski, V.A.; Degtyarev, A.V.; Kazheva, O.N.; Alexandrov, G.G.; Dyanchenko, O.A. Organometallic synthesis, molecular structure, and coloration of 2,7-disubstituted 1,8-bis(dimethylamino)naphthalenes. How significant is the influence of "buttressing effect" on their basicity? *J. Org. Chem.* **2003**, *69*, 10109–10122. [CrossRef] [PubMed]

87. Reuben, J. Isotopic Multiplets in the Carbon-13 NMR Spectra of Aniline Derivatives and Nucleosides with Partially Deuterated Amino Groups: Effects of Intra- and Intermolecular Hydrogen Bonding. *J. Am. Chem. Soc.* **1987**, *109*, 316–321. [CrossRef]

88. Schaefer, T. A Relationship between Hydroxy Proton Chemical Shifts and Torsional Frequencies in Some ortho-Substituted Phenol Derivatives. *J. Phys. Chem.* **1975**, *79*, 1888–1890. [CrossRef]

89. Chiara, J.L.; Gomez-Sanchez, A.; Bellanato, J. Spectral properties and isomerism of nitroenamines. Part 4. β-Amino-α-nitro-α,β-unsaturated ketones. *J. Chem. Soc. Perkin Trans.* **1998**, *2*, 1797–1806. [CrossRef]

90. Tupikina, E.Y.; Sigalov, M.; Shenderovich, I.G.; Mulloyarova, V.V.; Denisov, G.S.; Tolstoy, P.M. Correlations of NHN hydrogen bond energy with geometry and ^{1}H NMR chemical shift difference of NH protons for aniline complexes. *J. Chem. Phys.* **2019**, *150*, 114305. [CrossRef] [PubMed]

91. Cuma, M.; Scheiner, S.; Kar, T. Competition between rotamerization and proton transfer in *o*-hydroxybenzaldehyde. *J. Am. Chem. Soc.* **1998**, *120*, 10497–10503. [CrossRef]

92. Jablonski, M.; Kaczmarek, A.; Sadlej, S.J. Estimates of the Energy of Intramolecular Hydrogen Bonds. *J. Phys. Chem. A* **2006**, *110*, 10890–10898. [CrossRef]

93. Seyedkatouli, S.; Vakili, M.; Tayyari, S.F.; Hansen, P.E.; Kamounah, F. SMolecular structure, vibrational analysis, and intramolecular hydrogen bond strength (IHBs) of 3-methyl-4-amino-3-penten-2-one and the effect of N-methyl and N-phenyl substitutions on the IHBs. An Experimental and Theoretical approach. *J. Mol. Struct.* **2019**, *1184*, 233–245. [CrossRef]

94. Gholipur, A.; Neyband, R.S.; Farhadi, S. NMR investigation of substituent effects on strength the intramolecular hydrogen bonding interaction in X-phenylhydrazones switches: A theoretical study. *Chem. Phys. Lett.* **2017**, *676*, 6–11. [CrossRef]

95. Rozas, I.; Alkorta, I.; Elguero, J. Behavior of Ylides Containing N, O, and C Atoms as Hydrogen Bond Acceptors. *J. Am. Chem. Soc.* **2000**, *122*, 11154–11161. [CrossRef]

96. Hansen, P.E.; Tüchsen, E. Deuterium Isotope Effects on Carbonyl Chemical Shifts of BPTI. Hydrogen-bonding and Structure Determination in Proteins. *Acta Chem. Scand.* **1989**, *43*, 710–712. [CrossRef]

97. Tüchsen, E.; Hansen, P.E. Hydrogen bonding monitored by deuterium isotope effects on carbonyl 13C chemical shift in BPTI: Intra-residue hydrogen bonds in antiparallel β-sheet. *Int. J. Biol. Macromol.* **1991**, *13*, 2–8. [CrossRef]

98. Hansen, P.E.; Kamounah, F.S.; Saeed, B.A.; MacLachlan, M.J.; Spanget-Larsen, J. Intramolecular Hydrogen Bonds in Normal and Sterically Compressed *o* Hydroxy Aromatic Aldehydes. Isotope Effects on Chemical Shifts and Hydrogen Bond Strength. *Molecules* **2019**, *24*, 4533. [CrossRef]

99. Becke, A.D. A new mixing of Hartree-Fock and local density-functional theories. *J. Chem. Phys.* **1993**, *98*, 1372–1377. [CrossRef]

100. Bader, R.W.F. *Atoms in Molecules: A Quantum Theory*; Oxford University Press: New York, NY, USA, 1990.

101. Kieth, T.A. *AIMAll (Version 16.10.31)*; TK Gristmill Software: Overland Park, KS, USA, 2016.

102. Hansen, P.E.; Bolvig, S.; Duus, F.; Petrova, M.V.; Kawecki, R.; Kozerski, L. Deuterium Isotope Effects on [13]C Chemical Shifts of Intramolecularly Hydrogen-bonded Olefins. *Magn. Reson. Chem.* **1995**, *33*, 621–631. [CrossRef]

103. Chiara, J.L.; Gomez-Sanchez, A.; Sanchez Marcos, E. Spectral Properties and Isomerism of Nitro Enamines. Part 2. 3-Amino-2-nitrocrotonic Esters. *J. Chem. Soc. Perkin Trans.* **1990**, *2*, 385–392. [CrossRef]

104. Gilli, P.; Pretto, L.; Bertolasi, V.; Gilli, G. Predicting Hydrogen-Bond Strength from Acid-Base Molecular Properties. The pKa Slide Rule: Toward the Solution of a Long-Lasting Problem. *Acc. Chem. Res.* **2009**, *42*, 33–44. [CrossRef] [PubMed]

105. Gorobets, N.Y.; Yermolayev, S.A.; Gurley, T.; Gurinov, A.A.; Tolstoy, P.M.; Shenderovich, I.G.; Leadbeater, N.E. Difference between 1H NMR signals of primary amide protons as a simple spectral index of the amide intramolecular hydrogen bond strength. *J. Phys. Org. Chem.* **2012**, *25*, 287–295. [CrossRef]

106. Arunan, E.; Desiraju, G.R.; Klein, R.A.; Sadlej, J.; Scheiner, S.; Alkorta, I.; Clary, D.C.; Crabtree, R.H.; Dannenberg, J.J.; Hobza, P.; et al. Definition of the hydrogen bond (IUPAC Recommendations 2011). *Pure Appl. Chem.* **2011**, *83*, 1637–1641. [CrossRef]

107. Martyniak, A.; Majerz, I.; Filarowski, A. Pecularities of quasi-aromatic hydrogen bonding. *RSC Adv.* **2012**, *2*, 8135–8144. [CrossRef]

108. Adhikary, R.; Zimmermann, J.; Liu, J.; Forrest, R.P.; Janicki, T.D.; Dawson, P.E.; Corcelli, S.A.; Romesberg, F.E. Evidence of an unusual N-H . . . N Hydrogen Bond in Proteins. *J. Am. Chem. Soc.* **2014**, *136*, 13474–13477. [CrossRef]
109. Novak, A. Hydrogen Bonding in Solids. Correlation of Spectroscopic and Crystallographic Data. *Struct. Bond.* **1974**, *18*, 177–216
110. Grech, E.; Malarski, Z.; Sobczyk, L. Isotope Effects in NH.N Hydrogen bonds. *Chem. Phys. Lett.* **1986**, *128*, 259–263. [CrossRef]

Review

Modern Theoretical Approaches to Modeling the Excited-State Intramolecular Proton Transfer: An Overview

Joanna Jankowska [1,*] and Andrzej L. Sobolewski [2]

1 Faculty of Chemistry, University of Warsaw, 02-093 Warsaw, Poland
2 Institute of Physics, Polish Academy of Sciences, 02-668 Warsaw, Poland; sobola@ifpan.edu.pl
* Correspondence: jjankowska@chem.uw.edu.pl

Abstract: The excited-state intramolecular proton transfer (ESIPT) phenomenon is nowadays widely acknowledged to play a crucial role in many photobiological and photochemical processes. It is an extremely fast transformation, often taking place at sub-100 fs timescales. While its experimental characterization can be highly challenging, a rich manifold of theoretical approaches at different levels is nowadays available to support and guide experimental investigations. In this perspective, we summarize the state-of-the-art quantum-chemical methods, as well as molecular- and quantum-dynamics tools successfully applied in ESIPT process studies, focusing on a critical comparison of their specific properties.

Keywords: excited-state intramolecular proton transfer; photochemistry; photobiology; quantum chemistry; molecular dynamics; ultrafast processes

Citation: Jankowska, J.; Sobolewski, A.L. Modern Theoretical Approaches to Modeling the Excited-State Intramolecular Proton Transfer: An Overview. *Molecules* **2021**, *26*, 5140. https://doi.org/10.3390/molecules 26175140

Academic Editor: Mirosław Jabłoński

Received: 15 July 2021
Accepted: 23 August 2021
Published: 25 August 2021

Publisher's Note: MDPI stays neutral with regard to jurisdictional claims in published maps and institutional affiliations.

1. Introduction

Photochemistry of organic molecular systems is an extremely rich and exciting field of research, continuously growing and, thus, pushing forward the frontiers of our understanding of light–matter interactions. Among many chemical processes induced with photon absorption, the excited-state intramolecular proton transfer (ESIPT) stands out with its ultrashort timescale and strong impact on the molecular electronic structure, which is often manifested with large emission Stokes shifts. Relying on a proton exchange between two electronegative centers along a pre-existing intramolecular hydrogen bond, the ESIPT process is recognized to provide a mechanism of excellent photostability to natural and artificial molecular systems [1–3], finds applications in fluorescent probes and imaging agents [4–6], governs characteristic emission of the green fluorescent protein and its analogs [7–9], and opens rich possibilities for multicolor emission in organic light-emitting diodes (OLEDs) [10–14]. Last but not least, ESIPT may also activate other excited-state reaction channels, facilitating the design of complex molecular photo-devices [15–19].

In its typical arrangement, ESIPT occurs upon photoexcitation of a molecular system including two moieties connected on the one side by an intramolecular hydrogen bond and with an electronically conjugated network of covalent bonds on the other side, as shown in Figure 1. The reaction occurs in an excited electronic state and is usually being parameterized by the distance between the proton-donor (D) atom (most commonly oxygen or nitrogen [15,20,21]) and the transferring proton. The proton-accepting (A) moiety consists of another electronegative center, often including a carbonyl or an imine group [11,17], which, in order for the ESIPT process to be efficient, should exhibit stronger basicity in the excited state than the proton donor. After the proton transfer, the system undergoes further electronic relaxation—either radiative or nonradiative in nature. In the former case, the characteristic strongly red-shifted fluorescence is nowadays regarded as the hallmark of ESIPT. The latter scenario requires the presence of an independent nonradiative deactivation channel, induced, for instance, by a *cis/trans* isomerization reaction [22]. While the ultrafast ESIPT process is often reported to have ballistic nature (that is, barrierless excited-state potential-energy (PE) landscape in Figure 1), it may also involve passage through an energy barrier or include nonadiabatic transition/intersystem crossing between different electronic states. Similarly, after the relaxation to the ground electronic state, the system may reach a local PT minimum or may undergo a spontaneous back-transfer to the initial D–H bonded isomer. This final reaction-cycle closing transformation is sometimes referred to as a ground-state intramolecular proton transfer (GSIPT).

In the context of the following discussion, it is also important to underline a distinction between the ESIPT reaction investigated herein and similar processes, especially the proton-coupled electron transfer (PCET) reaction [23–26]. The latter phenomenon, often of nonadiabatic character, has a generally much more complex nature and may involve ground and excited-state reactions, such as intra- and intermolecular, concerted, and stepwise processes. Under certain conditions, ESIPT may play the role of an elementary step in a complex PCET reaction.

In this review, we identify and discuss three fundamental families of theoretical approaches to modeling the ESIPT process: (i) the static methods, (ii) the mixed quantum–classical molecular dynamics, and (iii) the quantum dynamics methods. In the following sections, we briefly outline their theoretical assumptions, comment on the scope of their applicability and performance in ESIPT studies, and highlight recent achievements in each field, focusing on the most illustrative results from the last 5 years. For a broader view of ESIPT-focused research, including also experimental insights, the interested reader is referred to other up-to-date reviews [4,6,21,27,28] and monographs [29–32] that have been published on the subject.

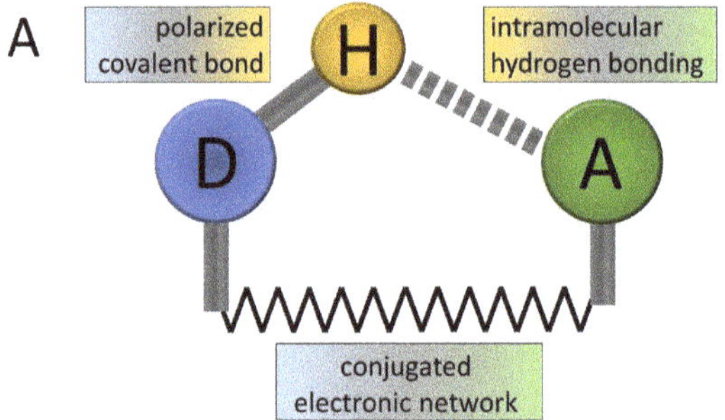

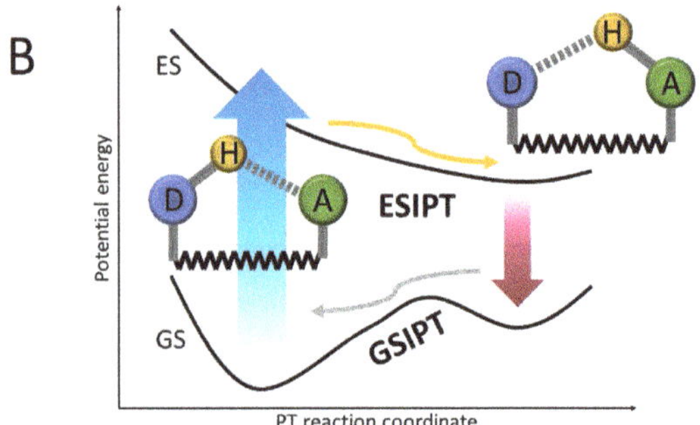

Figure 1. Schematic representation of the ESIPT mechanism: (**A**) initial atomic arrangement of an ESIPT system; (**B**) typical (most basic) potential energy landscape along the ESIPT reaction coordinate. D—proton donor; A—proton acceptor; GS—ground electronic state; ES—excited electronic state; blue arrow—initial photoabsorption; red arrow—Stokes-shifted fluorescence.

2. Static Investigation Approach

2.1. Objective of the Static Calculations

The most straightforward approach to the theoretical characterization of a new photo-chemical system relies on a "static" quantum-chemical investigation, which itself might be a target research strategy or an initial step of a more complex protocol. The static ESIPT investigation is primarily oriented toward providing high-quality absorption and emission optical energies, as well as the topographical description of the investigated system's PE landscape. It might also be considered a lower-cost computational option for large polyatomic molecules as this protocol, compared to the dynamic ones, usually involves a relatively limited number of demanding energy-gradient calculations and facilitates further savings by allowing calculations under fixed system symmetry, if such is present. Typical outcomes of the static approach are absorption and emission vertical electronic energies [33–36], upper-bound estimations for possible energy barriers in the ground and electronically excited states [36–38], and detailed characterization of these states in terms of symmetry and orbital configuration [38,39]. Moreover, the number of other molecular features complementing the experimental ESIPT characterization can be determined, including, e.g., tautomers relative energies [35,40], atomic charges [41–43], and vibrational modes attribution [44,45].

2.2. Typical Investigation Workflow

A typical workflow scheme of the static protocol is presented in Figure 2. In the first step, a search for stationary points on the ground-state (GS) potential energy surface (PES) is performed, followed by vertical electronic excitation energies calculations. At this stage, the electronic structure of the GS and the character of the relevant excited states (ES) need to be carefully evaluated, with a special focus on expected requirements for the excited-state methods to be applied in the following steps. Afterward, an analysis of ES relaxed properties is conducted, and the barrierless/barrier-restricted character of ESIPT is determined by excited-state geometry optimization of the relevant isomers, along with predictions for the energy and intensity of the Stokes-shifted fluorescence [38,46]. In the final step, adiabatic potential energy profiles (PEPs) [47,48] or PESs [3,33] may be calculated, if one or multiple reaction coordinates, respectively, need to be explicitly considered. In certain cases, the energy profiles of linearly interpolated reaction paths might also efficiently support the static ESIPT analysis [49]. However, in accordance with the static protocol name, it should be stressed that neither dynamic nor kinetic effects beyond the zero-point energy (ZPE) corrections to electronic stationary-point energies are included at this level of theory.

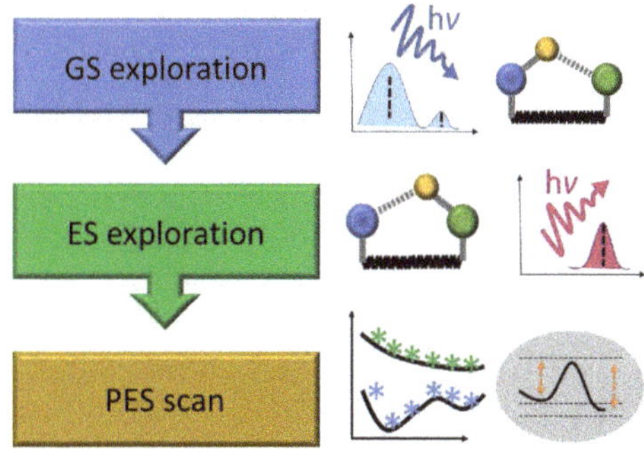

Figure 2. Typical workflow of the static ESIPT investigation protocol.

2.3. Applied Tools and Methods

In principle, the static ESIPT investigation can be performed with any kind of electronic structure method capable of treating the electronic structure of excited states in a relaxed manner. Typically, single-reference electronic structure methods can be trusted to reproduce accurately the topography of the involved electronic states, with the exception of an anti-Kasha ESIPT [50] or systems with low-lying doubly excited states [42,51]. The choice of the optimum electronic structure method for a particular ESIPT study is usually dictated, on the one hand, by the size of the molecular system and, on the other, by its specific electronic structure features. An important general observation is that the correct characterization of the PES topography of the proton-transfer process necessarily requires the inclusion of dynamic electronic correlation effects, as artificial or overestimated reaction barriers have been reported otherwise [52–54].

2.3.1. Ab Initio Wave Function Approaches

For the smaller molecular systems (generally, up to 50 heavy atoms), coupled-cluster electronic structure methods, such as the simplified version of singles and doubles, CC2 [55,56], and the algebraic diagrammatic construction method (ADC(2)) [57,58], have been the methods of choice for a long time [3,34,46,49,59]. The reason is their universality [60] and the availability of well-tested and efficient implementations in widely distributed quantum-chemical software packages. While the CC2 method yields overall slightly more accurate electronic excitation energies [58], the virtue of the ADC(2) approach lies in better numerical stability near-electronic excited-state crossings [61–63].

At the same time, regarding the recent reports of previously unrecognized troubles of the CC2 and ADC(2) methods in predicting accurate excited-state PES beyond the Franck–Condon vicinity [64,65], spin-component scaled CC2 (SCS-CC2 [66]) and scaled opposite spin CC2 (SOS-CC2 [67,68]) approaches have been found particularly promising in the context of ESIPT studies. In this direction, one of us recently employed both these protocols in combination with the ADC(2) method to model photophysical transformations in several salicylaldimine derivatives [34], observing indeed their improved performance for ESIPT-driven fluorescence energy calculations; similar results have also been reported by Kielesinski et al. for coumarins [69]. In this latter work, performed quantum-chemical investigation yielded correct predictions of solvatochromic effects in a series of compounds, studied both by experimental and theoretical means. Moreover, a direct explanation for single- and multicolor emission observed experimentally in closely related coumarin systems has been formulated on the grounds of a detailed computational analysis of the lowest-energy electronic excited states' properties.

2.3.2. Density Functional Theory Methods

The second widely applied family of electronic structure methods for ESIPT investigations is time-dependent density functional theory (TD-DFT [70]), in its original design and within the Tamm–Dancoff approximation (TDA-DFT [71]). Abundant ESIPT studies at this level of theory [33,40,46,48,72,73] take advantage of the favorable scaling of DFT with the system size. At the same time, due to known difficulties of TD-DFT with the description of charge-transfer states, and more recent findings on its troubles with the proper determination of state orders in inverted singlet/triplet systems [74,75], the choice of the exchange-correlation functional and method validation usually need to be carefully conducted before meaningful conclusions can be formulated [36,59,76].

In recent years, many functionals of different types have been employed in ESIPT studies [59]. In particular, the popular Becke three-parameter Lee–Yang–Parr (B3LYP) [77,78] functional was found to perform well for systems exhibiting small or no charge-transfer effect in the excited states involved in the ESIPT reaction [3,33,37,72,73,76]. Other recently applied and promising functionals include hybrid meta M06-2X [38,40,46,79], and long-range and dispersion-corrected ωB97X-D [40,80,81]. Among other reported possibilities, the Coulomb-attenuated hybrid functional CAM-B3LYP [82] has also recently gained a relatively trusted position as a tool for ESIPT investigations [36,73,76,81]. At the same time, none of these functional choices appear to be fully universal as of today [59]. As for the TD-DFT relation to TDA-DFT, the latter shows generally higher stability at the interstate crossings, including improved performance in the vicinity of conical intersections [83], even those involving the reference electronic state, and allows for some additional computational-time savings [81,84], appearing particularly attractive in the context of ESIPT dynamics simulations.

Finally, due to known DFT deficiencies in describing dispersion interactions [85], it is worth noting the role of these effects in ESIPT modeling at the TD-DFT level. It is observed that a suitable correction, such as D3 or D4 as proposed by Grimme et al. [86,87] or direct application of a dispersion-corrected functional (e.g., ωB97X-D) is typically required for proper treatment of microsolvated, supramolecular, or condensed-phase (e.g., crystal) systems, in which explicit interactions between the core molecule and the environment have to be included [88–90]. On the other hand, in most other cases, the omission of the dispersion part of interaction energy does not seem to play a significant role, as revealed by the generally good performance of common uncorrected exchange-correlation functionals [59,76].

2.3.3. Basis-Set Choice

Practical application of the methods discussed above requires making the additional choice of a basis set for the wave-function expansion, which has a direct impact on the quality of the results. In this case, again one needs to make a compromise between the computational cost and desired accuracy. Most common recent choices in ESIPT studies seem to be favoring the cc-pVTZ [91] basis set from the Dunning family on the one hand [3,33,39,40,48], and different variants of the Pople 6-311 G(d,p) [92] basis set, on the other [36,72,73]. The latter direction finds its support in a general study by Laurent et al. [93], in which the basis-set effect on vertical excitation energy calculations was investigated. On the grounds of reported results, however, it is not easy to make definite ESIPT-targeted recommendations for the basis-set choice since both system and ES-specific effects come into play [93].

2.3.4. Solvent Effects

Finally, a brief discussion of the environmental effects is appropriate, as ESIPT systems are usually investigated in solution or in other complex condensed-phase environments. In particular, the polarity of the surrounding medium has been observed to have a strong impact on the ESIPT reaction efficiency [73,76].

Thus far, several different approaches have been employed to tackle environmental effects on ESIPT, including microsolvation [43,69], the conductor-like screening model (COSMO) [48,94,95], the polarizable continuum model (PCM) [76,96,97], the solvent model density (SMD) method [46,98], and the integral equation formalism version of PCM (IEF-PCM) [99–101]. The latter approach, particularly popular recently [33,37,73,102], has been applied e.g. by Wang et al. to the BTS system [39] in methylene chloride, yielding very high accuracy predictions for excitation and emission wavelengths, with divergence from the experimental values measured in just a few nm. In addition to the general purpose methods listed above, state-specific PCM treatments of correlated linear response (cLR) [103] and the vertical excitation model within the unrelaxed density approximation (VEM-UD) [104,105] have been successfully applied to study ESIPT by Vérité et al. [40], who pointed out the advantages that these approaches bring for the description of charge-transfer states in ESIPT reactions. Nevertheless, the explicit inclusion of (typically few) solvent molecules is necessary in certain cases, especially for protic solvents and solvents exhibiting proton-accepting properties, since resulting competition between intra- and intermolecular hydrogen bond formation may drastically affect the ESIPT reaction yield [88,90,106].

2.4. Summary of the Static ESIPT Investigation Methods

To summarize the section dedicated to the static ESIPT investigation protocol, we again underline its strengths as being a relatively affordable and yet informative approach, designed to provide a fundamental characterization of ESIPT, including the system's absorption and emission properties, as well as information on the topography of GS and ES PESs over pre-selected reaction coordinates. Due to its inherent compatibility with a great variety of electronic structure methods, this protocol allows researchers to take advantage of new developments in electronic structure theory and, thus, constantly provides opportunities for cutting-edge studies of ESIPT in all types of molecular systems.

At the same time, it should be noted that, under certain circumstances, the investigation of static ESIPT paths may not be sufficient. In particular, systems undergoing multiple PT reactions are typically challenging to be accurately studied with this protocol due to the large computational cost of multi-dimensional PES scans, on the one hand, and the critical role of the sequence of individual processes missed at this level, on the other hand. Another situation, in which special precautions should be taken, is when ESIPT occurs within a dense manifold of electronic states, such as in situations in which a competition between various photochemical transformations is to be expected; in these cases, one may need to explicitly determine the relative efficiency of each channel, which usually requires the inclusion of nuclear-dynamic effects.

3. Nonadiabatic Molecular Dynamic Approaches

New opportunities of delving deeper into the course of the ESIPT reaction open when one turns toward dynamic approaches. In the most general view, this refers to a large (and growing) number of methods allowing for real-time simulations of molecular systems' evolution in terms of their electronic and nuclear structure, beyond the static picture.

The time-dependent Schrödinger equation is the core and starting point for the dynamic methods, yielding two broad families of approaches, differed by the level of the applied approximations. The first family consists of fully quantum dynamic (QD) methods, in which electronic and nuclear degrees of freedom are treated both at the quantum-mechanical level. The second family is built on nonadiabatic mixed quantum–classical (NA-MQC) dynamics methods, in which the nuclei, which are much slower than the electrons, are treated at the classical or semi-classical level, propagating under the Newton equation of motion. The quantum–electronic and classical–nuclear subsystems are in this picture connected by the nonadiabatic coupling, ensuring the self-consistency of the description of the total molecular system. We start our discussion of ESIPT nonadiabatic molecular dynamics studies with the NA-MQC methods, which will be subsequently followed by the analysis of the QD performance, in line with the increasing level of the method exactness.

3.1. Mixed Quantum–Classical Dynamic Calculations in ESIPT Studies

The NA-MQC methods, recently summarized in an excellent review by Crespo-Otero and Barbatti [107], can be divided into several groups, out of which the trajectory surface hopping (TSH) [108–111], ab initio multiple spawning (AIMS) [112,113], and, most recently, the nuclear–electronic orbital Ehrenfest (NEO-Ehrenfest) [114] methods are, to the best of our knowledge, the ones that have been successfully employed in dynamic ESIPT investigations thus far. Below, a brief characterization of these approaches is provided, along with an illustration of their performance for the description of the ESIPT process. Additionally, a schematic representation of their underlying mechanisms is presented in Figure 3.

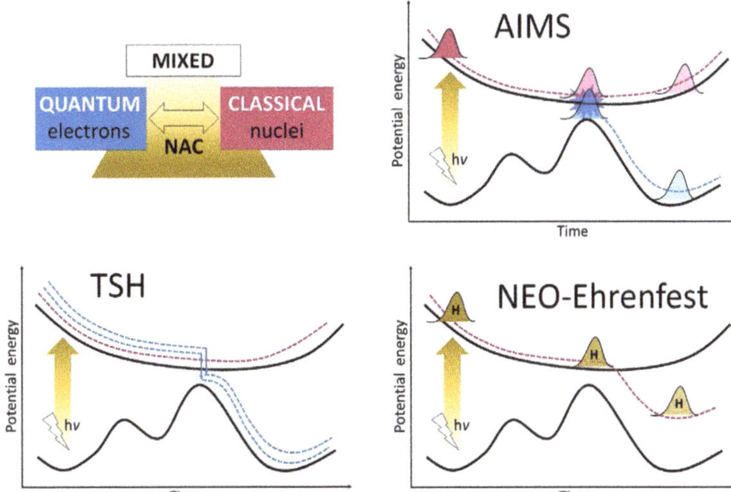

Figure 3. Schematic illustration of the mechanisms of the discussed NA-MQC dynamic approaches: trajectory surface hopping (TSH), ab inito multiple spawning (AIMS), and nuclear–electronic orbital Ehrenfest (NEO-Ehrenfest).

3.1.1. Trajectory Surface Hopping Approach

The trajectory surface hopping method, especially under Tully's fewest-switches (FSSH) [108] algorithm and under other algorithms based on the Landau–Zener (LZ) model [115,116], is the most widely used NA-MQC approach in ESIPT studies thus far. TSH relies on the modeling of the real-time evolution of the molecular system by a set of independent classical trajectories, which together are assumed to represent a nuclear wave packet in an approximate (statistical) way. While the trajectories are propagated on individual Born–Oppenheimer adiabatic PESs, nonadiabatic transitions between these surfaces are possible in regions characterized by large interstate nonadiabatic couplings (NACs). The interstate transitions are controlled by a stochastic algorithm (FSSH) or induced in minimum-energy-gap regions (LZ methods), with the "hopping" probability proportional to the NAC. Importantly, the TSH method can be implemented as an "on-the-fly" approach [117], which means that the actual PES, on which the system is propagating, does not have to be known in advance, and electronic properties, such as energies or gradients, are calculated along the NA-MQC path as needed. It should be noted, however, that usually, many trajectories are required for reliable and converged TSH results [107].

As of today, the TSH method has been implemented in a number of dedicated software packages, including Newton X [118,119], Shark [120], Jade [121], etc. [107]. In all cases, the NA-MQC dynamics protocol has to be paired with an electronic structure method, and its choice needs to be made with great care, as it directly impacts the quality of the results, as well as the simulation cost. One needs to be aware, however, that due to the inherent mixed-classical nature of the TSH approach, certain effects that may play an important role in the ESIPT reaction cannot be reproduced at the TSH level of theory. This includes all phenomena stemming from the nonlocality of the true nuclear wave function, which is reduced to a single point on adiabatic PES within the TSH picture. In particular, proton tunneling, wave-packet interference, and decoherence effects are not included, the latter being partially restored in the TSH simulations via the introduction of various kinds of decoherence corrections [122].

In terms of recent applications of the TSH methodology to particular ESIPT studies, Li et al. reported interesting results explaining 3-hydroxyflavone dual fluorescence in solvents containing protic contamination with a competition between intra- and inter-molecular excited-state PT reactions [90], discussing an effect of the number of explicitly included water molecules on the simulation outcomes. In this case, the applied FSSH/TD-DFT methodology allowed for high-quality predictions of the electronic excitation energies (typical error below 0.2 eV) and also yielded ESIPT timescale in very good agreement with available experimental data, with a deviation of less than 10 fs. Another challenging aspect of dealing with a large number of possible photo-reaction products has been tackled by Tuna et al., who employed a robust multiconfiguration interaction variant of the orthogonalization-corrected semi-empirical OM2 approach to model ESIPT-driven photochemistry of urocanic acid [123]. The same method has also been applied by Xia et al. to study relaxation mechanisms in the isolated benzodiazepinone molecule, in which several interconnected relaxation channels come to play [124]. Furthermore, we recently performed a TSH study at the TDA-DFT level to analyze the impact of the character of the lowest excited state on the ESIPT process efficiency [84], eventually confirming the important role of the $\pi\pi^*$ states.

3.1.2. Ab Initio Multiple Spawning Approach

Another NA-MQC approach that has found applications in time-resolved ESIPT simulations is the ab initio multiple spawning method. AIMS originates from the formally exact full multiple spawning methodology [125,126]. Its core concept relies on representing the nuclear wave function with partially coupled traveling Gaussian functions, having a finite width both in position and momentum coordinates and interacting during the dynamics. Importantly, the total number of the "on-the-fly" propagated Gaussian functions changes in time since on each passage through a PES region characterized with strong NAC, a new Gaussian is spawned (hence, the S in AIMS).

Similar to the TSH case, AIMS simulations require combining the particular MS protocol with a suitable electronic structure method. As for the AIMS code itself, as of today, it is available within several software packages, including GAMESS [127,128], MOL-PRO [129,130], and MOPAC [131,132]. Technically, AIMS involves a higher computational cost than TSH, yet it should be considered a superior approach, inherently including decoherence effects, and yielding a correct description of some non-local phenomena. At the same time, due to certain intrinsic limitations of the AIMS approach, the tunneling effect, although theoretically possible to be covered through the intrastate spawning procedure [112,125], is not reproduced at this level of theory [113].

Turning to recent interesting applications of the AIMS in ESIPT studies, Pijeau et al. investigated the photophysics of the paradigmatic salicylideneaniline (SA) system [133], focusing on the effect of nonplanarity on ESIPT and on the total deactivation mechanism. In this study, the AIMS protocol has been connected with the floating occupation molecular orbital complete active space configuration interaction (FOMO-CASCI) method, with further wave function-in-DFT embedding. The same group also tackled the hydroxyphenyl benzothiazole (HBT) system at this level of theory, obtaining very good agreement with the experimental results [134].

3.1.3. Nuclear–Electronic Orbital Ehrenfest Approach

Recently, a new NA-MQC dynamic approach, NEO-Ehrenfest, aiming toward the further enhanced recovery of nonlocal effects, has been developed. Within this method, protons are treated quantum mechanically on an equal footing with the electrons, yielding automatic inclusion of the ZPE, quantized vibrational levels, and tunneling effects associated with these species [114]. The NEO-Ehrenfest approach, specifically tailored to provide a high-level description of the ESIPT and PCET processes [135], is built on the concept of semi-classical traveling proton basis functions, which, on the one hand, provide means for the quantum-mechanical representation of protons, as has been demonstrated before for the time-independent case [136], and, on the other hand, enable the description of its long-range displacements.

In a recent pioneering NEO-Ehrenfest study by Zhao et al. the ESIPT process in o-hydroxy-benzaldehyde has been investigated [135]. Upon comparison of results obtained using the NEO-Ehrenfest and the traditional Ehrenfest approach [137] with all-classical nuclei, the proton transfer reaction acceleration in the quantum case has been observed, which has been ascribed to the delocalization of the proton wave function, resulting in a smaller necessary displacement of the proton-accepting and proton-donating centers. Moreover, the kinetic isotope effect upon deuterium substitution has been reproduced at this level of theory.

3.1.4. Summary of the NA-MQC Dynamic ESIPT Simulations

To summarize the section dedicated to mixed quantum–classical ESIPT studies, we again highlight the great contributions of the NA-MQD dynamic methods for the field. By allowing real-time picturing of the proton transfer process, characteristic timescales and unforeseen reaction mechanisms can be modeled at this level of theoretical description. While the methods share the mixed quantum–classical nature, they also still bear important differences, making them possible methods of choice for different conditions. In particular, TSH is a robust and probably most universal tool, reliable for the modeling of barrierless ESIPT, including complex situations, in which multiple PTs or competition from other photoreaction channels needs to be taken into account. The AIMS approach, formally more exact, may also be generally applied to this class of processes, as long as it does not become prohibitively expensive due to the extended molecular system size. At the same time, when nuclear quantum effects of protons are expected to play a role, such as in the barrier-restricted ESIPT case, the NEO-Ehrenfest method may be considered a good choice.

3.2. Quantum Dynamics Methods for ESIPT Simulations

Despite many useful conclusions on the ESIPT reaction course that may be taken from the NA-MQC dynamics, there are situations in which one needs to advance even further with the level of the system's dynamic description, up to the point of full quantum treatment of all the species, including nuclei. As has been already pointed out, the most typical reason of adopting this approach is when tunneling through an energy barrier along the ESIPT path needs to be included, i.e., when highly accurate rates or proton-transfer equilibrium have to be characterized. Another situation calling for the QD treatment is when a strongly nonadiabatic ESIPT mechanism is expected, e.g., when trivial interstate crossings are present, potentially threatening the correct NA-MQC dynamics performance. The latter problem, however, has been, in recent years, partially resolved by the successful design of correction strategies to several NA-MQC protocols [138–140].

Multiconfiguration Time-Dependent Hartree Method

Among the most robust approaches to solving the time-dependent Schrödinger equation that retain the quantum character of all the molecular system's components, the multiconfiguration time-dependent Hartree (MCTDH) method plays, up to date, the most prominent role [141]. Relying on the Born–Huang expansion, the MCTDH allows for propagating a wave-packet in time with the wave function of the system represented by the sum of products of so-called single-particle functions describing individual nuclear degrees of freedom (DOF), which are typically associated with the molecular normal vibrational modes. The MCTDH method employs model Hamiltonians, constructed individually for each system. In the case of the ESIPT studies, usually, a vibronic Hamiltonian is employed [142–144], including a pre-selected number of electronic PESs and nuclear DOFs. It should be noted that MCTDH requires the determination of the PESs prior to the MCTDH calculation. This is typically achieved by combining quantum-chemical probing of the PES regions expected to play the most important role in the investigated process with the application of various interpolation models to approximate the remaining PES areas.

In practical terms, the original MCTDH method can nowadays cover in a general case up to ca. 20 DOFs, but in recent years, new flavors of MCTDH have been developed, such as the multilayer multiconfiguration time-dependent Hartree (ML-MCTDH) method, which pushes this limit even up to several thousand DOFs [145]. The performance boost stems, in this case, from a tree-like (layered) representation of the nuclear wave function, in which the traditional SPFs are further expanded themselves in the MCTDH spirit. The eventual efficiency gain, however, depends strongly on the system's nature and size [145]. As of today, different variants of the MCTDHF methods are available in a few dedicated software packages, such as the Heidelberg MCTDH [146], or Quantics [147].

Moving to the MCTDHF applications to simulating the ESIPT process, interesting results on the photophysics of hydroxychromones have been reported by Perveaux et al. [148] and Anand et al. [149]. In the former case, the full-dimensional (48 DOFs) ML-MCTDH method was applied to analyze the interplay between the ESIPT reaction and the out-of-plane hydrogen torsion in 3-hydroxychromone, while the latter study comprised analogical simulations for the 3-hydroxychromone and 5-hydroxychromone systems, performed at the multimode MCTDH level with the inclusion of 25 DOFs. Both investigations led to similar conclusions on a critical role of a conical intersection between bright S_1 and dark S_2 states, of respective $\pi\pi^*$ and $n\pi^*$ character, which was interpreted as the reason for observation of two ESIPT rate constants for these molecules in the experiment. Anand et al. applied the same methodology to study ESIPT also in similar 3-hydroxyflavone [150] and 3-hydroxypyran-4-one [151] systems, confirming the important role of the S_2 state in their photorelaxation. Finally, recent thorough work by Cao et al. provided theoretical insights on ESIPT-driven mechanism and quantum dynamics of thermally activated delayed fluorescence in triquinolonobenzene [152], in which singlet-state ultrafast proton transfer occurs within a dense manifold of low-lying triplet states.

4. Summary and Future Outlook

In summary, in the present review, we gathered and discussed key features of the modern theoretical approaches employed in ESIPT investigations, with a special focus on their complementary capabilities and critical limitations. Depending on the particular research focus, e.g., manifested by the need for detailed knowledge of ES topography, equilibrium populations of different molecular isomers, or characterization of time-resolved effects, and the system-specific challenges, such as the isolated or band-like arrangement of the active excited states, presence of barrier-restricted or barrierless PT, the necessity of taking the intersystem crossing into account, etc., a proper theoretical approach in each case can be proposed. To this end, we hope that this sometimes challenging choice will be facilitated with the provided insights.

Looking toward future developments that would further strengthen the field, support from machine learning techniques should definitely be considered a promising direction for the QD efficiency enhancement, with first results already emerging [153,154], so as to improve the performance of other dynamic approaches [155]. Moreover, linking solvent-dependent optical properties with nonadiabatic ESIPT dynamics within the fully quantum framework could also provide powerful new tool to the existing set [151], opening new-level possibilities, e.g., for describing the competition between intra- and intermolecular excited-state proton transfer reactions on equal footing.

Funding: This research received no external funding.

Institutional Review Board Statement: Not applicable.

Informed Consent Statement: Not applicable.

Data Availability Statement: All data are available within the article and the included references.

Acknowledgments: The Authors would both like to thank Wolfgang Domcke for the great support and fruitful discussion on the preparation of the manuscript.

Conflicts of Interest: The authors declare no conflict of interest.

References

1. Gong, Y.; Wang, Z.; Zhang, S.; Luo, Z.; Gao, F.; Li, H. New ESIPT-inspired photostabilizers of two-photon absorption coumarin–benzotriazole dyads: From experiments to molecular modeling. *Ind. Eng. Chem.* **2016**, *55*, 5223–5230. [CrossRef]
2. Maity, S.; Chatterjee, A.; Chakraborty, N.; Ganguly, J. A dynamic sugar based bio-inspired, self-healing hydrogel exhibiting ESIPT. *New J. Chem.* **2018**, *42*, 5946–5954. [CrossRef]
3. Berenbeim, J.A.; Boldissar, S.; Owens, S.; Haggmark, M.R.; Gate, G.; Siouri, F.M.; Cohen, T.; Rode, M.F.; Schmidt Patterson, C.; De Vries, M.S. Excited state intramolecular proton transfer in hydroxyanthraquinones: Toward predicting fading of organic red colorants in art. *Sci. Adv.* **2019**, *5*, eaaw5227. [CrossRef]
4. Sedgwick, A.C.; Wu, L.; Han, H.H.; Bull, S.D.; He, X.P.; James, T.D.; Sessler, J.L.; Tang, B.Z.; Tian, H.; Yoon, J. Excited-state intramolecular proton-transfer (ESIPT) based fluorescence sensors and imaging agents. *Chem. Soc. Rev.* **2018**, *47*, 8842–8880. [CrossRef] [PubMed]
5. Long, R.; Tang, C.; Xu, J.; Li, T.; Tong, C.; Guo, Y.; Shi, S.; Wang, D. Novel natural myricetin with AIE and ESIPT characteristics for selective detection and imaging of superoxide anions in vitro and in vivo. *Chem. Comm.* **2019**, *55*, 10912–10915. [CrossRef]
6. Li, Y.; Dahal, D.; Abeywickrama, C.S.; Pang, Y. Progress in tuning emission of the excited-state intramolecular proton transfer (ESIPT)-based fluorescent probes. *ACS Omega* **2021**, *6*, 6547–6553. [CrossRef]
7. Hsieh, C.C.; Chou, P.T.; Shih, C.W.; Chuang, W.T.; Chung, M.W.; Lee, J.; Joo, T. Comprehensive studies on an overall proton transfer cycle of the ortho-green fluorescent protein chromophore. *JACS* **2011**, *133*, 2932–2943. [CrossRef] [PubMed]
8. Vegh, R.B.; Bloch, D.A.; Bommarius, A.S.; Verkhovsky, M.; Pletnev, S.; Iwaï, H.; Bochenkova, A.V.; Solntsev, K.M. Hidden photoinduced reactivity of the blue fluorescent protein mKalama1. *Phys. Chem. Chem. Phys.* **2015**, *17*, 12472–12485. [CrossRef]
9. Chatterjee, T.; Lacombat, F.; Yadav, D.; Mandal, M.; Plaza, P.; Espagne, A.; Mandal, P.K. Ultrafast dynamics of a green fluorescent protein chromophore analogue: Competition between excited-state proton transfer and torsional relaxation. *J. Phys. Chem. B* **2016**, *5*, 9716–9722 [CrossRef]
10. Tang, K.C.; Chang, M.J.; Lin, T.Y.; Pan, H.A.; Fang, T.C.; Chen, K.Y.; Hung, W.Y.; Hsu, Y.H.; Chou, P.T. Fine tuning the energetics of excited-state intramolecular proton transfer (ESIPT): White light generation in a single ESIPT system. *JACS* **2011**, *133*, 17738–17745. [CrossRef]
11. Wu, K.; Zhang, T.; Wang, Z.; Wang, L.; Zhan, L.; Gong, S.; Zhong, C.; Lu, Z.H.; Zhang, S.; Yang, C. De novo design of excited-state intramolecular proton transfer emitters via a thermally activated delayed fluorescence channel. *JACS* **2018**, *140*, 8877–8886. [CrossRef]
12. Duarte, L.G.T.A.; Germino, J.C.; Berbigier, J.F.; Barboza, C.A.; Faleiros, M.M.; de Alencar Simoni, D.; Galante, M.T.; de Holanda, M.S.; Rodembusch, F.S.; Atvars, T.D.Z. White-light generation from all-solution-processed OLEDs using a benzothiazole–salophen derivative reactive to the ESIPT process. *Phys. Chem. Chem. Phys.* **2019**, *21*, 1172–1182. [CrossRef]
13. Czernel, G.; Budziak, I.; Oniszczuk, A.; Karcz, D.; Pustuła, K.; Górecki, A.; Matwijczuk, A.; Gładyszewska, B.; Gagoś, M.; Niewiadomy, A.; et al. ESIPT-related origin of dual fluorescence in the selected model 1,3,4-thiadiazole derivatives. *Molecules* **2020**, *25*, 4168. [CrossRef]
14. Trannoy, V.; Léaustic, A.; Gadan, S.; Guillot, R.; Allain, C.; Clavier, G.; Mazerat, S.; Geffroy, B.; Yu, P. A highly efficient solution and solid state ESIPT fluorophore and its OLED application. *New J. Chem.* **2021**, *45*, 3014–3021. [CrossRef]
15. Rode, M.F.; Sobolewski, A.L. Effect of chemical substituents on the energetical landscape of a molecular photoswitch: An ab initio study. *J. Phys. Chem. A* **2010**, *114*, 11879–11889. [CrossRef]
16. Jankowska, J.; Rode, M.F.; Sadlej, J.; Sobolewski, A.L. Photophysics of Schiff bases: Theoretical study of salicylidene methylamine. *ChemPhysChem* **2012**, *13*, 4287–4294. [CrossRef] [PubMed]
17. Böhnke, H.; Bahrenburg, J.; Ma, X.; Röttger, K.; Näther, C.; Rode, M.F.; Sobolewski, A.L.; Temps, F. Ultrafast dynamics of the ESIPT photoswitch N -(3-pyridinyl)-2-pyridinecarboxamide. *Phys. Chem. Chem. Phys.* **2018**, *20*, 2646–2655. [CrossRef] [PubMed]
18. Georgiev, A.; Todorov, P.; Dimov, D. Excited state proton transfer and E/Z photoswitching performance of 2-hydroxy-1-naphthalene and 1-naphthalene 5,5'-dimethyl- and 5,5'-diphenylhydantoin Schiff bases. *J. Photochem. Photobiol. A* **2020**, *386*, 112143. [CrossRef]
19. Huang, A.; Hu, J.; Han, M.; Wang, K.; Xia, J.L.; Song, J.; Fu, X.; Chang, K.; Deng, X.; Liu, S.; et al. Tunable photocontrolled motions of anil-poly(ethylene terephthalate) systems through excited-state intramolecular proton transfer and trans–cis isomerization. *Adv. Mater.* **2021**, *33*, 2005249. [CrossRef]
20. Hsieh, C.C.; Jiang, C.M.; Chou, P.T. Recent experimental advances on excited-state intramolecular proton coupled electron transfer reaction. *Acc. Chem. Res.* **2010**, *43*, 1364–1374. [CrossRef]
21. Joshi, H.C.; Antonov, L. Excited-state intramolecular proton transfer: A short introductory review. *Molecules* **2021**, *26*, 1475. [CrossRef]
22. Sobolewski, A.L.; Domcke, W. Photophysics of intramolecularly hydrogen-bonded aromatic systems: Ab initio exploration of the excited-state deactivation mechanisms of salicylic acid. *Phys. Chem. Chem. Phys.* **2006**, *8*, 3410. [CrossRef]
23. Hammes-Schiffer, S. Theory of proton-coupled electron transfer in energy conversion processes. *Acc. Chem. Res.* **2009**, *42*, 1881–1889. [CrossRef]
24. Weinberg, D.R.; Gagliardi, C.J.; Hull, J.F.; Murphy, C.F.; Kent, C.A.; Westlake, B.C.; Paul, A.; Ess, D.H.; McCafferty, D.G.; Meyer, T.J. Proton-coupled electron transfer. *Chem. Rev.* **2012**, *112*, 4016–4093. [CrossRef] [PubMed]

25. Domcke, W.; Sobolewski, A.L.; Schlenker, C.W. Photooxidation of water with heptazine-based molecular photocatalysts: Insights from spectroscopy and computational chemistry. *J. Chem. Phys.* **2020**, *153*, 100902. [CrossRef]

26. Tyburski, R.; Liu, T.; Glover, S.D.; Hammarström, L. Proton-coupled electron transfer guidelines, fair and square. *JACS* **2021**, *143*, 560–576. [CrossRef]

27. Serdiuk, I.E.; Roshal, A.D. Exploring double proton transfer: A review on photochemical features of compounds with two proton-transfer sites. *Dyes Pigm.* **2017**, *138*, 223–244. [CrossRef]

28. Chen, L.; Fu, P.Y.; Wang, H.P.; Pan, M. Excited-state intramolecular proton transfer (ESIPT) for optical sensing in solid state. *Adv. Opt. Mater.* **2021**, *52*, 2001952. [CrossRef]

29. Demeter, A. Hydrogen bond basicity in the excited state: Concept and applications. In *Hydrogen Bonding and Transfer in the Excited State*; John Wiley & Sons, Ltd.: Hoboken, NJ, USA, 2010; Chapter 3, pp. 39–78. [CrossRef]

30. Tomin, V.I. Proton transfer reactions in the excited electronic state. In *Hydrogen Bonding and Transfer in the Excited State*; John Wiley & Sons, Ltd.: Hoboken, NJ, USA, 2010; Chapter 22, pp. 463–523. [CrossRef]

31. Stasyuk, A.J.; Solá, M. Excited state intramolecular proton transfer (ESIPT) process: A brief overview of computational aspects, conformational changes, polymorphism, and solvent effects. In *Theoretical and Quantum Chemistry at the Dawn of the 21st Century*; Taylor & Francis Group: London, UK, 2018; Chapter 24. Available online: https://www.taylorfrancis.com/books/edit/10.1201/9781351170963/theoretical-quantum-chemistry-dawn-21st-century-tanmoy-chakraborty-ramon-carbo-dorca (accessed on 3 July 2021). [CrossRef]

32. Tang, Z.; Wang, Y.; Zhao, N. Theoretical investigation of excited-state intramolecular proton transfer mechanism of flavonoid derivatives. In *Hydrogen-Bonding Research in Photochemistry, Photobiology, and Optoelectronic Materials*; World Scientific Publishing Co. Pte. Ltd.: Singapore, 2019; Chapter 8, pp. 179–213._0008. Available online: https://www.worldscientific.com/doi/pdf/10.1142/9781786346087_0008 (accessed on 5 July 2021). [CrossRef]

33. Zhang, M.; Zhou, Q.; Du, C.; Ding, Y.; Song, P. Detailed theoretical investigation on ESIPT process of pigment yellow 101. *RSC Adv.* **2016**, *6*, 59389–59394. [CrossRef]

34. Barboza, C.A.; Gawrys, P.; Banasiewicz, M.; Suwinska, K.; Sobolewski, A.L. Photophysical transformations induced by chemical substitution to salicylaldimines. *Phys. Chem. Chem. Phys.* **2020**, *22*, 6698–6705. [CrossRef]

35. Barboza, C.A.; Gawrys, P.; Banasiewicz, M.; Kozankiewicz, B.; Sobolewski, A.L. Substituent effects on the photophysical properties of tris(salicylideneanilines). *Phys. Chem. Chem. Phys.* **2021**, *23*, 1156–1164. [CrossRef]

36. Liang, X.; Fang, H. Theoretical insights into the directionality of ESIPT behavior of BTHMB molecule with two proton acceptors in solution. *Chem. Phys. Lett.* **2021**, *775*, 138670. [CrossRef]

37. Wang, L.; Wang, Y.; Zhang, Q.; Zhao, J. Theoretical exploration about the ESIPT mechanism and hydrogen bonding interaction for 2-(3,5-dichloro-2-hydroxy-phenyl)-benzoxazole-6-carboxylicacid. *J. Phys. Org. Chem.* **2020**, *33*, e4020. [CrossRef]

38. Liu, Z.Y.; Hu, J.W.; Huang, T.H.; Chen, K.Y.; Chou, P.T. Excited-state intramolecular proton transfer in the kinetic-control regime. *Phys. Chem. Chem. Phys.* **2020**, *22*, 22271–22278. [CrossRef] [PubMed]

39. Wang, J.; Liu, Q.; Yang, D. Theoretical insights into excited-state hydrogen bonding effects and intramolecular proton transfer (ESIPT) mechanism for BTS system. *Sci. Rep.* **2020**, *10*, 5119. [CrossRef]

40. Vérité, P.M.; Guido, C.A.; Jacquemin, D. First-principles investigation of the double ESIPT process in a thiophene-based dye. *Phys. Chem. Chem. Phys.* **2019**, *21*, 2307–2317. [CrossRef]

41. Tsai, H.H.G.; Sun, H.L.S.; Tan, C.J. TD-DFT Study of the Excited-State Potential Energy Surfaces of 2-(2′-Hydroxyphenyl)benzimidazole and its Amino Derivatives. *J. Phys. Chem. A* **2010**, *114*, 4065–4079. [CrossRef]

42. Houari, Y.; Chibani, S.; Jacquemin, D.; Laurent, A.D. TD-DFT assessment of the excited state intramolecular proton transfer in hydroxyphenylbenzimidazole (HBI) dyes. *J. Phys. Chem. B* **2015**, *119*, 2180–2192. [CrossRef]

43. Ramu, Y.L.; Jagadeesha, K.; Shivalingaswamy, T.; Ramegowda, M. Microsolvation, hydrogen bond dynamics and excited state hydrogen atom transfer mechanism of 2′,4′-dihydroxychalcone. *Chem. Phys. Lett.* **2020**, *739*, 137030. [CrossRef]

44. Zhao, G.J.; Han, K.L. Hydrogen bonding in the electronic excited state. *Acc. Chem. Res.* **2012**, *45*, 404–413. [CrossRef]

45. De Carvalho, F.; Coutinho Neto, M.; Bartoloni, F.; Homem-de Mello, P. Density Functional Theory Applied to Excited State Intramolecular Proton Transfer in Imidazole-, Oxazole-, and Thiazole-Based Systems. *Molecules* **2018**, *23*, 1231. [CrossRef]

46. Chrayteh, A.; Ewels, C.; Jacquemin, D. Dual fluorescence in strap ESIPT systems: A theoretical study. *Phys. Chem. Chem. Phys.* **2020**, *22*, 854–863. [CrossRef]

47. Wu, D.; Guo, W.W.; Liu, X.Y.; Cui, G. Excited-state intramolecular proton transfer in a blue fluorescence chromophore induces dual emission. *ChemPhysChem* **2016**, *17*, 2340–2347. [CrossRef]

48. Vivas, M.G.; Germino, J.C.; Barboza, C.A.; Simoni, D.D.A.; Vazquez, P.A.; De Boni, L.; Atvars, T.D.; Mendonça, C.R. Revealing the dynamic of excited state proton transfer of a π-conjugated salicylidene compound: An experimental and theoretical study. *J. Phys. Chem. C* **2017**, *121*, 1283–1290. [CrossRef]

49. Chaiwongwattana, S.; Škalamera, D.; Došlić, N.; Bohne, C.; Basarić, N. Substitution pattern on anthrol carbaldehydes: Excited state intramolecular proton transfer (ESIPT) with a lack of phototautomer fluorescence. *Phys. Chem. Chem. Phys.* **2017**, *19*, 28439–28449. [CrossRef]

50. Demchenko, A.P.; Tomin, V.I.; Chou, P.T. Breaking the Kasha Rule for More Efficient Photochemistry. *Chem. Rev.* **2017**, *117*, 13353–13381. [CrossRef] [PubMed]

51. Coe, J.D.; Martínez, T.J. Ab initio molecular dynamics of excited-state intramolecular proton transfer around a three-state conical intersection in malonaldehyde. *J. Phys. Chem. A* **2006**, *110*, 618–630. [CrossRef] [PubMed]

52. Sobolewski, A.L.; Domcke, W. Ab initio potential-energy functions for excited state intramolecular proton transfer: A comparative study of o-hydroxybenzaldehyde, salicylic acid and 7-hydroxy-1-indanone. *Phys. Chem. Chem. Phys.* **1999**, *1*, 3065–3072. [CrossRef]

53. Cembran, A.; Gao, J. Excited state intramolecular proton transfer in 1-(trifluoroacetylamino) naphthaquinone: A CASPT2//CASSCF computational study. *Mol. Phys.* **2006**, *104*, 943–955. [CrossRef]

54. Li, Y.; Wang, L.; Guo, X.; Zhang, J. A CASSCF/CASPT2 insight into excited-state intramolecular proton transfer of four imidazole derivatives. *J. Comput. Chem.* **2015**, *36*, 2374–2380. [CrossRef]

55. Hättig, C.; Weigend, F. CC2 Excitation Energy Calculations on Large Molecules Using the Resolution of the Identity Approximation. *J. Chem. Phys.* **2000**, *113*, 5154–5161. [CrossRef]

56. Hättig, C. Geometry optimizations with the coupled-cluster model CC2 using the resolution-of-the-identity approximation. *J. Chem. Phys.* **2003**, *118*, 7751–7761. [CrossRef]

57. Schirmer, J.; Trofimov, A.B. Intermediate state representation approach to physical properties of electronically excited molecules. *J. Chem. Phys.* **2004**, *120*, 11449–11464. [CrossRef]

58. Dreuw, A.; Wormit, M. The algebraic diagrammatic construction scheme for the polarization propagator for the calculation of excited states. *WIRES Comput. Mol. Sci.* **2015**, *5*, 82–95. [CrossRef]

59. Louant, O.; Champagne, B.; Liégeois, V. Investigation of the Electronic Excited-State Equilibrium Geometries of Three Molecules Undergoing ESIPT: A RI-CC2 and TDDFT Study. *J. Phys. Chem. A* **2018**, *122*, 972–984. [CrossRef] [PubMed]

60. Jacquemin, D.; Duchemin, I.; Blase, X. 0–0 Energies Using Hybrid Schemes: Benchmarks of TD-DFT, CIS(D), ADC(2), CC2, and BSE/GW formalisms for 80 Real-Life Compounds. *J. Chem. Theory Comput.* **2015**, *11*, 5340–5359. [CrossRef]

61. Hättig, C. Structure optimizations for excited states with correlated second-order methods: CC2 and ADC(2). *Adv. Quantum Chem.* **2005**, *50*, 37–60. [CrossRef]

62. Plasser, F.; Crespo-Otero, R.; Pederzoli, M.; Pittner, J.; Lischka, H.; Barbatti, M. Surface hopping dynamics with correlated single-reference methods: 9H-adenine as a case study. *J. Chem. Theory Comput.* **2014**, *10*, 1395–1405. [CrossRef]

63. Tuna, D.; Lefrancois, D.; Wolański, Ł.; Gozem, S.; Schapiro, I.; Andruniów, T.; Dreuw, A.; Olivucci, M. Assessment of approximate coupled-cluster and algebraic-diagrammatic-construction methods for ground- and excited-state reaction paths and the conical-intersection seam of a retinal-chromophore model. *J. Chem. Theory Comput.* **2015**, *11*, 5758–5781. [CrossRef]

64. Tajti, A.; Stanton, J.F.; Matthews, D.A.; Szalay, P.G. Accuracy of coupled cluster excited state potential energy surfaces. *J. Chem. Theory Comput.* **2018**, *14*, 5859–5869. [CrossRef]

65. Tajti, A.; Tulipán, L.; Szalay, P.G. Accuracy of spin-component ccaled ADC(2) excitation energies and potential energy surfaces. *J. Chem. Theory Comput.* **2020**, *16*, 468–474. [CrossRef]

66. Hellweg, A.; Grün, S.A.; Hättig, C. Benchmarking the performance of spin-component scaled CC2 in ground and electronically excited states. *Phys. Chem. Chem. Phys.* **2008**, *10*, 4119. [CrossRef]

67. Grimme, S. Improved second-order Møller-Plesset perturbation theory by separate scaling of parallel- and antiparallel-spin pair correlation energies. *J. Chem. Phys.* **2003**, *118*, 9095–9102. [CrossRef]

68. Jung, Y.; Lochan, R.C.; Dutoi, A.D.; Head-Gordon, M. Scaled opposite-spin second order Moller-Plesset correlation energy: An economical electronic structure method. *J. Chem. Phys.* **2004**, *121*, 9793–9802. [CrossRef]

69. Kielesiński, Ł.; Morawski, O.W.; Barboza, C.A.; Gryko, D.T. Polarized helical coumarins: [1,5] sigmatropic rearrangement and excited-state intramolecular proton transfer. *J. Org. Chem.* **2021**, *86*, 6148–6159. [CrossRef]

70. Li, T.C.; Tong, P.Q. Time-dependent density-functional theory for multicomponent systems. *Phys. Rev. A* **1986**, *34*, 529–532. [CrossRef]

71. Hirata, S.; Head-Gordon, M. Time-dependent density functional theory within the Tamm–Dancoff approximation. *Chem. Phys. Lett.* **1999**, *314*, 291–299. [CrossRef]

72. Sun, C.; Li, H.; Yin, H.; Li, Y.; Shi, Y. Effects of the cyano substitution at different positions on the ESIPT properties of alizarin: A DFT/TD-DFT investigation. *J. Mol. Liq.* **2018**, *269*, 650–656. [CrossRef]

73. Muriel, W.A.; Morales-Cueto, R.; Rodríguez-Córdoba, W. Unravelling the solvent polarity effect on the excited state intramolecular proton transfer mechanism of the 1- and 2-salicylideneanthrylamine. A TD-DFT case study. *Phys. Chem. Chem. Phys.* **2019**, *21*, 915–928. [CrossRef]

74. Ehrmaier, J.; Rabe, E.J.; Pristash, S.R.; Corp, K.L.; Schlenker, C.W.; Sobolewski, A.L.; Domcke, W. Singlet–triplet inversion in heptazine and in polymeric carbon nitrides. *J. Phys. Chem. A* **2019**, *123*, 8099–8108. [CrossRef]

75. De Silva, P. Inverted singlet-triplet gaps and their relevance to thermally activated delayed fluorescence. *J. Phys. Chem. Lett.* **2019**, *10*, 5674–5679. [CrossRef]

76. Savarese, M.; Raucci, U.; Fukuda, R.; Adamo, C.; Ehara, M.; Rega, N.; Ciofini, I. Comparing the performance of TD-DFT and SAC-CI methods in the description of excited states potential energy surfaces: An excited state proton transfer reaction as case study. *J. Comput. Chem.* **2017**, *38*, 1084–1092. [CrossRef]

77. Lee, C.; Yang, W.; Parr, R.G. Development of the Colle-Salvetti correlation-energy formula into a functional of the electron density. *Phys. Rev. B* **1988**, *37*, 785–789. [CrossRef]

78. Becke, A.D. Density-functional thermochemistry. III. The role of exact exchange. *J. Chem. Phys.* **1993**, *98*, 5648–5652. [CrossRef]

79. Zhao, Y.; Truhlar, D.G. The M06 suite of density functionals for main group thermochemistry, thermochemical kinetics, noncovalent interactions, excited states, and transition elements: Two new functionals and systematic testing of four M06-class functionals and 12 other function. *Theor. Chem. Acc.* **2008**, *120*, 215–241. [CrossRef]
80. Chai, J.D.; Head-Gordon, M. Long-range corrected hybrid density functionals with damped atom–atom dispersion corrections. *Phys. Chem. Chem. Phys.* **2008**, *10*, 6615. [CrossRef]
81. Tasheh, N.S.; Nkungli, N.K.; Ghogomu, J.N. A DFT and TD-DFT study of ESIPT-mediated NLO switching and UV absorption by 2-(2′-hydroxy-5′-methylphenyl)benzotriazole. *Theor. Chem. Acc.* **2019**, *138*, 100. [CrossRef]
82. Yanai, T.; Tew, D.P.; Handy, N.C. A new hybrid exchange–correlation functional using the Coulomb-attenuating method (CAM-B3LYP). *Chem. Phys. Lett.* **2004**, *393*, 51–57. [CrossRef]
83. Domcke, W.; Yarkony, D.R.; Köppel, H. *Conical Intersections*; World Scientific Publishing Co. Pte. Ltd.: Singapore, 2004. [CrossRef]
84. Jankowska, J.; Barbatti, M.; Sadlej, J.; Sobolewski, A.L. Tailoring the Schiff base photoswitching-a non-adiabatic molecular dynamics study of substituent effect on excited state proton transfer. *Phys. Chem. Chem. Phys.* **2017**, *19*, 5318–5325. [CrossRef]
85. Grimme, S. Density functional theory with London dispersion corrections. *WIREs Comput. Mol. Sci.* **2011**, *1*, 211–228. [CrossRef]
86. Grimme, S.; Antony, J.; Ehrlich, S.; Krieg, H. A consistent and accurate ab initio parametrization of density functional dispersion correction (DFT-D) for the 94 elements H-Pu. *J. Chem. Phys.* **2010**, *132*, 154104. [CrossRef]
87. Caldeweyher, E.; Ehlert, S.; Hansen, A.; Neugebauer, H.; Spicher, S.; Bannwarth, C.; Grimme, S. A generally applicable atomic-charge dependent London dispersion correction. *J. Chem. Phys.* **2019**, *150*, 154122. [CrossRef]
88. Naka, K.; Sato, H.; Higashi, M. Theoretical study of the mechanism of the solvent dependency of ESIPT in HBT. *Phys. Chem. Chem. Phys.* **2021** [CrossRef] [PubMed]
89. Laner, J.N.; Silva Junior, H.D.C.; Rodembusch, F.S.; Moreira, E.C. New insights on the ESIPT process based on solid-state data and state-of-the-art computational methods. *Phys. Chem. Chem. Phys.* **2021**, *23*, 1146–1155. [CrossRef] [PubMed]
90. Li, Y.; Siddique, F.; Aquino, A.J.A.; Lischka, H. Molecular dynamics simulation of the excited-state proton transfer mechanism in 3-hydroxyflavone using explicit hydration models. *J. Phys. Chem. A* **2021**, *125*, 5765–5778. [CrossRef]
91. Dunning, T.H. Gaussian basis sets for use in correlated molecular calculations. I. The atoms boron through neon and hydrogen. *J. Chem. Phys.* **1989**, *90*, 1007–1023. [CrossRef]
92. Self-consistent molecular orbital methods. XX. A basis set for correlated wave functions. *J. Chem. Phys.* **1980**, *72*, 650–654. [CrossRef]
93. Laurent, A.D.; Blondel, A.; Jacquemin, D. Choosing an atomic basis set for TD-DFT, SOPPA, ADC(2), CIS(D), CC2 and EOM-CCSD calculations of low-lying excited states of organic dyes. *Theor. Chem. Acc.* **2015**, *134*, 76. [CrossRef]
94. Sinnecker, S.; Rajendran, A.; Klamt, A.; Diedenhofen, M.; Neese, F. Calculation of solvent shifts on electronic g-tensors with the conductor-like screening model (COSMO) and its self-consistent generalization to real solvents (direct COSMO-RS). *J. Phys. Chem. A* **2006**, *110*, 2235–2245. [CrossRef]
95. Yang, Y.; Zhao, J.; Li, Y. Theoretical study of the ESIPT process for a new natural product quercetin. *Sci. Rep.* **2016**, *6*, 32152. [CrossRef]
96. Scalmani, G.; Frisch, M.J.; Mennucci, B.; Tomasi, J.; Cammi, R.; Barone, V. Geometries and properties of excited states in the gas phase and in solution: Theory and application of a time-dependent density functional theory polarizable continuum model. *J. Chem. Phys.* **2006**, *124*, 094107. [CrossRef] [PubMed]
97. Corni, S.; Cammi, R.; Mennucci, B.; Tomasi, J. Electronic excitation energies of molecules in solution within continuum solvation models: Investigating the discrepancy between state-specific and linear-response methods. *J. Chem. Phys.* **2005**, *123*, 134512. [CrossRef]
98. Marenich, A.V.; Cramer, C.J.; Truhlar, D.G. Universal solvation model based on solute electron density and on a continuum model of the solvent defined by the bulk dielectric constant and atomic surface tensions. *J. Phys. Chem. B* **2009**, *113*, 6378–6396. [CrossRef] [PubMed]
99. Cammi, R.; Tomasi, J. Remarks on the use of the apparent surface charges (ASC) methods in solvation problems: Iterative versus matrix-inversion procedures and the renormalization of the apparent charges. *J. Comput. Chem.* **1995**, *16*, 1449–1458. [CrossRef]
100. Cancès, E.; Mennucci, B.; Tomasi, J. A new integral equation formalism for the polarizable continuum model: Theoretical background and applications to isotropic and anisotropic dielectrics. *J. Chem. Phys.* **1997**, *107*, 3032–3041. [CrossRef]
101. Mennucci, B.; Cancès, E.; Tomasi, J. Evaluation of solvent effects in isotropic and anisotropic dielectrics and in ionic solutions with a unified integral equation method: Theoretical bases, computational implementation, and numerical applications. *J. Phys. Chem. B* **1997**, *101*, 10506–10517. [CrossRef]
102. Yuan, H.; Guo, X.; Zhang, J. Ab initio insights into the mechanism of excited-state intramolecular proton transfer triggered by the second excited singlet state of a fluorescent dye: An anti-Kasha behavior. *Mater. Chem. Front.* **2019**, *3*, 1225–1230. [CrossRef]
103. Caricato, M.; Mennucci, B.; Tomasi, J.; Ingrosso, F.; Cammi, R.; Corni, S.; Scalmani, G. Formation and relaxation of excited states in solution: A new time dependent polarizable continuum model based on time dependent density functional theory. *J. Chem. Phys.* **2006**, *124*, 124520. [CrossRef] [PubMed]
104. Marenich, A.V.; Cramer, C.J.; Truhlar, D.G.; Guido, C.A.; Mennucci, B.; Scalmani, G.; Frisch, M.J. Practical computation of electronic excitation in solution: Vertical excitation model. *Chem. Sci.* **2011**, *2*, 2143. [CrossRef]
105. Guido, C.A.; Scalmani, G.; Mennucci, B.; Jacquemin, D. Excited state gradients for a state-specific continuum solvation approach: The vertical excitation model within a Lagrangian TDDFT formulation. *J. Chem. Phys.* **2017**, *146*, 204106. [CrossRef]

106. Peng, Y.; Ye, Y.; Xiu, X.; Sun, S. Mechanism of Excited-State Intramolecular Proton Transfer for 1,2-Dihydroxyanthraquinone: Effect of Water on the ESIPT. *J. Phys. Chem. A* **2017**, *121*, 5625–5634. [CrossRef]
107. Crespo-Otero, R.; Barbatti, M. Recent advances and perspectives on nonadiabatic mixed quantum-classical dynamics. *Chem. Rev.* **2018**, *118*, 7026–7068. [CrossRef]
108. Tully, J.C. Molecular dynamics with electronic transitions. *J. Chem. Phys.* **1990**, *93*, 1061. [CrossRef]
109. Tully, J.C. Mixed quantum–classical dynamics. *Faraday Discuss.* **1998**, *110*, 407–419. [CrossRef]
110. Barbatti, M. Nonadiabatic dynamics with trajectory surface hopping method. *WIRES Comput. Mol. Sci.* **2011**, *1*, 620–633. [CrossRef]
111. Yarkony, D.R. Nonadiabatic quantum chemistry—Past, present, and future. *Chem. Rev.* **2012**, *112*, 481–498. [CrossRef] [PubMed]
112. Ben-Nun, M.; Quenneville, J.; Martínez, T.J. Ab initio multiple spawning: Photochemistry from first principles quantum molecular dynamics. *J. Phys. Chem. A* **2000**, *104*, 5161–5175. [CrossRef]
113. Curchod, B.F.E.; Martínez, T.J. Ab initio nonadiabatic quantum molecular dynamics. *Chem. Rev.* **2018**, *118*, 3305–3336. [CrossRef]
114. Zhao, L.; Wildman, A.; Tao, Z.; Schneider, P.; Hammes-Schiffer, S.; Li, X. Nuclear–electronic orbital Ehrenfest dynamics. *J. Chem. Phys.* **2020**, *153*, 224111. [CrossRef] [PubMed]
115. Belyaev, A.K.; Lasser, C.; Trigila, G. Landau–Zener type surface hopping algorithms. *J. Chem. Phys.* **2014**, *140*, 224108. [CrossRef]
116. Suchan, J.; Janoš, J.; Slavíček, P. Pragmatic approach to photodynamics: mixed Landau-Zener surface hopping with intersystem crossing. *J. Chem. Theory Comput.* **2020**, *16*, 5809–5820. [CrossRef]
117. Pratihar, S.; Ma, X.; Homayoon, Z.; Barnes, G.L.; Hase, W.L. Direct chemical dynamics simulations. *JACS* **2017**, *139*, 3570–3590. [CrossRef]
118. Barbatti, M.; Granucci, G.; Persico, M.; Ruckenbauer, M.; Vazdar, M.; Eckert-Maksić, M.; Lischka, H. The on-the-fly surface-hopping program system Newton-X: Application to ab initio simulation of the nonadiabatic photodynamics of benchmark systems. *J. Photochem. Photobiol. A* **2007**, *190*, 228–240. [CrossRef]
119. Barbatti, M.; Ruckenbauer, M.; Plasser, F.; Pittner, J.; Granucci, G.; Persico, M.; Lischka, H. Newton-X: A surface-hopping program for nonadiabatic molecular dynamics. *WIRES Comput. Mol. Sci.* **2014**, *4*, 26–33. [CrossRef]
120. Richter, M.; Marquetand, P.; González-Vázquez, J.; Sola, I.; González, L. SHARC: Ab initio molecular dynamics with surface hopping in the adiabatic representation including arbitrary couplings. *J. Chem. Theory Comput.* **2011**, *7*, 1253–1258. [CrossRef] [PubMed]
121. Du, L.; Lan, Z. An on-the-fly surface-hopping program JADE for nonadiabatic molecular dynamics of polyatomic systems: Implementation and applications. *J. Chem. Theory Comput.* **2015**, *11*, 1360–1374. [CrossRef] [PubMed]
122. Understanding the surface hopping view of electronic transitions and decoherence. *Annu. Rev. Phys. Chem.* **2016**, *67*, 387–417. [CrossRef] [PubMed]
123. Tuna, D.; Spörkel, L.; Barbatti, M.; Thiel, W. Nonadiabatic dynamics simulations of photoexcited urocanic acid. *Chem. Phys.* **2018**, *515*, 521–534. [CrossRef]
124. Xia, S.H.; Che, M.; Liu, Y.; Zhang, Y.; Cui, G. Photochemical mechanism of 1,5-benzodiazepin-2-one: Electronic structure calculations and nonadiabatic surface-hopping dynamics simulations. *Phys. Chem. Chem. Phys.* **2019**, *21*, 10086–10094. [CrossRef] [PubMed]
125. Ben-Nun, M.; Martínez, T.J. Ab initio quantum molecular dynamics. In *Advances in Chemical Physics*; John Wiley & Sons, Ltd.: Hoboken, NJ, USA, 2002; pp. 439–512. [CrossRef]
126. Leticia, G.; Roland, L. *Quantum Chemistry and Dynamics of Excited States: Methods and Applications*; Wiley: Hoboken, NJ, USA, 2021.
127. Fedorov, D.A.; Pruitt, S.R.; Keipert, K.; Gordon, M.S.; Varganov, S.A. Ab initio multiple spawning method for intersystem crossing dynamics: Spin-forbidden transitions between 3B1 and 1A1 States of GeH2. *J. Phys. Chem. A* **2016**, *120*, 2911–2919. [CrossRef]
128. Barca, G.M.J.; Bertoni, C.; Carrington, L.; Datta, D.; De Silva, N.; Deustua, J.E.; Fedorov, D.G.; Gour, J.R.; Gunina, A.O.; Guidez, E.; et al. Recent developments in the general atomic and molecular electronic structure system. *J. Chem. Phys.* **2020**, *152*, 154102. [CrossRef]
129. Levine, B.G.; Coe, J.D.; Virshup, A.M.; Martínez, T.J. Implementation of ab initio multiple spawning in the Molpro quantum chemistry package. *Chem. Phys.* **2008**, *347*, 3–16. [CrossRef]
130. Werner, H.J.; Knowles, P.J.; Knizia, G.; Manby, F.R.; Schütz, M. Molpro: A general-purpose quantum chemistry program package. *WIRES Comput. Mol. Sci.* **2012**, *2*, 242–253. [CrossRef]
131. Granucci, G.; Persico, M.; Toniolo, A. Direct semiclassical simulation of photochemical processes with semiempirical wave functions. *J. Chem. Phys.* **2001**, *114*, 10608–10615. [CrossRef]
132. Stewart, J. MOPAC: A semiempirical molecular orbital program. *J. Comput. Aided* **1990**, *4*, 1–103. [CrossRef] [PubMed]
133. Pijeau, S.; Foster, D.; Hohenstein, E.G. Effect of nonplanarity on excited-state proton transfer and internal conversion in salicylideneaniline. *J. Phys. Chem. A* **2018**, *122*, 5555–5562. [CrossRef] [PubMed]
134. Pijeau, S.; Foster, D.; Hohenstein, E.G. Excited-state dynamics of 2-(2′-hydroxyphenyl)benzothiazole: Ultrafast proton transfer and internal conversion. *J. Phys. Chem. A* **2017**, *121*, 4595–4605. [CrossRef] [PubMed]
135. Zhao, L.; Wildman, A.; Pavošević, F.; Tully, J.C.; Hammes-Schiffer, S.; Li, X. Excited state intramolecular proton transfer with nuclear-electronic orbital Ehrenfest dynamics. *J. Phys. Chem. Lett.* **2021**, *12*, 3497–3502. [CrossRef]
136. Webb, S.P.; Iordanov, T.; Hammes-Schiffer, S. Multiconfigurational nuclear-electronic orbital approach: Incorporation of nuclear quantum effects in electronic structure calculations. *J. Chem. Phys.* **2002**, *117*, 4106–4118. [CrossRef]

137. Li, X.; Tully, J.C.; Schlegel, H.B.; Frisch, M.J. Ab initio Ehrenfest dynamics. *J. Chem. Phys.* **2005**, *123*, 084106. [CrossRef]
138. Wang, L.; Prezhdo, O.V. A simple solution to the trivial crossing problem in surface hopping. *J. Phys. Chem. Lett.* **2014**, *5*, 713–719. [CrossRef]
139. Sun, Z.; Li, S.; Xie, S.; An, Z. Solution for the trivial crossing problem in surface hopping simulations by the classification on excited states. *J. Phys. Chem. C* **2018**, *122*, 8058–8064. [CrossRef]
140. Qiu, J.; Bai, X.; Wang, L. Crossing classified and corrected fewest switches surface hopping. *J. Phys. Chem. Lett.* **2018**, *9*, 4319–4325. [CrossRef] [PubMed]
141. Beck, M.; Jäckle, A.; Worth, G.A.; Meter, H.D. The multiconfiguration time-dependent Hartree (MCTDH) method: A highly efficient algorithm for propagating wavepackets. *Phys. Rep.* **2000**, *324*, 1–105. [CrossRef]
142. Cederbaum, L.S.; Köppel, H.; Domcke, W. Multimode vibronic coupling effects in molecules. *Int. J. Quantum Chem.* **1981**, *15*, 251–267. [CrossRef]
143. Cederbaum, L.S. The multistate vibronic coupling problem. *J. Chem. Phys.* **1983**, *78*, 5714–5728. [CrossRef]
144. Worth, G.A.; Meyer, H.D.; Köppel, H.; Cederbaum, L.S.; Burghardt, I. Using the MCTDH wavepacket propagation method to describe multimode non-adiabatic dynamics. *Int. Rev. Phys. Chem.* **2008**, *27*, 569–606. [CrossRef]
145. Vendrell, O.; Meyer, H.D. Multilayer multiconfiguration time-dependent Hartree method: Implementation and applications to a Henon–Heiles Hamiltonian and to pyrazine. *J. Chem. Phys.* **2011**, *134*, 044135. [CrossRef]
146. Worth, G.A.; Beck, M.H.; Jäckle, A.; Meyer, H.D. *The Heidelberg MCTDH Package*; University of Heidelberg: Heidelberg, Germany, 2000.
147. Worth, G.A.; Giri, K.; Richings, G.W.; Burghardt, I.; Beck, M.H.; Jäckle, A.; Meyer, H.D. *The QUANTICS Package*; University of Birmingham: Birmingham, UK, 2015.
148. Perveaux, A.; Lorphelin, M.; Lasorne, B.; Lauvergnat, D. Fast and slow excited-state intramolecular proton transfer in 3-hydroxychromone: A two-state story? *Phys. Chem. Chem. Phys.* **2017**, *19*, 6579–6593. [CrossRef]
149. Anand, N.; Isukapalli, S.V.K.; Vennapusa, S.R. Excited-state intramolecular proton transfer driven by conical intersection in hydroxychromones. *J. Comput. Chem.* **2020**, *41*, 1068–1080. [CrossRef]
150. Anand, N.; Welke, K.; Irle, S.; Vennapusa, S.R. Nonadiabatic excited-state intramolecular proton transfer in 3-hydroxyflavone: S 2 state involvement via multi-mode effect. *J. Chem. Phys.* **2019**, *151*, 214304. [CrossRef]
151. Anand, N.; Nag, P.; Kanaparthi, R.K.; Vennapusa, S.R. O-H vibrational motions promote sub-50 fs nonadiabatic dynamics in 3-hydroxypyran-4-one: Interplay between internal conversion and ESIPT. *Phys. Chem. Chem. Phys.* **2020**, *22*, 8745–8756. [CrossRef]
152. Cao, Y.; Eng, J.; Penfold, T.J. Excited state intramolecular proton transfer dynamics for triplet harvesting in organic molecules. *J. Phys. Chem. A* **2019**, *123*, 2640–2649. [CrossRef] [PubMed]
153. Richings, G.W.; Robertson, C.; Habershon, S. Can we use on-the-fly quantum simulations to connect molecular structure and sunscreen action? *Faraday Discuss.* **2019**, *216*, 476–493. [CrossRef] [PubMed]
154. Westermayr, J.; Marquetand, P. Machine learning and excited-state molecular dynamics. *MLST* **2020**, *1*, 043001. [CrossRef]
155. Westermayr, J.; Gastegger, M.; Marquetand, P. Combining SchNet and SHARC: The SchNarc machine learning approach for excited-state dynamics. *J. Phys. Chem. Lett.* **2020**, *11*, 3828–3834. [CrossRef]

molecules

Article

Polymorphism and Conformational Equilibrium of Nitro-Acetophenone in Solid State and under Matrix Conditions

Łukasz Hetmańczyk [1], Przemysław Szklarz [2], Agnieszka Kwocz [2], Maria Wierzejewska [2], Magdalena Pagacz-Kostrzewa [2], Mikhail Ya. Melnikov [3], Peter M. Tolstoy [4] and Aleksander Filarowski [2,5,*]

[1] Faculty of Chemistry, Jagiellonian University, Gronostajowa 2, 30-387 Kraków, Poland; hetmancz@chemia.uj.edu.pl

[2] Faculty of Chemistry, Wrocław University, I14 F. Joliot-Curie st., 50-383 Wrocław, Poland; przemyslaw.szklarz@chem.uni.wroc.pl (P.S.); agnieszka.kwocz@chem.uni.wroc.pl (A.K.); maria.wierzejewska@chem.uni.wroc.pl (M.W.); magdalena.pagacz-kostrzewa@chem.uni.wroc.pl (M.P.-K.)

[3] Department of Chemistry, Moscow State University, F. Joliot-Curie 14, 119991 Moscow, Russia; melnikov46@mail.ru

[4] Institute of Chemistry, St. Petersburg State University, Universitetskij pr. 26, 198504 St. Petersburg, Russia; peter.tolstoy@spbu.ru

[5] Frank Laboratory of Neutron Physics, Joint Institute of Nuclear Research, 141980 Dubna, Russia

* Correspondence: aleksander.filarowski@chem.uni.wroc.pl; Tel.: +48-71-3757283

Citation: Hetmańczyk, Ł.; Szklarz, P.; Kwocz, A.; Wierzejewska, M.; Pagacz-Kostrzewa, M.; Melnikov, M.Y.; Tolstoy, P.M.; Filarowski, A. Polymorphism and Conformational Equilibrium of Nitro-Acetophenone in Solid State and under Matrix Conditions. *Molecules* **2021**, *26*, 3109. https://doi.org/10.3390/molecules 26113109

Academic Editor: Mirosław Jabłoński

Received: 11 May 2021
Accepted: 21 May 2021
Published: 22 May 2021

Publisher's Note: MDPI stays neutral with regard to jurisdictional claims in published maps and institutional affiliations.

Abstract: Conformational and polymorphic states in the nitro-derivative of *o*-hydroxy acetophenone have been studied by experimental and theoretical methods. The potential energy curves for the rotation of the nitro group and isomerization of the hydroxyl group have been calculated by density functional theory (DFT) to estimate the barriers of the conformational changes. Two polymorphic forms of the studied compound were obtained by the slow and fast evaporation of polar and non-polar solutions, respectively. Both of the polymorphs were investigated by Infrared-Red (IR) and Raman spectroscopy, Incoherent Inelastic Neutron Scattering (IINS), X-ray diffraction, nuclear quadrupole resonance spectroscopy (NQR), differential scanning calorimetry (DSC) and density functional theory (DFT) methods. In one of the polymorphs, the existence of a phase transition was shown. The position of the nitro group and its impact on the crystal cell of the studied compound were analyzed. The conformational equilibrium determined by the reorientation of the hydroxyl group was observed under argon matrix isolation. An analysis of vibrational spectra was achieved for the interpretation of conformational equilibrium. The infrared spectra were measured in a wide temperature range to reveal the spectral bands that were the most sensitive to the phase transition and conformational equilibrium. The results showed the interrelations between intramolecular processes and macroscopic phenomena in the studied compound.

Keywords: polymorphism; isomerization; phase transition; nitro group; matrix isolation; IINS; FT-IR; Raman; X-ray; NQR; DSC; DFT

1. Introduction

The aim of this paper is to investigate polymorphic and conformational states of the nitro-derivative of *o*-hydroxy acetophenone. The study explains the influence of the intramolecular hydrogen bond on the phase transition and conformational equilibrium. The expected intramolecular dynamic processes in the *o*-hydroxy acetophenone molecule are the rotation of the nitro group and isomerization of the hydroxyl group (Scheme 1). These processes define such macroscopic phenomena as polymorphism, phase transition and the existence of stable isomers under different conditions. The study of conformational isomerism is a very important issue in modern chemistry for determining and modelling the physical–chemical properties of new materials [1–4] and pharmaceutical compounds [5–8].

It must be stressed that an intramolecular hydrogen bond strongly influences the isomerization equilibrium [9,10]. For deeper insight into the hydrogen-bonding effect on the polymorphic states and isomerization we applied a wide variety of research methods (DFT, X-ray, DSC, NQR, IINS, IR and Raman) in different environments and over a wide temperature range. It is worth noting that the method of matrix isolation was unique in its ability to trace metastable states and help interpret complex reactions [11–14]. Moreover, investigations of objects with hydrogen bonding by IINS [15–18] and NQR [19,20] techniques are useful.

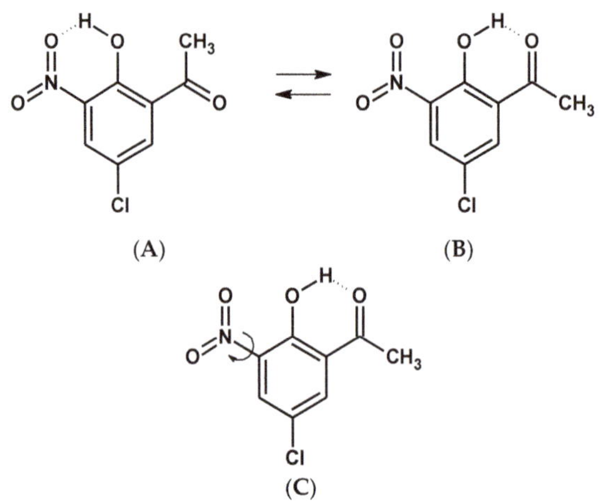

Scheme 1. Schemes of intramolecular hydrogen bonding isomerization (**A,B**) and nitro group rotation (**C**) for 5-chloro-3-nitro-2-hydroxyacetophenone (**CNK**).

In this paper the research methodology followed the following sequence: First, DFT calculations were performed to detect stable and metastable states of the studied molecule, 5-chloro-3-nitro-2-hydroxyacetophenone (**CNK**). The first stage of the studies predicted that **CNK** would crystallize into two polymorphic forms; therefore, its structural properties were analyzed by X-ray diffraction at different temperatures, and both polymorphs were studied by the DSC method to detect phase transition, which was found in one of the polymorphs and verified by ^{35}Cl NQR measurement. Spectral properties of both polymorphs were investigated by IINS, IR and Raman, and computational (DFT) methods in different states to obtain exhaustive information about which vibrational bands (as well as their assignments) were the most sensitive to the phase transition. The assignments of the spectral bands were accomplished on the basis of the H/D substitution of bridge hydrogen (OH → OD) and Potential Energy Distribution (PED) analysis. Spectral infrared studies of the two isotopologues (OH and OD) showed the presence of two conformers (A and B, Scheme 1) under the argon matrix condition.

2. Results and Discussion

At first, the studies dwelt on the DFT calculations of the potential curves to detect stable and metastable states of the molecule according to the concepts presented in the review paper by Bernstein [2]. The information about the energy difference between the global and local minima as well as the height of energy barriers made it possible to predict the presence of a particular conformer or polymorph depending on the environment. For the studied molecule it was logical to consider two possible intramolecular processes—reorientation of the hydroxyl group and rotation of the nitro group (Scheme 1). Therefore, the dependencies of the potential energy on the turning angle of the hydroxyl and acetyl groups were obtained by DFT calculations, which were performed under step-by-step changes to the torsional angle while all other geometric parameters of the molecule were optimized. The calculation of potential energy dependence on the Θ angle of turning the nitro group is given by the equation $\Delta E = f(\Theta)$, where $\Delta E = E_{min} - E_i$; E_{min} is the minimal energy of the system; and E_i is the energy of the system for each fixed Θ angle). The result demonstrated that the energy barrier was rather small and equaled 0.78 kcal/mol when the Θ angle equaled 0 degrees (Figure 1).

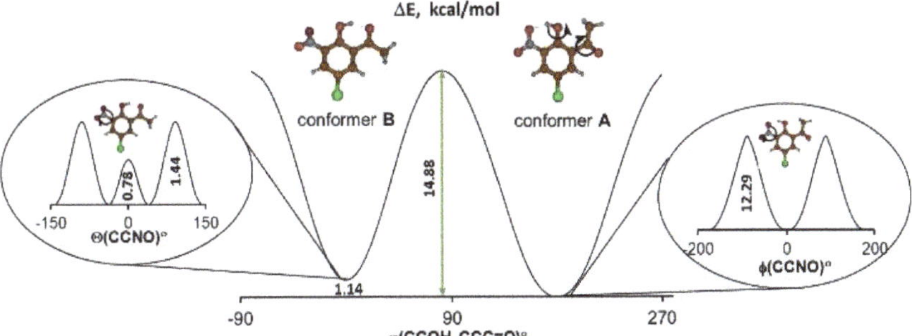

Figure 1. The calculated (B3LYP/6-311++G(2d,2p)) potential energy curves for the intramolecular hydrogen bond transformation (picture in the centre) and the nitro group rotation (picture in the left related to conformer B and picture in the right related to conformer A).

Such a result may seem unconvincing because the torsional angle at 0 degrees (a flat molecule) is usually characterized by minimal potential energy because of π-electronic coupling between the nitro group and phenyl moieties. However, the studied molecule displayed a significant steric repulsion between the oxygen of the nitro group and the oxygen of the hydroxyl group, which counteracts the π-electronic coupling. Such steric repulsion results in the appearance of an energy barrier ($\Delta E = 0.78$ kcal/moL) at $\Theta = 0°$ (Figure 1). Notably, this barrier is not large; thus, the nitro group can easily change its position with respect to the phenyl ring. The process of reorientation of the hydroxyl group leading to the transition from conformer A to conformer B is more complicated and is presented in Supplementary Materials (Supplementary Materials, Figure S1). Judging from the calculated energy barrier of 14–15 kcal/mol, one expected the existence of both conformational forms at a reasonable temperature [21]. Taking these findings into account, we assumed that the studied compound featured polymorphism brought about by the rotation of the nitro group, or by a significant conformational change introduced by the reorientation of the hydroxyl group.

We performed comprehensive studies to trace polymorphs, phase transitions and conformational change. Two polymorphs of the **CNK** compound were obtained by slow recrystallization from methanol (polymorph **I**) and fast re-crystallization from chloroform (polymorph **II**). X-ray studies of polymorphs **I** and **II** showed that they crystallized in a Pccn (T = 200 K) and P21/c (100 K) space group, respectively. The comparison of the structures of the polymorphs clearly showed the difference in the position of the nitro group (Figure 2 and SM Figures S2 and S3). The disoriented position of this group in polymorph **I** points to significant dynamics in the solid state. The crystal cell of polymorph **II** is characterized by a more defined orientation of the nitro group. The nitro groups of polymorph **I** are able to rotate, but the nitro groups of polymorph **II** are not since in the crystal cell of polymorph **II** the molecules are packed in a way that turns the nitro groups and blocks the rotation. The studies of both polymorphs were carried out by DSC to detect a phase transition. The DSC measurements of both polymorphs showed the presence of a phase transition for polymorph **I** at 109.8 K (cooling)/114.5 K (heating) (Figure S4, SM) and the absence of a phase transition for polymorph **II**. The phase transition is reversible and the transformation is enantiotropic.

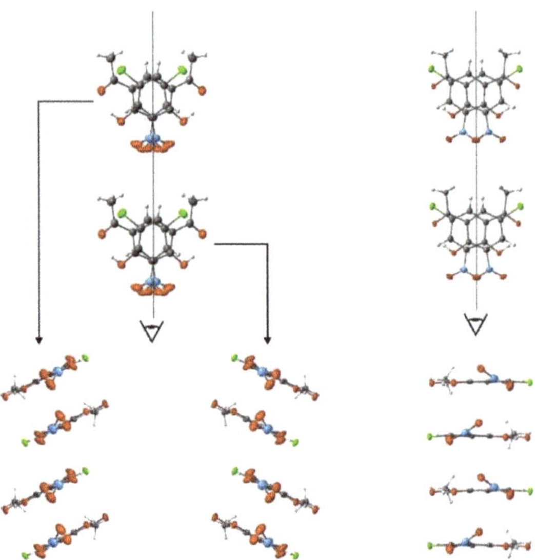

Figure 2. Selected view of the molecular packing of polymorph **I** (**left side**) and polymorph **II** (**right side**).

The comparison of the structural data and the packing of the molecules in the crystal cell of both polymorphs made it possible to conclude that the phase transition was conditioned by the following phenomenon. A decrease in temperature tightened the packing of the molecules in the cell and, therefore, caused a stronger interaction between the nitro groups. Polymorph **II** featured "jagged" nitro groups, which evoked the stable position of molecules in the crystal cell upon temperature decrease. Polymorph **I** lacked this phenomenon, having the possibility of a looser nitro-group rotation at room temperature, which stopped when the temperature was lowered. The decreasing temperature led to stronger interactions between nitro groups and shifted the molecules towards each other, triggering the phase transition. Such a conclusion was verified by comparing the packaging of **CNK** to that of the structurally similar 5-methyl-3-nitro-2-hydroxyacetophenone (**MNK**) [22]. For **MNK,** the nitro groups were oppositely directed and not able to interact strongly, so they did not provoke tensions in the crystal cell. Therefore, the structurally close **MNK** did not exhibit phase transition and polymorphism (cf. Figure S5, SM).

The phase transition in **CNK** was detected on the basis of NQR measurements. NQR was successfully used in the research of compounds with intermolecular [23] and intramolecular hydrogen bonds [24]. NQR studies revealed that the ^{35}Cl signal shifted to high frequencies (from 35.85 to 36.5 MHz) when the temperature fell from 300 to 120 K (the temperature of phase transition), whereas the signal was quite stable at temperatures below the transition state (Figure 3). A similar trend was observed for 1,3-diazinium hydrogen chloranilate monohydrate [25], morpholinium hydrogen chloranilate [26] and *p*-dichlorobenzene [27].

After comparing the structural data and the crystal packing of both polymorphs we concluded that the polymorphism and phase transition were conditioned by the position of the nitro group. Based on experimental data, this conclusion was in accordance with theoretical predictions.

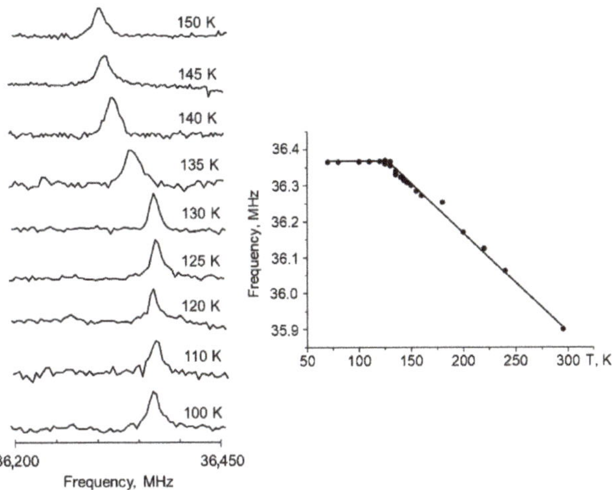

Figure 3. The ^{35}Cl NQR spectra (**left side**) and the temperature dependence of the ^{35}Cl NQR frequency (**right side**) of 5-chloro-3-nitro-2-hydroxyacetophenone.

2.1. Detection of Polymorphs in the Solid State and Isomers under the Matrix Condition by Spectroscopic Methods

To detect the spectral bands that are the most sensitive to polymorphic and conformational changes as well as to phase transition, we performed an analysis of the vibrational spectra measured in solid state and under the matrix condition. To that end, IR, Raman and IINS spectra of the studied compound and its deuterated derivative (OH → OD) were measured in the wide spectral and temperature ranges (50–4000 cm^{-1}, 300–5 K, Figures 4 and 5 and SM Figures S6–S8). The analysis of the spectra and the assignments of the bands were based on DFT and PED calculations (Tables S2–S5, SM). Below, the description of the spectra measured in the solid state and the matrix isolation condition is presented, on the basis of which the isomeric equilibrium analysis was carried out.

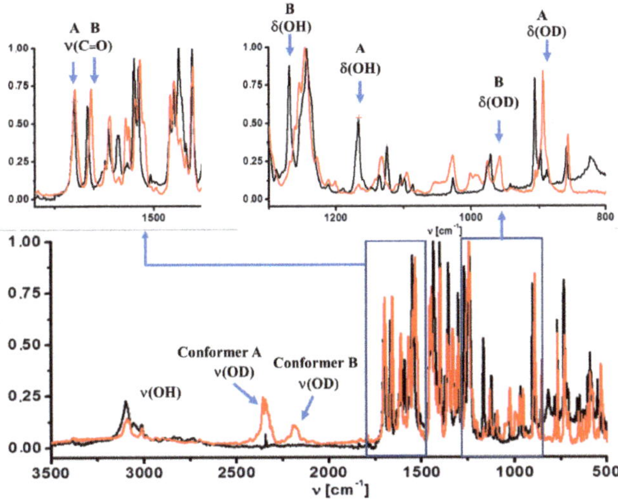

Figure 4. Matrix isolation IR spectra (in Ar) of 5-chloro-3-nitro-2-hyroxyacetophenone (black line) and its deuterated derivative (red line).

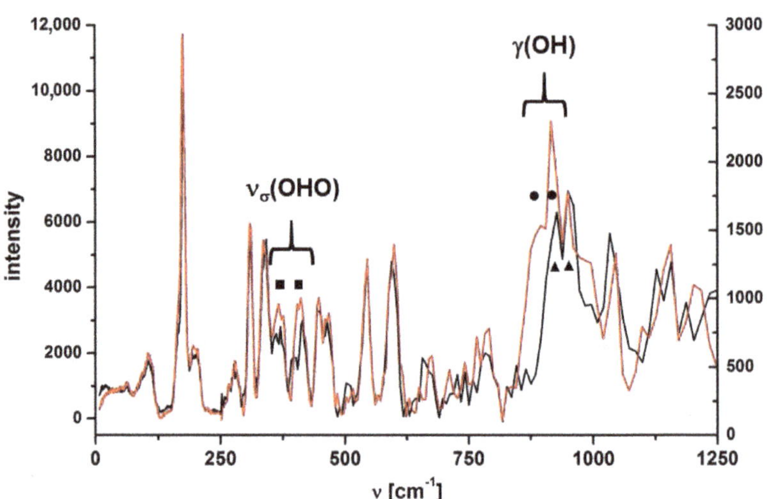

Figure 5. IINS spectra of **CNK** polymorph **I** (red line) and its deuterated analogue (black line). The IINS spectra were aligned and normalized in 0–250 cm^{-1} and 250–1250 cm^{-1} regions, separately.

2.1.1. The ν(OH) Stretching Mode

The adequate assignments of the bands of stretching (ν(OH)), in-plane (δ(OH)) and out-of-plane (γ(OH)) bending modes of the hydroxyl group, as well as the isotope effects of these vibrations caused by the replacement of the bridged hydrogen by deuterium (OH → OD) was completed.

Solid state. When comparing the infrared spectra of deutero-(**CNK-OD**) and non-deutero-(**CNK**) derivatives of the studied compound, the band at 3000–2100 cm^{-1} in the IR spectra recorded in the solid state was assigned to the ν(OH) stretching mode due to its shift to 2100–2000 cm^{-1} after deuteration (Figure S6, SM). It was necessary to underline that the measured infrared spectra in the solid state did not reveal the intense band at 3100 cm^{-1} that was assigned to the stretching vibration of the hydroxyl group, which is hydrogen bonded with the nitro group (cf. spectra of **CNK** and *o*-nitro-phenol [28]). This fact confirmed the absence of conformational form **A** of the studied compound in the solid state. The result agreed with the presented X-ray study.

Matrix condition. The analysis of the ν(OH) band in the infrared spectrum measured under the matrix conditions does not provide a clear proof for the presence of two conformers. The reason is the overlapping of ν(OH) bands of both conformers and ν(CH) bands. However, two bands at 2185 cm^{-1} and 2350 cm^{-1} appear in the spectrum of **CNK-OD**; they are assigned to the ν(OD) vibration of conformers **A** and **B** (Figure 4), respectively. This result manifests the presence two conformational forms under the matrix condition.

2.1.2. The δ(OH) and γ(OH) Bending Modes

Solid state. The in-plane bending mode (δ(OH)) was hard to analyze because of an uncharacteristic vibration in the solid state. According to PED analysis of the spectra calculated by the DFT method, this vibration was strongly coupled to the stretching vibrations of the phenyl ring (1563 and 1433 cm^{-1}, Table S2, SM).

A broad band at 860 cm^{-1} for polymorph **I** and the band at 838 cm^{-1} for polymorph **II** (Table S2 and Figure S7, SM) were assigned to the out-of-plane mode of the hydroxyl group (γ(OH)) in IR spectra measured in the solid state. The bands at 860 cm^{-1} and 838 cm^{-1} narrowed down drastically under the deutero replacement, and new bands arose at 628 cm^{-1} and 624 cm^{-1} in the **CNK-OD** spectra (Figure S6, ESI). These bands were assigned to the γ(OD) mode according to ISR = 1.35 (ISR—isotopic spectroscopic ratio).

Matrix condition. As for in-plane and out-of-plane bending modes, a more distinct picture (due to the absence of the overlapping bands) was observed in IR spectra measured under the matrix condition. In these spectra, obvious changes to two bands at 1269 cm^{-1} and 1166 cm^{-1} appeared after deuteration. They completely disappeared after deuteration and new bands appeared at 959 cm^{-1} and 895 cm^{-1} in the **CNK-OD** spectrum (Figure 4). According to the observed changes and the calculated isotopic ratio (ISR = 1.32), these two bands were assigned to the δ(OH) and δ(OD) modes of conformer **B** and conformer **A**, respectively (Table S3, SM). Rather broad and weak bands at 823 cm^{-1} and 719 cm^{-1} completely vanished and turn up at 619 cm^{-1} and 534 cm^{-1} after deuteration (ISR = 1.33), the bands having been assigned to the γ(OH) and γ(OD) modes of conformer **B** and **A**, respectively. The results presented above are in agreement with the data based on PED analysis (Table S3, SM) as well as the data obtained earlier for 5-methyl-3-nitro-2-hydroxyacetophenone [29].

2.1.3. The ν(C = O) Stretching Mode

It is noteworthy that the stretching vibration of the carbonyl group (ν (C = O)) was the most sensitive to the conformational equilibrium [30,31]. According to IR and Raman spectra measured in the solid state (Figure S7, SM), only one band was observed in the 1800–1600 cm^{-1} range at 1650 cm^{-1} and was assigned to ν (C = O) mode (PED analysis, Table S2, SM). Interestingly, the ν (C = O) bands of both polymorphs were nearly the same, and therefore, demonstrated very little sensitivity to polymorphic changes. However, IR spectra under the matrix condition were characterized by two bands at 1700 cm^{-1} and 1667 cm^{-1}, which were assigned to the ν (C = O) vibrations of conformers **A** and **B** (Figure 4), correspondingly. This statement is in accordance with our previous studies of the methyl derivative of *o*-hydroxy acetophenone [29].

2.1.4. Nitro Group Mode

Solid state. The bands most sensitive to polymorphic states were assigned to the nitro group vibrations. The bands at 1533 cm^{-1} (ν^{as} (NO$_2$)), 1346 cm^{-1} (ν^{sym} (NO$_2$)), 900 cm^{-1} (δ(NO$_2$)) in the Raman spectrum and at 899 cm^{-1} (δ(NO$_2$) in the IR spectrum shifted noticeably upon transitioning from polymorph **I** to polymorph **II** (1540 cm^{-1}, 1357 cm^{-1}, 881 cm^{-1} in Raman and 894 cm^{-1} in IR spectra, Figure S7, SM). A large increase in intensity of the band at 1346 cm^{-1} was also detected in the Raman spectrum upon transition from polymorph **I** to polymorph **II**.

Matrix condition. Two intense bands at 1550 cm^{-1} and 1539 cm^{-1} as well as two bands at 1356 cm^{-1} and 1303 cm^{-1} observed in IR spectrum registered under the matrix condition (Figure 4) were assigned to asymmetric and symmetric vibrations of the nitro group of conformers **B** and **A**, respectively. As for the solid state, the bands assigned to conformer **B** (cf. IR spectra obtained under the matrix condition and in the solid state, Figure 4 and Figure S7, SM) were absent in IR and Raman spectra.

2.2. Spectral Changes under Phase Transition in the Solid State

Regarding phase transition, no significant band splitting during the transition from one phase to the other was observed in IR spectra measured in the middle spectral range. However, the IR spectrum in the far-infrared range demonstrated visible changes (Figure S8, SM). At higher temperatures (300–120 K) there were single bands at 454, 409, 350 and 182 cm^{-1}, which split into doublets at the temperature below the phase transition. The IR spectra of polymorph **II** were measured to verify the results, which did not reveal the splitting of the abovementioned bands in the far-infrared range for the polymorph without the phase transition. The PED analysis showed that bands at 454, 350 and 182 cm^{-1} belonged to multicomponent modes (Table S2, SM). The intense band at 409 cm^{-1} in the IR spectrum was assigned to ν_σ (OHO) vibration due to a strong reduction of the intensity upon deutero replacement. The reliability of the assignment was also proved by IINS studies (Figure 5).

For a deeper insight into the vibrational spectra, isotopic effect and conformational polymorphism we studied the IINS spectra of **CNK** and **CNK-OD** within the 1200–0 cm^{-1} range in the solid state (Figure 5). These spectra showed the presence of two polymorphic forms at low temperature and verified the correctness of the assignment of the vibrations of the hydroxyl group and the hydrogen bridge (ν_σ (OHO)).

Most bands of the measured IINS spectra of both isotopologues were doublets (Figure 5; in the figure, the doublets are marked by filled squares, circles and triangles). In the papers by Tomkinson [32] and Margues [33], it is stated that the split bands in the IINS spectra occurred as a result of two crystallographically and energetically non-equivalent modes. The doubling of bands in the IINS spectra was also supposed to result from the Davydov effect [34], which is the separation of energy levels ascribed to the same vibration due to the presence of several interacting molecular entities in the unit cell. Therefore, the splitting of the bands in the IINS spectra in both isotopologues was due to the presence of two polymorphic forms at 10 K. Three bands at 949, 916 and 895 cm^{-1} were observed within the range of 1000–850 cm^{-1}, three of which were the result of th overlapping of two doublets: 949/916 cm^{-1} (circles) and 916/895 cm^{-1} (triangles). The doublet at 895/916 cm^{-1} disappeared upon deuteration (cf. spectra **CNK** and **CNK-OD**, Figure 5) and appeared as a very small doublet at 656 and 676 cm^{-1}. According to the calculated ISR coefficient (ISR = 1.35) and the PED analysis (Table S5, SM), these bands were assigned to the γ(OH) and γ(OD) out-of-plane bending modes. The values of the γ(OH) bands of IR and IINS spectra were compared in a qualitative way with the values obtained from the correlation R(OO) = 3.01 + 0.0044 × 10^{-4} γ(OH) (for OHO hydrogen bridges longer than 2.4 Å) presented in reference [35]. The position of the γ(OH) band shifted towards the high frequencies alongside the strengthening of the hydrogen bond (823 cm^{-1} in matrix condition <860 cm^{-1} in the solid state). This fact supports the tendency shown in the review by Novak [36] for medium-strong intramolecular hydrogen bonding.

3. Materials and Methods

3.1. Chemicals

Compounds and solvents were purchased from Sigma-Aldrich and used without purification. High-purity argon gas (N60 = 99.99990%) was obtained from Air Liquide. The deuterated sample was prepared by dissolving the product in deuterated methanol (CH$_3$OD). The solution was heated to 60 °C and refluxed for 20 min. Then the methanol was removed by evaporation under reduced pressure. This procedure was repeated three times. The deuteration degree was estimated to be ca. 80–90%.

3.2. Computational Details

Quantum-mechanical calculations using the B3LYP functional [37,38] with the 6-311++G(2d,2p) [39,40] basis set were performed with the GAUSSIAN 16 program [41]. The non-adiabatic approach was used to calculate $\Delta E = f(\mathrm{d}(\Theta))$ dependence. Structural parameters were optimized for each fixed angle, changing gradually by 10°. The potential energy distribution (PED) of the normal modes for each obtained equilibrium geometry was calculated using the internal coordinates by the GAR2PED program [42].

3.3. Thermal Measurements

Differential scanning calorimetry (DSC) experiments were recorded with PerkinElmer Model 8500 differential scanning calorimeter on the polycrystalline material in the temperature range of 100–300 K under a nitrogen atmosphere in hermetically sealed Al pans. Calibration was performed with *n*-heptane and indium as standards.

3.4. NQR Measurements

The powdered sample without further preparation was placed in a standard 5 mm glass sample tube, which was subsequently placed inside the probe of the NQR spectrometer. ^{35}Cl NQR spectra were recorded using NMR/NQR Tecmaq Redstone spectrometer by the single pulse method in the frequency range 35–37 MHz. The pulse duration was 3 μs, followed by 1 ms acquisition and subsequent 400 ms delay time. The number of scans was ca. 3000. The temperature was controlled using an Oxford Instrument cryostat and stabilized with the precision better than ± 1 K.

3.5. Crystallographic Details

The intensity data were collected at 100 K using a Kuma KM4CCD diffractometer and graphite-monochromated MoKα (0.71073 Å) radiation generated from an X-ray tube operating at 50 kV and 35 mA. The images were indexed, integrated, and scaled using the Oxford Diffraction data reduction package [43]. The experimental details together with crystallographic data are given in Table S5. An absorption correction was omitted. The structure was solved by direct methods using SHELXS97 [44] and refined by the full-matrix least-squares method on all F^2 data (SHELXL97) [45]. Non-hydrogen atoms were refined with anisotropic thermal parameters; hydrogen atoms were included from $\Delta\rho$ maps and refined isotropically. The supplementary crystallographic data for this paper is deposited to Cambridge Crystallographic Data Centre (CCDC) under no. 1937047-1937049 for 5-chloro-3-nitro-2-hydroxyacetophenone. These data can be obtained free of charge via www.ccdc.cam.ac.uk/conts/retrieving.html (accessed on 11 May 2021) (or from the Cambridge Crystallographic Data Centre, 12 Union Road, Cambridge CB2 IEZ, UK: fax: (+44) 1123-336-033. e-mail: deposit@ccdc.cam.ac.uk). The molecular structures with atom labeling are shown in Figure S5 (SM).

3.6. IR and Raman Measurements

The standard infrared and Raman spectra were measured using Bruker IFS 66 FT-IR and Nicolet Magna 860 FT-Raman spectrometers in the solid state with a 2 cm^{-1} resolution, respectively. The In–Ga–Ar laser line at 1064 nm was employed for the Raman excitation measurements. To obtain matrices containing **CNK**, the crystalline sample was allowed to sublimate from a small electric oven located inside the vacuum vessel of the cryostat. The **CNK** vapours, mixed with a large excess of matrix gas (argon), were deposited onto a CsI window kept at 15 K in a closed cycle helium cryostat (APD-Cryogenics). The sample temperature was maintained with a temperature controller (Scientific Instruments 9700) equipped with a silicone diode and a resistive heater. Infrared spectra were recorded at 11 K between 4000 and 50 cm^{-1} with a resolution of 0.5 cm^{-1} by means of a Fourier transform IR spectrometer (Bruker IFS 66) equipped with a liquid N$_2$ cooled MCT detector.

3.7. Incoherent Inelastic Neutron Scattering (IINS) Measurement

Neutron scattering data were collected at the pulsed IBR–2M reactor at the Joint Institute of Nuclear Research (Dubna, Russia) using the time-of-flight inverted geometry spectrometer NERA-PR at 10 K. To avoid misunderstanding in the comparison of the IINS spectra, the samples of the studied compounds were measured as a powered substance.

4. Conclusions

To summarize, the paper showed an application of quantum-mechanical calculations for the detection of particular physical-chemical processes in a molecule. DFT calculations enabled the prediction of the possible observation of two phenomena for the studied compound: polymorphism in the solid state and the conformational equilibrium under the condition approaching to the gas phase. The first of the predicted phenomena—the existence of two polymorphs of the studied compound and the phase transition between them-was revealed by DSC and NQR techniques. The non-equivalency of the structures in crystal cells was reflected in the splitting of the bands in the IINS spectrum, and the bands in IR spectrum registered in far-infrared region.

The second of the predicted phenomena was the conformational equilibrium between two hydrogen bonds (N = O$\cdots$H–O and O–H$\cdots$O = C(CH$_3$)). Two conformational forms detected in the matrix isolation conditions by infrared spectroscopy unambiguously proved

Molecules **2021**, *26*, 3109

this prediction. Complete assignments of spectral bands based on the PED analysis and isotopic substitution showed the presence of two conformers under the matrix condition and two polymorph forms in the solid state. The studies confirmed the agreement of the theoretical and experimental results.

Supplementary Materials: The following are available online, Figures S1–S8, Table S1–S5.

Author Contributions: Conceptualization, A.F.; methodology, Ł.H., P.S., A.K. and M.P.-K.; software, Ł.H., P.S., M.P.-K. and A.K.; validation, A.F.; formal analysis, all Authors; investigation, Ł.H., P.S., A.K., M.W., M.Y.M., P.M.T., M.P.-K. and A.F.; resources, P.M.T. and A.F.; data curation, Ł.H., P.S., A.K., M.W., M.Y.M., P.M.T., M.P.-K. and A.F.; writing—original draft preparation, Ł.H., P.S., A.K., M.W., M.Y.M., P.M.T., M.P.-K. and A.F.; writing—review and editing, A.F.; visualization, Ł.H., P.S., A.K., M.W., M.Y.M., P.M.T., M.P.-K. and A.F.; supervision, A.F.; project administration, A.F.; funding acquisition, M.Y.M., P.M.T. and A.F. All authors have read and agreed to the published version of the manuscript.

Funding: This research was funded by RSF, grant number no: 18-13-00050.

Institutional Review Board Statement: Not applicable.

Informed Consent Statement: Not applicable.

Data Availability Statement: All data associated with this article are included in Supplementary Materials.

Acknowledgments: This work was supported by RSF (no: 18-13-00050) grant. The Authors acknowledge the Wrocław Centre for Networking and Supercomputing Centres (WCSS) for providing computational time and facilities.

Conflicts of Interest: The authors declare no conflict of interest.

Sample Availability: Samples of the compound 5-chloro-3-nitro-2-hydroxyacetophenone are available from the authors.

References

1. Bernstein, J. *Polymorphism in Molecular Crystals*, 2nd ed.; Oxford University Press: Oxford, UK, 2008.
2. Cruz-Cabeza, A.J.; Bernstein, J. Conformational Polymorphism. *Chem. Rev.* **2014**, *114*, 2170–2191. [CrossRef] [PubMed]
3. Panda, M.K.; Centore, R.; Causà, M.; Tuzi, A.; Borbone, F.; Naumov, P. Strong and Anomalous Thermal Expansion Precedes the Thermosalient Effect in Dynamic Molecular Crystals. *Sci. Rep.* **2016**, *6*, 29610. [CrossRef] [PubMed]
4. Nangia, A. Conformational polymorphism in organic crystals. *Acc. Chem. Res.* **2008**, *41*, 595–604. [CrossRef] [PubMed]
5. Steed, K.M.; Steed, J.W. Packing Problems: High Z′ Crystal Structures and Their Relationship to Cocrystals, Inclusion Compounds, and Polymorphism. *Chem. Rev.* **2015**, *115*, 2895–2933. [CrossRef]
6. Hilfiker, R. *Polymorphism: In the Pharmaceutical Industry*; Wiley-VCH: Weinheim, Germany, 2006.
7. Brittain, H.G. *Polymorphism in Pharmaceutical Solids*, 1st ed.; Marcel Dekker Inc.: New York, NY, USA, 1999.
8. Yu, L. Polymorphism in Molecular Solids: An Extraordinary System of Red, Orange, and Yellow Crystals. *Acc. Chem. Res.* **2010**, *43*, 1257–1266. [CrossRef]
9. Schmidtmann, M.; Gutmann, M.J.; Middlemiss, D.S.; Wilson, C.C. Towards proton transfer in hydrogen bonded molecular complexes: Joint experimental and theoretical modelling and an energy scale for polymorphism. *CrystEngComm* **2007**, *9*, 743–745. [CrossRef]
10. Bilton, C.; Howard, J.A.K.; Laxmi Madhavi, N.N.; Nangia, A.; Desiraju, G.R.; Allen, F.H.; Wilson, C.C. When is a polymorph not a polymorph? Helical trimeric O–H···O synthons in *trans*-1,4-diethynylcyclohexane-1,4-diol. *Chem. Commun.* **1999**, 1675–1676. [CrossRef]
11. Barnes, A.J. Matrix isolation studies of hydrogen bonding—An historical perspective. *J. Mol. Struct.* **2018**, *1163*, 77–85. [CrossRef]
12. Barnes, A.J.; Orville-Thomas, W.J.; Müller, A.; Gaufrès, R. (Eds.) *Matrix Isolation Spectroscopy*; Springer Publishing Company: Dordrecht, The Netherlands, 1981.
13. Nogueira, B.A.; Ildiz, G.O.; Canotilho, J.; Eusebio, M.E.S.; Henriques, M.S.C.; Paixao, J.A.; Fausto, R. 5-Methylhydantoin: From Isolated Molecules in a Low-Temperature Argon Matrix to Solid State Polymorphs Characterization. *J. Phys. Chem. A* **2017**, *121*, 5267–5279. [CrossRef]
14. Avadanei, M.; Kus', N.; Cozan, V.; Fausto, R. Structure and Photochemistry of *N*-Salicylidene-*p*-carboxyaniline Isolated in Solid Argon. *J. Phys. Chem. A* **2015**, *119*, 9121–9132. [CrossRef]
15. Mitchell, P.C.H.; Parker, S.F.; Ramirez-Cuesta, A.J.; Tomkinson, J. *Vibrational Spectroscopy with Neutrons, with Applications in Chemistry, Biology, Materials Science and Catalysis*; World Scientific: Singapore, 2005.

16. Marques, M.P.M.; Valero, R.; Parker, S.F.; Tomkinson, J.; Batista de Carvalho, L.A.E. Polymorphism in Cisplatin Anticancer Drug. *J. Phys. Chem. B* **2013**, *117*, 6421–6429. [CrossRef] [PubMed]
17. Rivera, S.A.; Allis, D.G.; Hudson, B.S. Importance of Vibrational Zero-Point Energy Contribution to the Relative Polymorph Energies of Hydrogen-Bonded Species. *Cryst. Growth Des.* **2008**, *8*, 3905–3907. [CrossRef]
18. Johnson, M.R.; Trommsdorff, H.P. Vibrational spectra of crystalline formic and acetic acid isotopologues by inelastic neutron scattering and numerical simulations. *Chem. Phys.* **2009**, *355*, 118–122. [CrossRef]
19. Seliger, J.; Zagar, V. Hydrogen bonding and proton transfer in cocrystals of 4,4′-bipyridyl and organic acids studied using nuclear quadrupole resonance. *Phys. Chem. Chem. Phys.* **2014**, *16*, 18141–18147. [CrossRef] [PubMed]
20. Seliger, J.; Zagar, V.; Asaji, T.; Gotoh, K.; Ishida, H. A N-14 nuclear quadrupole resonance study of phase transitions and molecular dynamics in hydrogen bonded organic antiferroelectrics 55DMBP-H(2)ca and 1,5-NPD-H(2)ca. *Phys. Chem. Chem. Phys.* **2011**, *13*, 9165–9172. [CrossRef] [PubMed]
21. Orville-Thomas, W.J. *Internal Rotation in Molecules*; John Wiley & Sons, Inc.: New York, NY, USA, 1974.
22. Filarowski, A.; Kochel, A.; Koll, A.; Bator, G.; Mukjerhee, S. Phase transition and intramolecular hydrogen bonding in nitro derivatives of ortho-hydroxy acetophenones. *J. Mol. Struct.* **2006**, *785*, 7–13. [CrossRef]
23. Kalenik, J.; Majerz, I.; Sobczyk, L.; Grech, E.; Habeeb, M.M.M. ^{35}Cl nuclear quadrupole resonance and infrared studies of hydrogen-bonded adducts of 2-chloro-4-nitrobenzoic acid. *J. Chem. Soc. Faraday Trans. I* **1989**, *85*, 3187–3193. [CrossRef]
24. Apih, T.; Gregorovic, A.; Zagar, V.; Seliger, J. Nuclear Quadrupole Resonance Study of Proton and Deuteron Migration in Short Strong Hydrogen Bonds Formed in Molecular Complex 3,5-Dinitrobenzoic Acid-Nicotinic Acid and in Deuterated 3,5-Pyridinedicarboxylic Acid. *J. Phys. Chem. C* **2016**, *120*, 9992–10000. [CrossRef]
25. Asaji, T.; Hoshino, M.; Ishida, H.; Konnai, A.; Shinoda, Y.; Seliger, J.; Žagar, V. Phase transition and proton exchange in 1,3-diazinium hydrogen chloranilate monohydrate. *Hyperfine Interact.* **2010**, *198*, 85–91. [CrossRef]
26. Tobu, Y.; Ikeda, R.; Nihei, T.A.; Gotoh, K.; Ishida, H.; Asaji, T. Temperature dependence of one-dimensional hydrogen bonding in morpholinium hydrogen chloranilate studied by Cl-35 nuclear quadrupole resonance and multi-temperature X-ray diffraction. *Phys. Chem. Chem. Phys.* **2012**, *14*, 12347–12354. [CrossRef]
27. Dolinenkov, P.; Korneva, I.; Sinyavsky, N. The Distribution Change of Relaxation Times in Cl-35 NQR for Phase Transitions in p-Dichlorobenzene. *Appl. Magn. Reson.* **2015**, *46*, 17–24. [CrossRef]
28. Kovacs, A.; Izvekov, V.; Keresztury, G.; Pongor, G. Vibrational analysis of 2-nitrophenol. A joint FT-IR, FT-Raman and scaled quantum mechanical study. *Chem. Phys.* **1998**, *238*, 231–243. [CrossRef]
29. Pająk, J.; Maes, G.; De Borggraeve, W.M.; Boens, N.; Filarowski, A. Matrix-isolation FT-IR and theoretical investigation of the competitive intramolecular hydrogen bonding in 5-methyl-3-nitro-2-hydroxyacetophenone. *J. Mol. Struct.* **2008**, *880*, 86–96. [CrossRef]
30. Konopacka, A.; Filarowski, A.; Pawełka, Z. Solvent influence on the transformation of intramolecular hydrogen bonds in 2-hydroxy-5-methyl-3-nitroacetophenone. *J. Solution Chem.* **2005**, *34*, 929–945. [CrossRef]
31. Sobczyk, L.; Chudoba, D.M.; Tolstoy, T.M.; Filarowski, A. Some Brief Notes on Theoretical and Experimental Investigations of Intramolecular Hydrogen Bonding. *Molecules* **2016**, *21*, 1657. [CrossRef] [PubMed]
32. Tomkinson, J.; Parker, S.F.; Braden, D.A.; Hudson, B.S. Inelastic neutron scattering spectra of the transverse acoustic modes of the normal alkanes. *Phys. Chem. Chem. Phys.* **2002**, *4*, 716–721. [CrossRef]
33. Marques, M.P.M.; Batista de Carvalho, L.A.E.; Valero, R.; Machado, N.F.L.; Parker, S.F. An inelastic neutron scattering study of dietary phenolic acids. *Phys. Chem. Chem. Phys.* **2014**, *16*, 7491–7500. [CrossRef]
34. Davydov, A.S. *Teorya Molekularnyzh Eksitonov*; Nauka: Moscow, Russia, 1968.
35. Howard, J.; Tomkinson, J.; Eckert, J.; Goldstone, J.A.; Taylor, A.D. Inelastic neutron scattering studies of some intramolecular hydrogen bonded complexes: A new correlation of γ(OHO) vs R (OO). *J. Chem. Phys.* **1983**, *78*, 3150–3155. [CrossRef]
36. Novak, A. Hydrogen bonding in solids correlation of spectroscopic and crystallographic data. *Struct. Bonding* **1974**, *18*, 177–216.
37. Becke, A.D. Density-functional thermochemistry. III. The role of exact exchange. *J. Chem. Phys.* **1993**, *98*, 5648–5652. [CrossRef]
38. Lee, C.; Yang, W.; Parr, R.G. Development of the Colle-Salvetti correlation-energy formula into a functional of the electron density. *Phys. Rev. B* **1988**, *37*, 785–789. [CrossRef] [PubMed]
39. McLean, A.D.; Chandler, G.S. Contracted Gaussian basis sets for molecular calculations. I. Second row atoms, $Z = 11–18$. *J. Chem. Phys.* **1980**, *72*, 5639–5648. [CrossRef]
40. Frisch, M.J.; Pople, J.A.; Binkley, J.S. Self-consistent molecular orbital methods 25. Supplementary functions for Gaussian basis sets. *J. Chem. Phys.* **1984**, *80*, 3265–3269. [CrossRef]
41. Frisch, M.J.; Trucks, G.W.; Schlegel, H.B.; Scuseria, G.E.; Robb, M.A.; Cheeseman, J.R.; Scalmani, G.; Barone, V.; Petersson, G.A.; Nakatsuji, H.; et al. *Gaussian 16*; Revision B.01; Gaussian, Inc.: Wallingford, CT, USA, 2016.
42. Martin, J.M.L.; Van Alsenoy, C. *Gar2ped*; University of Antwerp: Antwerpen, Belgium, 1995.
43. Oxford Diffraction Poland Sp. *CryslAlis CCD*; CryslAlis RED Version 1.171.13 beta (Release 14.11.2003) Copyright 1995−2003; Oxford Diffraction Ltd.: Abingdon, UK, 2003.
44. Sheldrick, G.M. *SHELXS97 Program for Solution of Crystal Structure*; University of Goettingen: Goettingen, Germany, 1997.
45. Sheldrick, G.M. *SHELXl97 Program for Refinement of Crystal Structure*; University of Goettingen: Goettingen, Germany, 1997.

molecules

Article

Spectroscopic Identification of Hydrogen Bond Vibrations and Quasi-Isostructural Polymorphism in N-Salicylideneaniline

Łukasz Hetmańczyk [1], Eugene A. Goremychkin [2], Janusz Waliszewski [3], Mikhail V. Vener [4,5], Paweł Lipkowski [6], Peter M. Tolstoy [7,*] and Aleksander Filarowski [2,8,*]

[1] Faculty of Chemistry, Jagiellonian University, 2 Gronostajowa Str., 30-387 Cracow, Poland; lukasz.hetmanczyk@uj.edu.pl

[2] Frank Laboratory of Neutron Physics, Joint Institute for Nuclear Research 6 F. Joliot-Curie str., 141980 Dubna, Russia; goremychkin@jinr.ru

[3] Faculty of Physics, University of Bialystok, 1L Ciolkowskiego str., 15-245 Bialystok, Poland; j.waliszewski@uwb.edu.pl

[4] Quantum Chemistry Department, Mendeleev University of Chemical Technology, Miusskaya Square 9, 125047 Moscow, Russia; venermv@muctr.ru

[5] Kurnakov Institute of General and Inorganic Chemistry, Russian Academy of Sciences, Leninskii Prosp. 31, 119991 Moscow, Russia

[6] Department of Physical and Quantum Chemistry, Wrocław University of Science and Technology, Wybrzeże Wyspiańskiego 27, 50-370 Wrocław, Poland; pawel.lipkowski@pwr.edu.pl

[7] Institute of Chemistry, St. Petersburg State University, Universitetskij pr. 26, 198504 Saint Petersburg, Russia

[8] Faculty of Chemistry, University of Wrocław 14 F. Joliot-Curie str., 50-383 Wrocław, Poland

* Correspondence: peter.tolstoy@spbu.ru (P.M.T.); aleksander.filarowski@chem.uni.wroc.pl (A.F.); Tel.: +48-71-375-7229 (A.F.)

Citation: Hetmańczyk, Ł.; Goremychkin, E.A.; Waliszewski, J.; Vener, M.V.; Lipkowski, P.; Tolstoy, P.M.; Filarowski, A. Spectroscopic Identification of Hydrogen Bond Vibrations and Quasi-Isostructural Polymorphism in N-Salicylideneaniline. *Molecules* **2021**, *26*, 5043. https://doi.org/10.3390/molecules26165043

Academic Editor: Mirosław Jabłoński

Received: 31 July 2021
Accepted: 16 August 2021
Published: 20 August 2021

Abstract: The *ortho*-hydroxy aryl *Schiff* base 2-[(E)-(phenylimino)methyl]phenol and its deutero-derivative have been studied by the inelastic incoherent neutron scattering (IINS), infrared (IR) and Raman experimental methods, as well as by Density Functional Theory (DFT) and Density-Functional Perturbation Theory (DFPT) simulations. The assignments of vibrational modes within the 3500–50 cm^{-1} spectral region made it possible to state that the strong hydrogen bond in the studied compound can be classified as the so-called quasi-aromatic bond. The isotopic substitution supplemented by the results of DFT calculations allowed us to identify vibrational bands associated with all five major hydrogen bond vibrations. Quasi-isostructural polymorphism of 2-[(E)-(phenylimino)methyl]phenol (**SA**) and 2-[(E)-(phenyl-D$_5$-imino)methyl]phenol (**SA-C$_6$D$_5$**) has been studied by powder X-ray diffraction in the 20–320 K temperature range.

Keywords: *Schiff* bases; inelastic incoherent neutron scattering; hydrogen bond; isotopic effect

1. Introduction

This paper dwells on the studies of one of the most popular photo-thermochromic compounds, 2-[(E)-(phenylimino)methyl]phenol (N-salicylideneaniline), from the group of the ortho-hydroxy aryl Schiff bases. The first compounds of this type were synthesized in 1864 by Hugo Schiff [1] and have attracted attention ever since [2,3]. The ortho-hydroxy aryl Schiff bases, and materials based on them, demonstrate a number of interesting and useful characteristics. For example, they manifest polymorphic properties [4], ionic liquids' properties [5], elastic bending capability [6] and recognised anti-cancer properties [7]. The chiral aldimines possess the photochromic feature in the crystalline state, caused by photoinduced proton transfer in the intramolecular OHN hydrogen bond [8,9]. The chiroptical and optical anisotropic properties of photomechanical Schiff bases were previously studied under UV irradiation [10]. Such properties as ferroelectricity, piezoelectricity and second-order optical non-linearity were reported [11,12].

Many of the abovementioned features are linked to the existence of an intramolecular OHN hydrogen bond and to various conformational changes that ortho-hydroxy aryl Schiff bases could undergo. The conformational changes in the excited state were first studied in the year 1962 [13] and continue to attract attention to date. The development of new synthetic routes led to a significant increase in the number of available compounds and allowed novel types of liquid crystals and photo-optical switches to be obtained [14]. Such characteristics are usually conditioned by the isomerization of the imine fragment [15–18]. Besides, isomerization of an ortho-hydroxy aryl Schiff base attached to a BODIPY chromophore was used to tune its UV absorption spectra [19].

A long discussion takes place in the literature [20–30] on what intramolecular changes are responsible for the emergence and fine tuning of photo-thermochromic properties of ortho-hydroxy aryl Schiff bases: isomerization of the hydroxyl- and imine- groups (Chart **a**, Figure 1) or intramolecular proton transfer with the consequent isomerization of the amino group (Chart **b**, Figure 1) or aniline ring rotation (Chart **c**, Figure 1).

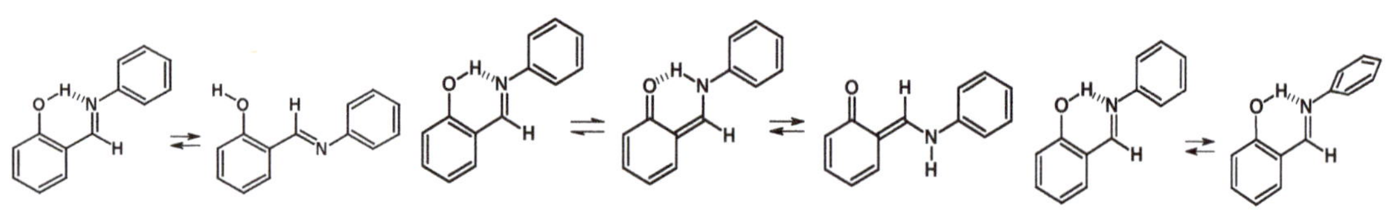

(a)

(b)

(c)

Figure 1. Conformational equilibria of 2-[(E)-(phenylimino)methyl]phenol.

The recent comprehensive studies seem to indicate that the proton transfer in the intramolecular OHN hydrogen bond and the isomerization of the aldimine fragment (Chart **b**, Figure 1) is in charge of the appearance of photo-thermochromic properties in ortho-hydroxy aryl Schiff bases [28]. Notably, the strength of the hydrogen bond and tautomeric equilibrium in it are linked to the observation of certain physical-chemical properties. In cases where the OH tautomeric form is prevailing, it is possible to observe OH isomerization, leading to the absence of photo-thermochromic properties, Chart **a**. Recently, Mielke et al. [29,30] have discovered the existence of the trans-OH form in the specific condition of matrix isolation. However, the prevailing of the NH tautomeric form makes it possible to observe the isomerization of the amino group (Chart **b**) and, therefore, the emergence of photo-thermochromic properties. It is important to underline the contribution by Ogawa et al. [24] who were the first to show the existence of trans-NH form in the solid state by X-ray method. The possibility of rotation of aldimine ring (Chart **c**) was first demonstrated by Cohen et al. [13,20], and later it was associated with the appearance of polymorphism of the studied compound.

The conformational changes, occurring in ortho-hydroxy aryl Schiff bases in different conditions and states, were investigated mostly by the methods of vibrational spectroscopy in the middle spectral range [29–36], while the low- and high-frequency hydrogen bridge vibrations have never been in the focus of the spectroscopic studies before. Therefore, in order to fill in the gap, we have synthesized 2-[(E)-(phenylimino)methyl]phenol (N-salicylideneaniline, **SA**) and its deutero-derivatives (**SA-OD** and **SA-C$_6$D$_5$**, see Figure 2) to analyse the vibrational spectra measured by IINS, Raman and IR methods. The quantum-mechanical DFT and DFPT calculations, as well as Potential Energy Distribution (PED) analysis, have been conducted to assign the spectral bands. The isotopic replacements of the bridged hydrogen for deuterium (O-H$\cdots$N $\rightarrow$ O-D$\cdots$N) and the hydrogen atoms for the deuterium atoms in aldimine ring (NC$_6$H$_5$ $\rightarrow$ NC$_6$D$_5$) were performed to additionally confirm the assignment. To that end, the Isotopic Spectral Ratios (ISRs) were used, which are defined for a given vibrational mode as the ratio of frequencies for non-deuterated and deuterated species. The ISR values are well studied for high-frequency stretching and bending proton vibrations [37], but poorly studied for the low-frequency hydrogen bridge vibrations.

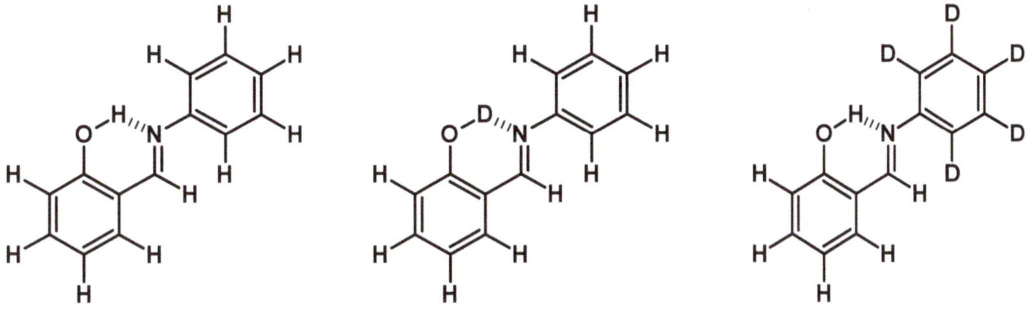

Figure 2. Structures of studied 2-[(E)-(phenylimino)methyl]phenol and its isotopologues.

Moreover, this paper delves into the spectroscopic manifestations of polymorphism of N-salicylideneaniline and its isotopologues in a wider temperature range in 20–320 K. N-salicylideneaniline was studied earlier in papers [27,28] in the 100–320 K temperature range.

2. Results and Discussion

2.1. Assignment of Hydrogen Bonding Vibrations

This part of the paper focuses on the assignments of the spectral bands to the vibrations involving the hydroxyl group (ν(OH), δ(OH) and γ(OH))) and the hydrogen bridge (ν_σ(ON) and ν_β(ON)), Scheme 1, continuing and expanding our studies started in Ref. [38].

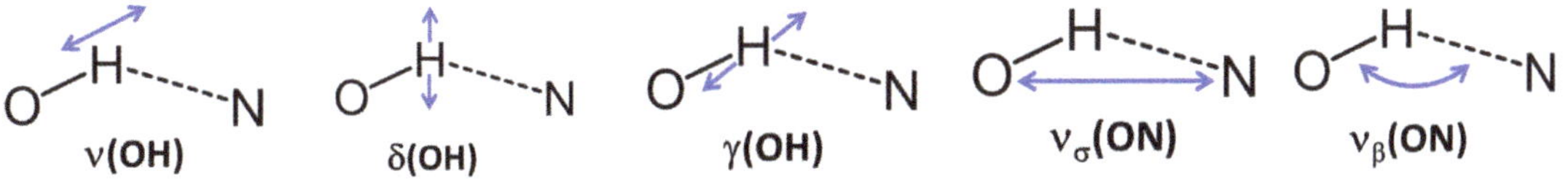

Scheme 1. The OHN hydrogen bond vibrations of the studied 2-[(E)-(phenylimino)methyl]phenols.

We start the discussion with the high-frequency region of vibrational spectra, which are presented in Figure S1 in the supporting information for brevity. The broad band located within the range of 2900–1500 cm^{-1} in the measured infrared spectrum is assigned to the stretching vibrations ν(OH) as a result of the comparison of the spectra of the non-deuterated and deuterated isopologues (ISR = 1.320). This band is characteristic for OH form of Schiff bases (enol-imine) and not to NH form (keto-amine). The ν(OH) band is rather wide and can be assigned to the so-called Zundel's continuum absorption [39]. This fact proves that the hydrogen bond in these compounds is strong. Moreover, the intensity of the ν(OH) band is rather weak despite its noticeable shift towards lower frequencies, which indicates that the hydrogen bond in **SA** is a quasi-aromatic one, which is also referred to as resonance-assisted hydrogen bond (RAHB) [40–48].

The low-frequency region of the IR, Raman and IINS spectra of non-deuterated (**SA**) and deuterated (**SA-OD**) compounds are shown in Figure 3. The experimental spectra were interpreted using the calculated vibrational spectra (DFT) and the results of the PED analysis (Tables S2–S4).

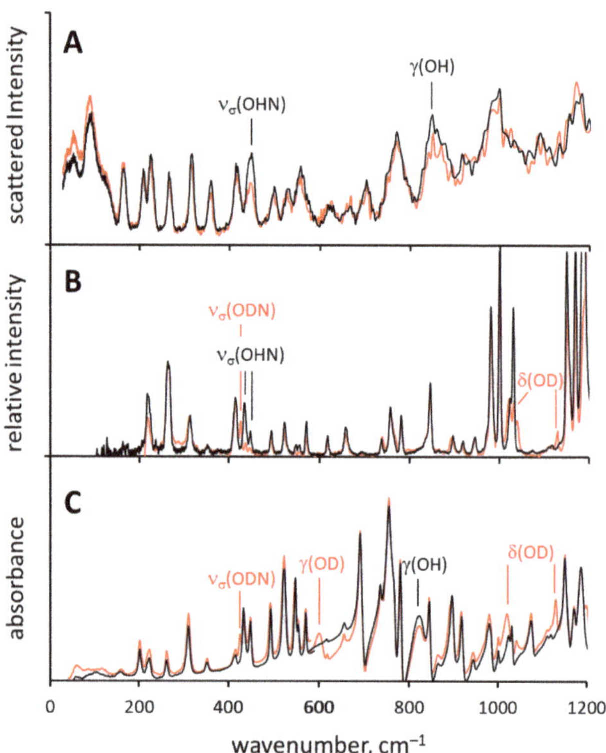

Figure 3. Normalized experimental IINS (**A**), Raman (**B**) and IR (**C**) spectra of **SA** (black line) and its deuterated derivative **SA-OD** (red lines).

When it comes to the in-plane deformational vibrations δ(OH), they are not characteristic for ortho-hydroxy Schiff bases [33,35,38]. According to PED analysis, δ(OH) vibration is mixed with the ν(C$_{alk}$C$_{alk}$) and γ(CH) vibrations. After the OH $\rightarrow$ OD substitution the bands located at 1571 and 1484 cm^{-1} (both IR and Raman active) decrease in intensity and shift slightly, while the more characteristic δ(OD) bands appear at ca. 1130 cm^{-1} and ca. 1020 cm^{-1} both in IR and Raman spectra of **SA-OD** (Figure 3).

As for the out-of-plane deformational vibration γ(OH), in the IR spectrum of **SA** the corresponding band is found in the range of 860–820 cm^{-1}, but it strongly overlaps with the bands of other vibrations and was identified with the help of DFT calculations. However, in the IR spectra of **SA-OD** this band is clearly visible at 601 cm^{-1}. The γ(OH) vibration of **SA** is practically Raman-inactive, as in the 860–820 cm^{-1} region of the Raman spectrum there are no clearly identifiable bands which would be sensitive to the OH $\rightarrow$ OD replacement. The band at 848 cm^{-1} in the IINS spectrum drops in intensity after the deuteration and thus it was assigned to the γ(OH) vibration. Note, that for IINS method the scattering cross-section for the deutron is much smaller than that of the proton and thus the contribution of the vibrations involving deutrons in IINS spectra is weak. The experimentally observed positions of γ(OH) bands and the H/D isotope effects on them are consistent with the literature data [49–51].

The stretching vibration of the hydrogen bridge (ν_σ(OHN)) gives rise to the isotope-sensitive bands at 449 cm^{-1} in the IINS spectrum and two bands at 448 and 434 cm^{-1} in the Raman spectrum of **SA** (Figure 3). The intensity of these bands is greatly decreased in the IINS and the Raman spectra of **SA-OD**, while a new band appears at 425 cm^{-1} in the Raman spectrum, assigned to the vibration of the ODN bridge, ν_σ(ODN). The

IINS spectrum does not show the band of the deformational vibration $\nu_\beta(\text{OHN})$ due to insignificant deformational motion of the bridged proton. This phenomenon was described earlier in Ref. [38]. To make the interpretation of the hydrogen bridge vibrations more accurate and reliable, the synthesis of **SA-C$_6$D$_5$** isotopologue was performed (deuteration in the aldimine ring, $NC_6H_5 \rightarrow NC_6D_5$). In Figure 4 the IINS spectra of **SA**, **SA-OD** and **SA-C$_6$D$_5$** are compared. Upon deuteration in the hydrogen bridge site (**SA-OD**), the intensity of the bands assigned to the $\gamma(\text{OH})$ and $\nu_\sigma(\text{OHN})$ vibrations (848 cm^{-1} and 449 cm^{-1}, respectively) is decreasing (Figure 4A), while the IINS spectrum of the **SA-C$_6$D$_5$** derivative features a more complicated picture (Figure 4B). Firstly, the IINS spectrum of **SA-C$_6$D$_5$** is characterized by the disappearance of a number of bands assigned to $\gamma(\text{CH})$ and $\tau(\text{CH})$ vibrations, which were located at 1183, 1175, 1153, 1089, 703 and 521 cm^{-1} in the spectrum of **SA** (see blue arrows in Figure 4). Secondly, the bands visible at 208, 264, 359 and 495 cm^{-1} in the spectrum of **SA** (see red arrows in Figure 4B) shift to lower frequencies, namely, to 202, 255, 345 and 491 cm^{-1}, respectively. The intensity of the first three bands is decreasing, while that of the fourth one is increasing slightly. Judging from Tables S2 and S4, these bands stem from the skeleton vibrations of the aldimine ring, where the deuteration occurs.

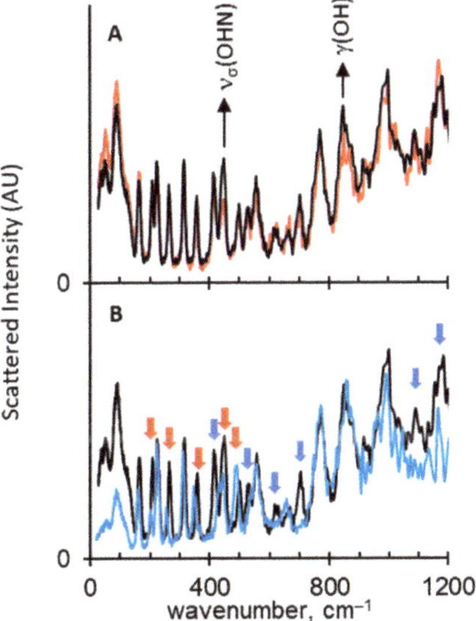

Figure 4. Normalized IINS spectra of **SA** (black traces), **SA-OD** (panel (**A**), red trace) and **SA-C$_6$D$_5$** (panel (**B**), blue trace). Blue arrows: bands of $\gamma(\text{CH})$ and $\tau(\text{CH})$ vibrations, which disappear after deuteration. Red arrows: bands of various vibrations, which shift after the deuteration. See text for more details.

The band at 449 cm^{-1}, which is assigned to $\nu_\sigma(\text{OHN})$ vibration (also marked by the red arrow in Figure 4B), is also sensitive to deuteration in the aldimine ring: the band broadens and its peak intensity decreases. Such a long-range H/D isotope effect on hydrogen bond vibrations is reported here for the first time. Below in Section 2.3 it is shown that **SA** exists as a mixture of two quasi-isostructural polymorphs and the shift of the $\nu_\sigma(\text{OHN})$ band might be associated with the change of the mole fractions of these polymorphs upon **SA** $\rightarrow$ **SA-C$_6$D$_5$** deuteration, though this question requires additional consideration which is beyond the scope of this work.

2.2. Spectral Manifestations of Polymorphism

The **SA** could crystallize in three polymorphic states—called α_1, α_2 and β—the crystal structures and cell packing of which were earlier published by F. Arod in papers [27,28]; for a visual representation of the cell packing see Figure 2 (polymorph β) and Figure 8 (polymorphs α_1 and α_2) in [27]. All three polymorphs exhibit enol-imine form with intramolecular OH$\cdots$N hydrogen bond. These polymorphs differ only slightly in molecular positions and conformations, representing "very close points in the crystal structure landscape" [52–54]; one of the larger differences between α and β states is the rotational conformation of the aldimine ring (Chart **c**, Figure 1). The polymorphs α_1 and α_2 are called quasi-isostructural. The structural mobility and polymorphism of different compounds [55–59], and Schiff bases in that number [60,61], was investigated by various methods. In this part of the paper our goal was to determine if vibrational spectra could be used to unambiguously identify the particular polymorphic state and which vibrational marker modes would be most informative. For that purpose, the crystallographic structures of polymorphs α_2 and β were optimized with CRYSTAL software, the IR spectra were calculated by DFPT method and compared with the experimental one. The result of the comparison is shown in Figure 5. There is a reasonably good agreement—both in bands positions and their relative intensities—between the experimental spectrum (Figure 5A) obtained at 295 K in which polymorph α_2 prevails and the calculated spectrum of polymorph α_2 (Figure 5B), while the agreement with that of polymorph β is significantly worse (Figure 5C). The most informative bands are marked by asterisks in Figure 5A,B. These bands are assigned to the following vibrations: ν(C=N), ν(C$_{ar}$C$_{ar}$), γ(CH), τ(CH) and τ(CC).

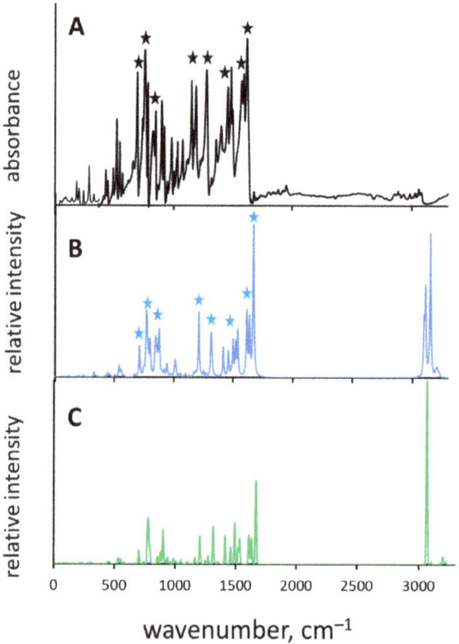

Figure 5. Normalized IR experimental ((**A**), black trace; measured at 295 K) and calculated by DFPT method spectra of polymorphs α_2 ((**B**), blue trace) and β ((**C**), green trace) of **SA**. Black and blue asterisks: the most informative bands, allowing one to identify polymorph α_2. See text for more details.

The abovementioned observations support the applicability of DFPT computational method for the research of polymorphic states.

2.3. X-ray Powder Diffraction (XPD) Study of Polymorphism in SA

X-ray Powder Diffraction measurements of **SA** and **SA-C$_6$D$_5$** were carried out in the 20–320 K temperature range. The X-ray diffraction pattern for **SA-OD** is not discussed here, because the results closely match those for **SA**. The experiments revealed that both **SA** and **SA-C$_6$D$_5$** crystallize in a triclinic form, which is in agreement with the single crystal X-ray data for polymorph α_1 obtained earlier [27]. For **SA**, several reflexes are observed as dual signals in the 20–295 K temperature range. As an example, in Figure 6 the reflexes, 002 and 0-11, for **SA** and **SA-C$_6$D$_5$** are shown. The positions and relative intensities of components of the dual signals are temperature dependent. Similar observations are valid for the deuterated derivative **SA-C$_6$D$_5$** (Figure 6C,D). Such behaviour is often attributed to the co-existence of two quasi-isostructural polymorphs, preserving the same crystal symmetry [52–56]. In case of **SA**, following the results of Refs. [27,28] we assign these polymorphs to α_1 and α_2 forms.

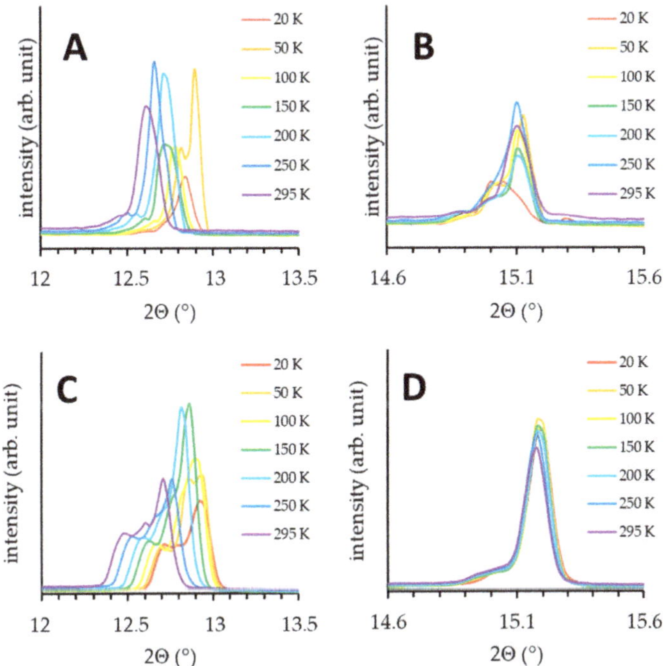

Figure 6. Temperature behaviour of (002) (**A,C**) and (0-11) (**B,D**) reflections of **SA** top (**A,B**) and **SA-C$_6$D$_5$** bottom (**C,D**) panels.

For a better understanding of this phenomenon we performed DFT calculations (in the gas phase) of the potential curves for the rotation of the aldimine fragment of **SA** in the enol-imine and keto-amine forms. The calculation confirmed that the structure of the keto-amine form is evidently flat and the twist of the aldimine fragment by up to 20° virtually does not change the potential energy (Figure 7, top). In contrast, the optimized geometry of the enol-imine form is not flat: torsional angle Θ(C=N-C=C) = 40° (Figure 7, bottom) and crossing of the phenol ring plane requires to overcome a 1 kcal/mol barrier. According to the postulate of Benstein et al. [62], such a barrier in a non-homogeneous environment of a crystal lattice could make it possible to obtain two polymorphic forms.

Though the keto-amine form is less stable than the enol-imine form by ca. 4.6 kcal/mol and unlikely to be present at room temperature, one could speculate see that the flat structure of the keto-amine form would not be prone to polymorphism.

The keto-amine if the keto-amine form would be,

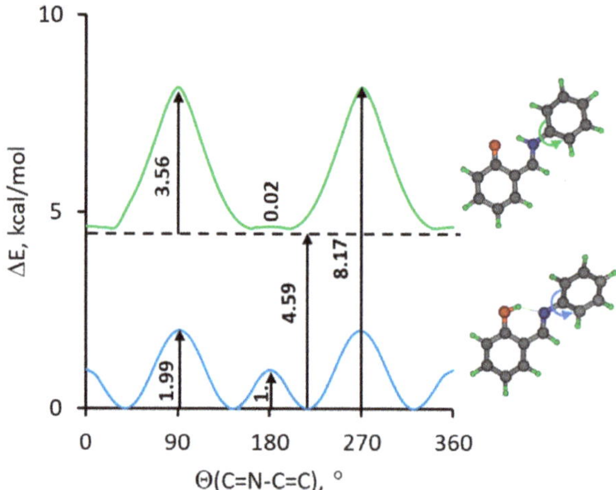

Figure 7. The calculated potential energy profile (B3LYP/6-311++G(d,p)) for a gradual rotation around the N-C bond for enol-imine (blue line) and keto-amine (green line) forms of studied Schiff base.

Upon further heating, a significant change of diffractograms of **SA** and **SA-C₆D₅** is observed at ca. 310 K (see the set of diffractograms in Figures S2 and S3). Based on the available XPD data, it is difficult to speculate which structural changes are responsible for this, but it is unlikely to be the $\alpha \leftrightarrow \beta$ transition, because the melting temperature of studied crystal was 325 K, coinciding with that previously reported for α_1 in Ref. [27].

3. Materials and Methods

3.1. Compounds and Deuteration

2-[(E)-(phenylimino)methyl]phenol (**SA**) and 2-[(E)-(phenyl-D₅-imino)methyl]phenol (**SA-C₆D₅**) were synthesized from stoichiometric mixtures of the salicylaldehyde with aniline or aniline-D₅ in refluxing methanol, respectively. The solvents were purchased from *Sigma-Aldrich* and used without further purification. For the deuteration in the mobile proton site, the solution of 2-[(E)-(phenylimino)methyl]phenol in methanol-OD was heated to 60 °C and refluxed during 30 min, then the methanol was removed by evaporation, leaving **SA-OD**. The deuteration degree was estimated to be ca. 90%.

3.2. Infrared, Raman and IINS Measurements

The infrared measurements were performed using a *Bruker Vertex 70v* spectrometer. The spectra were collected with a resolution of 2 cm⁻¹. The FT-FIR spectra (500–50 cm⁻¹) were collected for sample suspended in Apiezon N grease and placed on a polyethylene (PE) disc. The FT-MIR spectra were collected for sample in a KBr pellet. The Raman spectra of the analysed samples were obtained using an FT-Nicolet Magma 860 spectrophotometer. The In:Ga:Ar laser excitation at 1064 nm was employed for the Raman measurements. The spectra were recorded at the room temperature with the spectral resolution 4 cm⁻¹ and with 512 scans. Neutron scattering data were collected at the pulsed IBR-2 reactor in the Joint Institute of Nuclear Research (Dubna, Russia) using the time-of-flight inverted geometry spectrometer NERA at 10 K temperature. The experimental features are described in Ref. [50].

3.3. X-ray Powder Diffraction

The powder X-ray measurements were performed using EMPYREAN equipped with the water-cooled X-ray tube with Cu anode with the wavelengths of characteristic radiation $\lambda_{CuK\alpha1}$ = 1.54056 Å and $\lambda_{CuK\alpha2}$ = 1.54443 Å. In order to suppress the K_β radiation the Ni filter was used. The central part of the data acquisition system was the silicon strip detector PIXcel. The low temperature measurements were made using the closed cycle helium cryostat Phenix—Oxford Cryosystems.

3.4. Calculations

Quantum-mechanical calculations using the B3LYP functional (DFT, the three-parameter exchange hybrid functional of Becke [63], and gradient-corrected correlation functional of Lee, Yang and Parr) [64] with the 6-311++G(d,p) basis set [65–67] were performed for full geometry optimizations with the *Gaussian* 16 Rev. C01 suite of program [68]. The structures were visualized by the MOLDEN program [69]. The potential energy distribution (PED) of the normal modes was calculated in terms of natural internal coordinates using the Gar2ped program [70]. Static periodic (solid-state) DFT calculations (DFPT *Density-Functional Perturbation Theory*) were performed in the CRYSTAL09 [71,72] software package using the B3LYP functional with the Grimme D2 dispersion correction [73] and 6-31G** basis set [67]. The space groups and unit cell parameters of the considered crystals obtained in the experimental studies [27,28] were fixed and structural relaxations were limited to the position of atoms. Such approximation gives a reasonable description of the structure and spectroscopic features of intra- and intermolecular H-bonds of different types and strengths in molecular crystals [74–76]. The experimental atomic positions were used as the starting point in the periodic DFT computations. Periodic DFT computations were conducted for polymorphs α_2 and β. The disordered α_1-polymorph [27] is not suitable for periodic calculations. Since α_1 and α_2 polymorphs are quasi-isostructural, we assumed that their IR spectra are very close to each other. Therefore, only the IR spectrum of the second polymorph was considered in the present study. Periodic DFT computations of α_2-polymorph led to the appearance of imaginary frequencies. This problem is usually solved by reducing the space symmetry of a crystal [77,78]. Reducing the space symmetry group to P1 allowed us to get rid of imaginary frequencies of α_2-polymorph. An accurate interpretation of the experimental spectrum assumes knowledge of the relative stability of polymorphs α_2 and β. This requires the calculation of the sublimation enthalpy, which is very cumbersome in the case of conformationally mobile molecules [79]. In addition, the error in the calculated values can reach 15 kJ/mol [80]. Such an accuracy of calculations is hardly suitable for describing "very close points in the crystal structure landscape" of crystals with conformationally mobile molecules, e.g., see [81].

4. Conclusions

The paper presents the synthesis of 2-[(E)-(phenylimino)methyl]phenol (**SA**) and its two deutero-derivatives (**SA-OD** and **SA-C$_6$D$_5$**). The bands corresponding to hydrogen bond vibrational modes (see Scheme 1) were assigned in experimental IR, IINS and Raman spectra. The analysis of the obtained spectral results proved that in a crystal state, **SA** exists in enol-imine form with a so-called quasi-aromatic intramolecular OH···N hydrogen bond of medium strength. The measured IINS spectra **SA** and **SA-C$_6$D$_5$** made it possible to show the influence of deuteron replacement in the remote group (aldimine ring) on the hydrogen bridge stretching vibrations ν_σ(OHN). The calculations of structure and vibrational spectra of two polymorphs (earlier studied by X-ray method [27]) performed by static periodic (solid state) DFT method showed a good agreement between the measured and calculated spectra, allowing for spectroscopic distinction between polymorphs α_2 and β, as well as confirming the applicability of DFPT calculations for characterization of polymorphic states. Based on analysis of IR spectra, the **SA** in the studied samples was shown to exist in α-form. However, the XPD measurements carried out for **SA** and **SA-C$_6$D$_5$** showed dual signals

Molecules **2021**, *26*, 5043

for several reflexes and did not show a phase transition in the temperature range from 20 to 295 K. On the basis of these measurements two quasi-isostructural polymorphs of **SA** (most probably, so-called α_1 and α_2 forms) were assumed to co-exist in the mentioned temperature range.

Supplementary Materials: The following are available online, Figure S1: Normalized experimental Raman and IR spectra of **SA** and its deuterated derivative **SA-OD**. Figure S2: Temperature behaviour of XRD patterns for **SA** and **SA-C$_6$D$_5$**. Figure S3: Temperature dependence of X-ray powder diffraction patterns of **SA** and **SA-C6D5**. Table S1: Definitions of the internal coordinates used in the potential energy distribution (PED) analysis for the assignments of the vibrational spectra, Table S2: Experimental IR, Raman, IINS and calculated (B3LYP/6-311+G(d,p)) spectral data of **SA**, Table S3: Experimental IR, Raman and calculated (B3LYP/6-311+G(d,p)) spectral data of **SA-OD**, Table S4: Experimental IINS and calculated (B3LYP/6-311+G(d,p)) spectral data of **SA-C$_6$D$_5$** for the region 0–1200 cm^{-1}.

Author Contributions: Conceptualization, A.F.; methodology, A.F.; validation, all authors; formal analysis, A.F. and P.M.T.; investigation, Ł.H., E.A.G., J.W., M.V.V., P.L., P.M.T. and A.F.; resources, A.F. and P.M.T.; data curation, Ł.H., E.A.G., J.W., M.V.V., P.L., P.M.T. and A.F.; writing—original draft preparation, A.F.; writing—review and editing, A.F. and P.M.T.; visualization, Ł.H., E.A.G., J.W., M.V.V., P.L., P.M.T. and A.F.; supervision, A.F.; project administration, A.F.; funding acquisition, P.M.T. All authors have read and agreed to the published version of the manuscript.

Funding: This research was funded by the Russian Science Foundation (RFBR, grant no 18-13-00050) and Polish Government Plenipotentiary for JINR in Dubna (PWB/168-19/2021; data 11.03.2021).

Institutional Review Board Statement: Not applicable.

Informed Consent Statement: Not applicable.

Data Availability Statement: All data are available within the article or Supplementary Materials.

Acknowledgments: A.F. acknowledges the Wrocław Centre for Networking and Supercomputing Centres (WCSS) for providing computational time and facilities.

Conflicts of Interest: The authors declare no conflict of interest.

Sample Availability: Samples of the studied compounds are not available from the authors.

References

1. Schiff, H. Mittheilungen aus dem Universitätslaboratorium in Pisa Eine neue Reihe organischer Basen. *Justus Liebigs Ann. Chem.* **1864**, *131*, 118–119. [CrossRef]
2. Tidwell, T.T. Hugo (Ugo) Schiff, Schiff Bases, and a Century of β-Lactam Synthesis. *Angew. Chem.* **2008**, *47*, 1016–1020. [CrossRef]
3. Anselrnino, O. Isomere Schiff'sche Basen. *Ber. Dtsch. Chem. Ges.* **1905**, *38*, 3989–3997. [CrossRef]
4. Carletta, A.; Dubois, J.; Tilborg, A.; Wouters, J. Solid-state investigation on a new dimorphic substituted *N*-salicylidene compound: Insights into its thermochromic behaviour. *CrystEngComm* **2015**, *17*, 3509–3518. [CrossRef]
5. Ossowicz, P.; Janus, E.; Szady-Chelmieniecka, A.; Rozwadowski, Z. Influence of modification of the amino acids ionic liquids on their physico-chemical properties: Ionic liquids versus ionic liquids-supported Schiff bases. *J. Mol. Liq.* **2016**, *224*, 211–218. [CrossRef]
6. Liu, H.; Ye, K.; Zhang, Z.; Zhang, H. An Organic Crystal with High Elasticity at an Ultra-Low Temperature (77 K) and Shape ability at High Temperatures. *Angew. Chem. Int. Ed.* **2019**, *58*, 19081–19086. [CrossRef]
7. Napier, I.; Ponka, P.; Richardson, D.R. Iron trafficking in the mitochondrion: Novel pathways revealed by disease. *Blood* **2005**, *105*, 1867–1874. [CrossRef]
8. Kawato, T.; Koyama, H.; Kanatomi, H.; Shigemizu, H. Difference in the Rate of Photo-induced Unimolecular Motion of Chiral Salicylideneamines in the Chiral Crystal Environments. *Chem. Lett.* **1997**, *26*, 401–402. [CrossRef]
9. Koshima, H.; Matsuo, R.; Matsudomi, M.; Uemura, Y.; Shiro, M. Light-Driven Bending Crystals of Salicylidenephenylethylamines in Enantiomeric and Racemate Forms. *Cryst. Growth Des.* **2013**, *13*, 4330–4337. [CrossRef]
10. Takanabe, A.; Tanaka, M.; Johmoto, K.; Uekusa, H.; Mori, T.; Koshima, H.; Asahi, T. Optical Activity and Optical Anisotropy in Photomechanical Crystals of Chiral Salicylidenephenylethylamines. *J. Am. Chem. Soc.* **2016**, *45*, 15066–15077. [CrossRef]
11. Centore, R.; Jazbinsek, M.; Tuzi, A.; Roviello, A.; Capobianco, A.; Peluso, A. A series of compounds forming polar crystals and showing single-crystal-to-single-crystal transitions between polar phases. *CrystEngComm* **2012**, *14*, 2645–2653. [CrossRef]
12. Takanabe, A.; Katsufuji, T.; Johmoto, K.; Uekusa, H.; Shiro, M.; Koshima, H.; Asahi, T. Reversible Single-Crystal-to-Single-Crystal Phase Transition of Chiral Salicylidenephenylethylamine. *Crystals* **2017**, *7*, 7. [CrossRef]

13. Cohen, M.D.; Schmidt, G.M. Photochromy and thermochromy of anils. *J. Phys. Chem.* **1962**, *66*, 2442–2446. [CrossRef]
14. Riddle, J.A.; Bollinger, J.C.; Lee, D. Escape from a Nonporous Solid: Mechanically Coupled Biconcave Molecules. *Angew. Chem. Int. Ed.* **2005**, *44*, 6689–6693. [CrossRef]
15. Jia, Y.; Li, J. Molecular Assembly of Schiff Base Interactions: Construction and Application. *Chem. Rev.* **2015**, *115*, 1597–1621. [CrossRef]
16. Kandappa, S.K.; Valloli, L.K.; Ahuja, S.; Parthiban, J.; Sivaguru, J. Taming the excited state reactivity of imines—from non-radiative decay to aza Paternó–Büchi reaction. *Chem. Soc. Rev.* **2021**, *50*, 1617–1641. [CrossRef] [PubMed]
17. Muriel, W.A.; Botero-Cadavid, J.F.; Cardenas, C.; Rodriguez-Cordoba, W. A theoretical study of the photodynamics of salicylidene-2-anthrylamine in acetonitrile solution. *Phys. Chem. Chem. Phys.* **2018**, *20*, 29399–29411. [CrossRef] [PubMed]
18. Matamoros, E.; Cintas, P.; Light, M.E.; Palacios, J.C. Electronic effects in tautomeric equilibria: The case of chiral imines from d-glucamine and 2-hydroxyacetophenones. *Org. Biomol. Chem.* **2019**, *17*, 10209–10222. [CrossRef]
19. Filarowski, A.; Lopatkova, M.; Lipkowski, P.; Van der Auwerar, M.; Leen, V.; Dehaen, W. Solvatochromism of BODIPY-Schiff base Dye. *J. Phys. Chem. B* **2015**, *119*, 2576–2584. [CrossRef]
20. Cohen, M.D.; Schmidt, G.M.J.; Flavian, S. Topochemistry. Part VI. Experiments on photochromy and thermochromy of crystalline anils of salicylaldehydes. *J. Chem. Soc.* **1964**, 2041–2051. [CrossRef]
21. Andes, R.V.; Manikowski, D.M. Photochromism of Salicylidene Aniline. *Appl. Opt.* **1968**, *7*, 1179–1183. [CrossRef]
22. Destro, R.; Gavezzotti, A.; Simonetta, M. Salicylideneaniline. *Acta Crystallogr.* **1978**, *34*, 2867–2869. [CrossRef]
23. Hadjoudis, E.; Vittorakis, M.; Moustakali-Mavridis, I. Photochromism of organic compounds in the crystal state. *Tetrahedron* **1987**, *43*, 1345–1360. [CrossRef]
24. Ogawa, K.; Kasahara, Y.; Ohtani, Y.; Harada, J. Crystal Structure Change for the Thermochromy of N-Salicylideneanilines. The First Observation by X-ray Diffraction. *J. Am. Chem. Soc.* **1998**, *120*, 7107–7108.
25. Gao, A.-H.; Wang, M.-S. Nonadiabatic *ab initio* molecular dynamics study of photoisomerization in N-salicilydenemethylfurylamine (SMFA). *J. Chem. Phys.* **2017**, *146*, 124312. [CrossRef] [PubMed]
26. Rawat, M.S.M.; Mal, S.; Singh, P. Photochromism in Anils—A Review. *Open Chem. J.* **2015**, *2*, 7–19. [CrossRef]
27. Arod, F.; Pattison, P.; Schenk, K.J.; Chapuis, G. Polymorphism in N-Salicylideneaniline Reconsidered. *Cryst. Growth Des.* **2007**, *7*, 1679–1685. [CrossRef]
28. Arod, F.; Gardon, M.; Pattison, P.; Chapuis, G. The α2-polymorph of salicylideneaniline. *Acta Crystallogr.* **2005**, *61*, o317–o320. [CrossRef]
29. Grzegorzek, J.; Filarowski, A.; Mielke, Z. The photoinduced isomerization and its implication in the photo-dynamical processes in two simple Schiff bases isolated in solid argon. *Phys. Chem. Chem. Phys.* **2011**, *13*, 16596–16605. [CrossRef]
30. Grzegorzek, J.; Mielke, Z.; Filarowski, A. C=N–N=C conformational isomers of 2'-hydroxyacetophenone azine: FTIR matrix isolation and DFT study. *J. Mol. Struct.* **2010**, *976*, 371–376. [CrossRef]
31. Sıdır, İ.; Gülseven Sıdır, Y.; Góbi, S.; Berber, H.; Fausto, R. Structural Relevance of Intramolecular H-Bonding in *Ortho*-Hydroxyaryl Schiff Bases: The Case of 3-(5-bromo-2-hydroxybenzylideneamino) Phenol. *Molecules* **2021**, *26*, 2814. [CrossRef] [PubMed]
32. Majerz, I.; Pawlukojć, A.; Sobczyk, L.; Dziembowska, T.; Grech, E.; Szady-Chełmieniecka, A. The infrared, Raman and inelastic neutron scattering studies on 5-nitro-N-salicylideneethylamine. *J. Mol. Struct.* **2000**, *552*, 243–247. [CrossRef]
33. Filarowski, A.; Koll, A.; Lipkowski, P.; Pawlukojć, A. Inelastic neutron scattering and vibrational spectra of 2-(N-methyl-α-iminoethyl)-phenol and 2-(N-methyliminoethyl)-phenol: Experimental and theoretical approach. *J. Mol. Struct.* **2008**, *880*, 97–108. [CrossRef]
34. Moosavi-Tekyeh, Z.; Dastani, N. Intramolecular hydrogen bonding in N-salicylideneaniline: FT-IR spectrum and quantum chemical calculations. *J. Mol. Struct.* **2015**, *1102*, 314–322. [CrossRef]
35. Pajₐk, J.; Maes, G.; De Borggraeve, W.M.; Boens, N.; Filarowski, A. Matrix-isolation FT-IR and theoretical investigation of the vibrational properties of the sterically hindered *ortho* hydroxy acylaromatic *Schiff* bases. *J. Mol. Struct.* **2007**, *844–845*, 83–93. [CrossRef]
36. Dziembowska, T.; Szafran, M.; Katrusiak, A.; Rozwadowski, Z. Crystal structure of and solvent effect on tautomeric equilibrium in Schiff base derived from 2-hydroxy-1-naphthaldehyde and methylamine studied by X-ray diffraction, DFT, NMR and IR methods. *J. Mol. Struct.* **2009**, *929*, 32–42. [CrossRef]
37. Sobczyk, L.; Obrzud, M.; Filarowski, A. H/D Isotope Effects in Hydrogen Bonded Systems. *Molecules* **2013**, *18*, 4467–4476. [CrossRef]
38. Kwocz, A.; Panek, J.J.; Jezierska, A.; Hetmańczyk, Ł.; Pawlukojć, A.; Kochel, A.; Lipkowski, P.; Filarowski, A. A molecular roundabout: Triple cycle-arranged hydrogen bonds in light of experiment and theory. *New J. Chem.* **2018**, *42*, 19467–19477. [CrossRef]
39. Zundel, G. Easily Polarizable Hydrogen Bonds—Their Interactions with the Environment—IR Continuum and Anomalous Large Conductivity. In *The Hydrogen Bond: Recent Developments in Theory and Experiments*; Schuster, P., Zundel, G., Sandorfy, C., Eds.; North-Holland: Amsterdam, The Netherland, 1976; Volume 2, pp. 683–766.
40. Shigorin, D.N. *Vodorodnaya Svyaz' (Hydrogen Bond)*; Sokolov, N.D., Chulanovsky, A.D., Eds.; Nauka: Moscow, Russia, 1964; p. 195.
41. Hadži, D.; Bratos, S. Vibrational spectroscopy of the hydrogen bond. In *The Hydrogen Bond: Recent Developments in Theory and Experiments*; Schuster, P., Zundel, G., Sandorfy, C., Eds.; North-Holland: Amsterdam, The Netherlands, 1976; Volume 2, pp. 565–611.

42. Tsubomura, H. Nature of the Hydrogen Bond. III. The Measurement of the Infrared Absorption Intensities of Free and Hydrogen-Bonded OH Bands. Theory of the Increase of the Intensity Due to the Hydrogen Bond. *J. Chem. Soc.* **1956**, *24*, 927. [CrossRef]
43. Hadzi, D. Absorption spectra and structure of some solid hydroxyazo-compounds. *J. Chem. Soc.* **1956**, 2143–2150. [CrossRef]
44. Filarowski, A.; Koll, A. Intergrated intensity of ν_s(OH) absorption bands in bent hydrogen bonds in *ortho*-dialkylaminomethyl phenols. *Vib. Spectrosc.* **1996**, *12*, 15–24. [CrossRef]
45. Filarowski, A.; Koll, A. Specificity of the intramolecular hydrogen bond. The differences in spectroscopic characteristics of the intermolecular and intramolecular H-bonds. *Vib. Spectrosc.* **1998**, *17*, 123–131. [CrossRef]
46. Takasuka, M.; Matsui, Y. Experimental Observations and CN D0/2 Calculations for Hydroxy Stretching Frequency Shifts, Intensities, and Hydrogen Bond Energies of Intramolecular Hydrogen Bonds in ortho-Substituted Phenols. *J. Chem. Soc. Perkin II* **1979**, 1743–1750. [CrossRef]
47. Nguyen, Y.H.; Lampkin, B.J.; Venkatesh, A.; Ellern, A.; Rossini, A.J.; VanVeller, B. Open-Resonance-Assisted Hydrogen Bonds and Competing Quasiaromaticity. *J. Org. Chem.* **2018**, *83*, 9850–9857. [CrossRef]
48. Gilli, G.; Gilli, P. *The Nature of the Hydrogen Bond: Outline of a Comprehensive Hydrogen Bond Theory*; Oxford University Press: Oxford, UK, 2009.
49. Hetmańczyk, Ł.; Szklarz, P.; Kwocz, A.; Wierzejewska, M.; Pagacz-Kostrzewa, M.; Melnikov, M.Y.; Tolstoy, P.M.; Filarowski, A. Polymorphism and Conformational Equilibrium of Nitro-Acetophenone in Solid State and under Matrix Conditions. *Molecules* **2021**, *26*, 3109. [CrossRef]
50. Piękoś, P.; Jezierska, A.; Panek, J.J.; Goremychkin, E.A.; Pozharskii, A.F.; Antonov, A.S.; Tolstoy, P.M.; Filarowski, A. Symmetry/asymmetry of the NHN hydrogen bond in protonated 1,8-bis(dimethylamino)naphthalene. *Symmetry* **2020**, *12*, 1924. [CrossRef]
51. Jóźwiak, K.; Jezierska, A.; Panek, J.J.; Goremychkin, E.A.; Tolstoy, P.M.; Shenderovich, I.G.; Filarowski, A. Inter- *vs.* intra-molecular hydrogen bond patterns and proton dynamics in phthalic acid associates. *Molecules* **2020**, *25*, 4720. [CrossRef]
52. Prohens, R.; Barbas, R.; Font-Bardia, M. Morphotropism and "Quasi-Isostructurality" in the Three High Z′ Concomitant Polymorphs of Efinaconazole. *Cryst. Grow. Des.* **2020**, *20*, 4238–4242. [CrossRef]
53. Kálmán, A.; Párkányi, L.; Argay, G. Classification of the isostructurality of organic molecules in the crystalline state. *Acta Cryst.* **1993**, *49*, 1039–1049. [CrossRef]
54. Dey, D.; Thomas, S.P.; Spackman, M.A.; Chopra, D. Quasi-isostructural polymorphism' in molecular crystals: Inputs from interaction hierarchy and energy frameworks. *Chem. Commun.* **2016**, *52*, 2141–2144. [CrossRef]
55. Fischer, J.; Lima, J.A.; Freire, P.T.C.; Melo, F.E.A.; Havenith, R.W.A.; Filho, J.M.; Broer, R.; Eckert, J.; Bordallo, H.N. Molecular flexibility and structural instabilities in crystalline L-methionine. *Biophys. Chem.* **2013**, *180–181*, 76–85. [CrossRef]
56. De Souza, J.M.; Freire, P.T.C.; Argyriou, D.N.; Stride, J.A.; Barthòs, M.; Kalceff, W.; Bordallo, H.N. Raman and Neutron Scattering Study of Partially Deuterated l-Alanine: Evidence of a Solid-Solid Phase Transition. *ChemPhysChem* **2009**, *10*, 3337–3343. [CrossRef]
57. Hetmańczyk, J.; Hetmańczyk, Ł.; Bilski, P.; Kozak, A. Complex water dynamics in crystalline [Ca(H2O)2](ReO4)2, studied by the vibrational spectroscopy and proton magnetic resonance. *J. Mol. Struct.* **2020**, *1205*, 127610. [CrossRef]
58. Szostak, E.; Hetmańczyk, J.; Migdał-Mikuli, A. Inelastic and elastic neutron scattering studies of the vibrational and reorientational dynamics, crystal structure and solid–solid phase transition in [Mn(OS(CH3)2)6](ClO4)2 supported by theoretical (DFT) calculations. *Spectrochim. Acta A* **2015**, *145*, 368–375. [CrossRef] [PubMed]
59. Hetmańczyk, J.; Hetmańczyk, Ł.; Nowicka-Scheibe, J.; Pawlukojć, A.; Maurin, J.K.; Schilf, W. Structural, Thermal, and Vibrational Properties of N,N-Dimethylglycine–Chloranilic Acid—A New Co-Crystal Based on an Aliphatic Amino Acid. *Materials* **2021**, *14*, 3292. [CrossRef] [PubMed]
60. Shapenova, D.S.; Shiryaev, A.A.; Bolte, M.; Kukułka, M.; Szczepanik, D.W.; Hooper, J.; Babashkina, M.G.; Mahmoudi, G.; Mitoraj, M.P.; Safin, D.A. Resonance Assisted Hydrogen Bonding Phenomenon Unveiled from Both Experiment and Theory—An Example of New Family of Ethyl N-salicylideneglycinate Dyes. *Chem. Eur. J.* **2020**, *26*, 12987–12995. [CrossRef]
61. Safin, D.A.; Robeyns, K.; Babashkina, M.G.; Filinchuk, Y.; Rotaru, A.; Jureschi, C.; Mitoraj, M.P.; Hooper, J.; Brela, M.; Garcia, Y. Polymorphism driven optical properties of an anil dye. *CrystEngComm* **2016**, *18*, 7249–7259. [CrossRef]
62. Cruz-Cabeza, A.J.; Bernstein, J. Conformational Polymorphism. *Chem. Rev.* **2014**, *114*, 2170–2191. [CrossRef]
63. Becke, A.D. Density-functional thermochemistry. III. The role of exact exchange. *J. Chem. Phys.* **1993**, *98*, 5648–5652. [CrossRef]
64. Lee, C.; Yang, W.; Parr, R.G. Development of the Colle-Salvetti Correlation-Energy Formula into a Functional of the Electron Density. *Phys. Rev. B* **1988**, *37*, 785–789. [CrossRef] [PubMed]
65. McLean, A.D.; Chandler, G.S. Contracted Gaussian basis sets for molecular calculations. I. Second row atoms, Z=11-18. *J. Chem. Phys.* **1980**, *72*, 5639–5648. [CrossRef]
66. Krishnan, R.; Binkley, J.S.; Seeger, R.; Pople, J.A. Self-consistent molecular orbital methods. XX. A basis set for correlated wave functions. *J. Chem. Phys.* **1980**, *72*, 650–654. [CrossRef]
67. Frisch, M.J.; Pople, J.A.; Binkley, J.S. Self-consistent molecular orbital methods 25. Supplementary functions for Gaussian basis sets. *J. Chem. Phys.* **1984**, *80*, 3265–3269. [CrossRef]
68. Frisch, M.J.; Trucks, G.W.; Schlegel, H.B.; Scuseria, G.E.; Robb, M.A.; Cheeseman, J.R.; Scalmani, G.; Barone, V.; Petersson, G.A.; Nakatsuji, H.; et al. *Gaussian 16, Revision C.01*; Gaussian, Inc.: Wallingford, CT, USA, 2016.
69. Schaftenaar, G.; Noordik, J.H. Molden: A pre- and post-processing program for molecular and electronic structures. *J. Comput. Aided Mol. Des.* **2000**, *14*, 123–134. [CrossRef]

70. Martin, J.M.L.; Van Alsenoy, C. *GAR2PED*; University of Antwerp: Antwerpen, Belgium, 1995.
71. Dovesi, R.; Orlando, R.; Civalleri, B.; Roetti, C.; Saunders, V.R.; Zicovich-Wilson, C.M. CRYSTAL: A computational tool for the *ab initio* study of the electronic properties of crystals. *Z. Kristallogr.* **2005**, *220*, 571–573. [CrossRef]
72. Dovesi, R.; Saunders, V.R.; Roetti, C.; Orlando, R.; Zicovich-Wilson, C.M.; Pascale, F.; Civalleri, B.; Doll, K.; Harrison, N.M.; Bush, I.J.; et al. *CRYSTAL09 User's Manual*; University of Torino: Torino, Italy, 2009.
73. Grimme, S. Semiempirical GGA-type density functional constructed with a long-range dispersion correction. *J. Comput. Chem.* **2006**, *27*, 1787–1799. [CrossRef] [PubMed]
74. Surov, A.O.; Voronin, A.P.; Vener, M.V.; Churakov, A.V.; Perlovich, G.L. Specific features of supramolecular organisation and hydrogen bonding in proline cocrystals: A case study of fenamates and diclofenac. *CrystEngComm* **2018**, *20*, 6970–6981. [CrossRef]
75. Voronin, A.P.; Surov, A.O.; Churakov, A.V.; Parashchuk, O.D.; Rykounov, A.A.; Vener, M.V. Combined X-ray Crystallographic, IR/Raman Spectroscopic, and Periodic DFT Investigations of New Multicomponent Crystalline Forms of Anthelmintic Drugs: A Case Study of Carbendazim Maleate. *Molecules* **2020**, *25*, 2386. [CrossRef]
76. Surov, A.O.; Vasilev, N.A.; Churakov, A.V.; Parashchuk, O.D.; Artobolevskii, S.V.; Alatortsev, O.A.; Makhrov, D.E.; Vener, M.V. Two Faces of Water in the Formation and Stabilization of Multicomponent Crystals of Zwitterionic Drug-Like Compounds. *Symmetry* **2021**, *13*, 425. [CrossRef]
77. Vener, M.V.; Sauer, J. Environmental effects on vibrational proton dynamics in $H_5O_2^+$: DFT study on crystalline $H_5O_2^+ClO_4^-$. *Phys. Chem. Chem. Phys.* **2005**, *7*, 258–263. [CrossRef]
78. Sen, A.; Mitev, P.D.; Eriksson, A.; Hermansson, K. H-bond and electric field correlations for water in highly hydrated crystals. *Int. J. Quantum Chem.* **2016**, *116*, 67–80. [CrossRef]
79. Červinka, C.; Fulem, M. Cohesive properties of the crystalline phases of twenty proteinogenic α-aminoacids from first-principles calculations. *Phys. Chem. Chem. Phys.* **2019**, *21*, 18501–18515. [CrossRef] [PubMed]
80. Beran, G.J.O. Modeling Polymorphic Molecular Crystals with Electronic Structure Theory. *Chem. Rev.* **2016**, *116*, 5567–5613. [CrossRef] [PubMed]
81. Surov, A.O.; Manin, A.N.; Voronin, A.P.; Churakov, A.V.; Perlovich, G.L.; Vener, M.V. Weak Interactions Cause Packing Polymorphism in Pharmaceutical Two-Component crystals. The case study of the Salicylamide Cocrystal. *Cryst. Growth Des.* **2017**, *17*, 1425–1437. [CrossRef]

molecules

Article

A Family of Ethyl *N*-Salicylideneglycinate Dyes Stabilized by Intramolecular Hydrogen Bonding: Photophysical Properties and Computational Study

Larisa E. Alkhimova [1,*], **Maria G. Babashkina** [2] **and Damir A. Safin** [1,3,4,*]

1   Institute of Chemistry, University of Tyumen, Volodarskogo Str. 6, 625003 Tyumen, Russia
2   Institute of Condensed Matter and Nanosciences, Université Catholique de Louvain, Place L. Pasteur 1, 1348 Louvain-la-Neuve, Belgium; maria.babashkina@gmail.com
3   Kurgan State University, Sovetskaya Str. 63/4, 640020 Kurgan, Russia
4   Innovation Center for Chemical and Pharmaceutical Technologies, Ural Federal University Named after the First President of Russia B.N. Eltsin, Mira Str. 19, 620002 Ekaterinburg, Russia
*   Correspondence: l.e.alkhim@gmail.com (L.E.A.); damir.a.safin@gmail.com (D.A.S.)

Abstract: In this work we report solvatochromic and luminescent properties of ethyl *N*-salicylideneglycinate (**1**), ethyl *N*-(5-methoxysalicylidene)glycinate (**2**), ethyl *N*-(5-bromosalicylidene)glycinate (**3**), and ethyl *N*-(5-nitrosalicylidene)glycinate (**4**) dyes. **1–4** correspond to a class of *N*-salicylidene aniline derivatives, whose photophysical properties are dictated by the intramolecular proton transfer between the OH-function and the imine N-atom, affording tautomerization between the enol-imine and keto-enamine forms. Photophysical properties of **1–4** were studied in different pure non-polar and (a)protic polar solvents as well as upon gradual addition of NEt_3, NaOH, and CH_3SO_3H. The DFT calculations were performed to verify the structures of **1–4** as well as their electronic and optical properties.

Keywords: Schiff base; *N*-salicylidene aniline derivative; photophysical properties; solvatochromism; Hirshfeld surface analysis; DFT

Citation: Alkhimova, L.E.; Babashkina, M.G.; Safin, D.A. A Family of Ethyl *N*-Salicylideneglycinate Dyes Stabilized by Intramolecular Hydrogen Bonding: Photophysical Properties and Computational Study. *Molecules* **2021**, *26*, 3112. https://doi.org/10.3390/molecules26113112

Academic Editor: Mirosław Jabłoński

Received: 23 April 2021
Accepted: 21 May 2021
Published: 23 May 2021

Publisher's Note: MDPI stays neutral with regard to jurisdictional claims in published maps and institutional affiliations.

1. Introduction

Schiff base is a condensation product of aldehyde and primary amine. When salicylaldehyde derivative is used as an aldehyde and amine is a monoamine derivative, the condensation yields a NO-Schiff base compound that may produce complexes with a great variety of metal ions upon deprotonation. Thus, Schiff bases are still of great importance in coordination chemistry, although more than a century has passed since their discovery [1–3]. Schiff bases have been known to be the most actively used ligands thanks to their ease of synthesis, availability, structural versatility, and solubility in common solvents. Moreover, Schiff bases are not only efficient ligands to coordinate/chelate many elements but can stabilize elements in various and specific oxidation states. It is also noteworthy that Schiff bases, as well as their corresponding complexes, can possess pronounced bioactivity [4–6], including against coronavirus [7]. The latter is becoming even more important since these days humanity is in dire need of drugs for COVID-19 treatment.

Schiff bases fabricated from salicylaldehyde derivatives (*N*-salicylidene aniline derivatives) have been the main focus of our extensive studies due to their chromic properties and a broad color palette [8–18]. Particularly, these compounds are known to a possible intramolecular transfer of the OH proton to the imine N-atom affording tautomerization between the enol-imine and keto-enamine forms [19–28]. This phenomenon is usually responsible for the fascinating optical properties of these compounds. Notably, the enol-imine and keto-enamine forms can adopt either the *trans*- or *cis*-isomers, yielding a rich color palette from colorless to red through yellow (Scheme 1) [29,30]. These colors can be called by different external stimuli (light irradiation, temperature, solvent, pH, etc.). Thus, chromic properties are one of the main features of *N*-salicylidene aniline derivatives and are important for many practical applications [31–35].

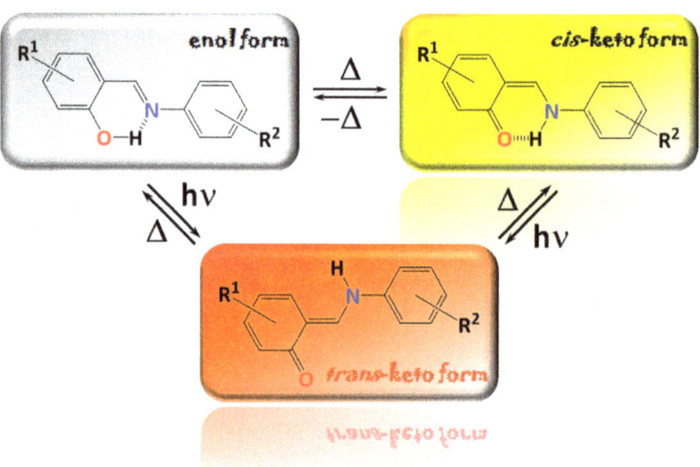

Scheme 1. Isomeric forms and the corresponding color palette of *N*-salicylidene aniline derivatives.

Notably, due to the abovementioned tautomerization, *N*-salicylidene aniline derivatives can also possess the so-called ESIPT (excited state intramolecular proton transfer) [36], which can afford an equilibrium between the excited keto-enamine* and enol-imine* forms [29,30,37,38]. The keto-enamine form usually, emits at a lower energy than the enol-imine tautomer leading to dual fluorescence, which can be observed as two (or more) emission bands [29,30,37,38]. The ratio of these bands can be influenced by the nature of a solvent, which allows tuning the resulting emission color.

It should be noticed that we have recently initiated comprehensive studies of closely related *N*-salicylidene aniline derivatives **1–4**, derived from the ethyl ester of glycine (Chart 1) [17,18]. We found that **1–4** in the solid state each exhibits an enol-imine form. Notably, the electron-withdrawing NO_2 substituent in **4** is mainly responsible for the formation of the *cis*-keto form isomer in EtOH, while dyes **1–3** adopt the enol-imine form in the same solvent. Additionally, only **4** is emissive under ambient conditions in the solid state, while **2** is emissive in EtOH.

Chart 1. Diagrams of the discussed dyes **1** (X = H), **2** (X = OMe), **3** (X = Br) and **4** (X = NO_2).

Intrigued by the obtained results, herein, we continue our comprehensive studies of the dyes **1–4**. Particularly, to probe solvatochromism, we focused on the optical properties of these compounds in different solvents, as well as shedding more light on their crystal structures using the Hirshfeld surface analysis to examine in-depth the non-covalent interactions responsible for crystal packing, which might be responsible for the observed photophysical properties of **1–4** in the solid state.

2. Results and Discussion

Dyes **1–4** were produced according to the described synthetic procedure [18]. We first applied a Hirshfeld surface analysis [39] to study the intermolecular interactions in the reported crystal structures of **1–4** [17]. As a result, a set of 2D fingerprint plots [40] were generated using CrystalExplorer 3.1 [41]. In order to estimate the propensity of two chemical species being in contact, we also calculated the enrichment ratios (E) [42] of the intermolecular contacts.

The intermolecular H···H, H···C and H···O contacts occupy a majority of a molecular surface of **1** (Table 1). There is a barely observed splitting of the shortest H···H fingerprint at $d_e + d_i \approx 2.4$ Å, which is due to the shortest contact being between three atoms, rather than for direct two-atom contact (Figure S1 in Supplementary Materials) [40]. The H···C and H···O shortest contacts are shown at $d_e + d_i \approx 2.7$ and 2.3 Å, respectively (Figure S1). The latter contacts in the corresponding fingerprint plot are shown as two spikes and mainly correspond to the contacts formed by the carbonyl oxygen atom. The molecular surface of **1** is also characterized by a negligible proportion of the H···N, C···O and O···O contacts, ranging from 0.6% to 2.4% (Table 1).

Table 1. Hirshfeld contact surfaces and derived "random contacts" and "enrichment ratios" for the crystal structures of **1–4**.

	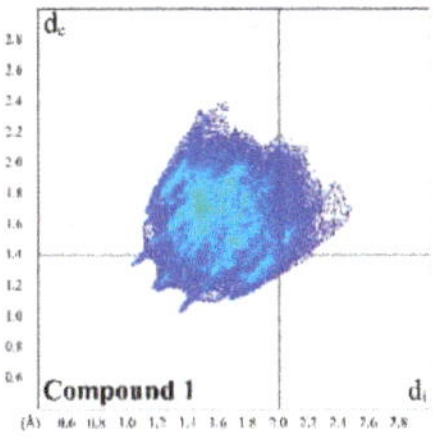 Compound 1				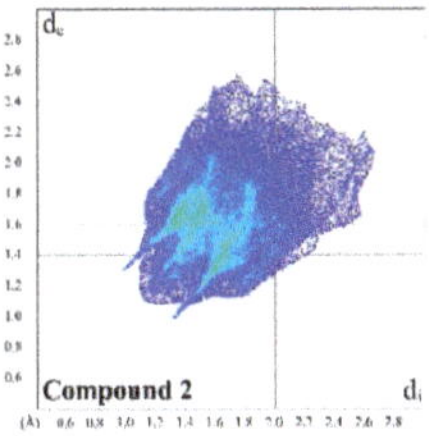 Compound 2				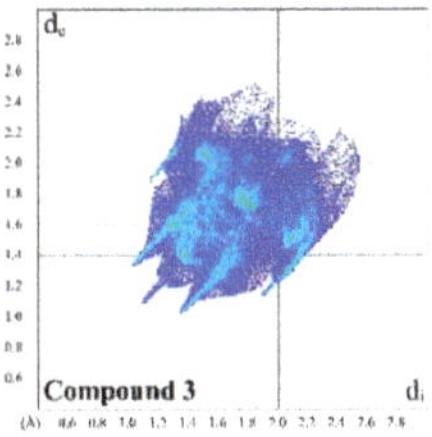 Compound 3					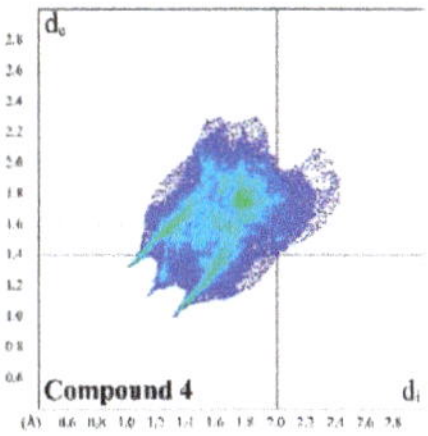 Compound 4			
	H	C	N	O	H	C	N	O	H	C	N	O	Br	H	C	N	O
Contacts (C, %) [1]																	
H	48.3	–	–	–	45.5	–	–	–	35.4	–	–	–	–	30.0	–	–	–
C	22.6	0.0	–	–	20.9	0.6	–	–	13.7	4.1	–	–	–	11.5	4.9	–	–
N	2.4	0.0	0.0	–	2.4	0.0	0.0	–	1.9	0.4	0.0	–	–	3.3	0.6	0.0	–
O	24.9	1.2	0.0	0.6	30.1	0.3	0.0	0.3	23.4	1.6	0.5	0.2	–	44.1	1.8	1.7	2.1
Br	–	–	–	–	–	–	–	–	17.4	0.5	0.0	0.0	0.9	–	–	–	–
Surface (S, %)																	
	73.3	11.9	1.2	13.7	72.2	11.2	1.2	15.5	63.6	12.2	1.4	13.0	9.9	59.5	11.9	2.8	25.9
Random contacts (R, %)																	
H	53.7	–	–	–	52.1	–	–	–	40.4	–	–	–	–	35.4	–	–	–
C	17.4	1.4	–	–	16.2	1.3	–	–	15.5	1.5	–	–	–	14.2	1.4	–	–
N	1.8	0.0	0.0	–	1.7	0.0	0.0	–	1.8	0.3	0.0	–	–	3.3	0.0	0.1	–
O	20.1	3.3	0.3	1.9	22.4	3.5	0.4	2.4	16.5	3.2	0.4	1.7	–	30.8	6.2	1.5	6.7
Br	–	–	–	–	–	–	–	–	12.6	2.4	0.3	2.6	1.0	–	–	–	–
Enrichment (E) [2]																	
H	0.90	–	–	–	0.87	–	–	–	0.88	–	–	–	–	0.85	–	–	–
C	1.30	0.0	–	–	1.29	0.46	–	–	0.88	2.73	–	–	–	0.81	3.50	–	–
N	1.33	–	–	–	1.41	–	–	–	1.06	–	–	–	–	1.00	–	–	–
O	1.24	0.36	–	0.32	1.34	0.09	–	0.13	1.42	0.50	–	0.12	–	1.43	0.29	1.13	0.31
Br	–	–	–	–	–	–	–	–	1.38	0.21	–	0.0	0.90	–	–	–	–

[1] Values are obtained from CrystalExplorer 3.1 [41]. [2] The "enrichment ratios" were not computed when the "random contacts" were lower than 0.9%, as they are not meaningful [42].

Substitution of the 5-hydrogen atom by the MeO function in **2** decreases a proportion of the H$\cdots$H and H$\cdots$C contacts by 2.8 and 1.7%, respectively, with the simultaneous increase of a proportion of the H$\cdots$O contacts up to 30.1% (Table 1). The H$\cdots$H and H$\cdots$C shortest contacts are now shown at $d_e + d_i \approx 2.2$ and 2.9 Å, respectively (Figure S2). The H$\cdots$O contacts in the corresponding 2D fingerprint plot are also shown as two spikes with the shortest contacts at $d_e + d_i \approx 2.3$ Å, also being formed by the carbonyl oxygen atom (Figure S2). The Hirshfeld molecular surface of **2** is further described by a very minor proportion of the H$\cdots$N, C$\cdots$C, C$\cdots$O and O$\cdots$O contacts, varying from 0.3 to 2.4% (Table 1).

In **3**, incorporation of the bromine instead of the hydrogen in **1** or the methoxy group in **2** significantly decreases a proportion of the H$\cdots$H and H$\cdots$C intermolecular contacts down to 35.4 and 13.7%, respectively. A proportion of the H$\cdots$O contact remains almost the same as in **1** (Table 1). The H$\cdots$H, H$\cdots$C and H$\cdots$O shortest contacts are shown in the corresponding 2D fingerprint plots at $d_e + d_i \approx 2.2$, 2.8, and 2.3 Å, respectively (Figure S3). The molecular surface of **3** is additionally characterized by a remarkable proportion of the H$\cdots$Br intermolecular contacts of 17.4%, with the shortest contacts at $d_e + d_i \approx 3.0$ Å (Figure S3). Notably, the molecular surface of **3** is further described by a distinct proportion of the C$\cdots$C contacts of 4.1%, which is due to intermolecular $\pi \cdots \pi$ stacking interactions between the phenylene rings [17]. These contacts are shown on the 2D plot as a typical area at $d_e = d_i \approx 1.7$–2.0 Å (Figure S3). Insignificant contributions from H$\cdots$N, C$\cdots$N, C$\cdots$O, C$\cdots$Br, N$\cdots$O, O$\cdots$O, and Br$\cdots$Br contacts in **3** have also been revealed (Table 1).

Dye **4** contains the NO$_2$ group in the same position as the corresponding substituent in **2** and **3** (Chart 1). The molecular surface of **4** is characterized by a dominant contribution from the H$\cdots$O contacts of 44.1% with the shortest contacts, shown in the corresponding 2D fingerprint plot at $d_e + d_i \approx 2.3$ Å mainly formed by both the NO$_2$ and carbonyl oxygen atoms (Figure S4). The H$\cdots$H and H$\cdots$C contacts occupy 30.0 and 11.5%, respectively. Notably, the structure of **4**, similar to **3**, also exhibits a distinct proportion of the C$\cdots$C contacts of 4.9% due to the formation of intermolecular $\pi \cdots \pi$ stacking interactions between the phenylene rings [17]. These contacts are also shown on the 2D plot as a typical area at $d_e = d_i \approx 1.7$–2.0 Å (Figure S4). Remaining contacts, namely H$\cdots$N, C$\cdots$N, C$\cdots$O, N$\cdots$O and O$\cdots$O contacts, occupy a negligible proportion of the molecular surface of **4** (Table 1).

All the H$\cdots$X (X = C, N, O) contacts are highly favored in **1** and **2** since the corresponding enrichment ratios E_{HX} are significantly higher than unity (Table 1). This is explained by a relatively higher proportion of these contacts on the total Hirshfeld surface area over a corresponding proportion of random contacts R_{HX} (Table 1). However, only H$\cdots$N and H$\cdots$O intermolecular contacts are favored in **3** and **4**, while the H$\cdots$C contacts on the surface of molecules are much less favored, which is related to a remarkably high enrichment of C$\cdots$C contacts in **3** and **4** (Table 1), formed in aromatic stacking. Notably, both **3** and **4** are additionally characterized by highly favored H$\cdots$Br and N$\cdots$O contacts, respectively, and **3** is also described by less favored Br$\cdots$Br contacts (Table 1). The H$\cdots$H intermolecular contacts on the molecular surfaces of all compounds are less favored since the corresponding enrichment ratios E_{HH} are less than unity ranging from 0.85 to 0.90 (Table 1). The remaining contacts are significantly impoverished with the corresponding enrichment ratios less than 0.50 (Table 1).

We have also examined the optical properties of **1–4** in different solvents. We first probed such solvents as cyclohexane (non-polar), THF (polar aprotic), CH$_3$CN (polar aprotic), and EtOH (polar protic) to study the enol-imine and keto-enamine tautomerization in the applied solvents. It is known that the dipole moments for the enol-imine isomers are smaller than those for the keto-enamine derivatives [43,44], which means that the keto-enamine tautomer is more prevalent in polar solvents.

The absorption spectra of **1–3** each contain bands corresponding to n→π* and π→π* transitions only in the UV region, regardless of the solvent (Figures S5–S7). Interestingly, the longest wavelength band in the spectra of **2** is remarkably red-shifted (Figures S5–S7). This is due to the electron-donating properties of the MeO fragment in comparison to the H and Br substituents in **1** and **3**, respectively. Furthermore, the spectra of **1–3** in EtOH exhibit an additional low-intense band, centered at about 400–420 nm (Figures S5–S7), which corresponds to the traces of the *cis*-keto-enamine form.

The absorption spectra of **4** in the same solvents, except cyclohexane, contain an intense band in the visible region, which arises from the *cis*-keto-enamine form (Figure 1). Notably, in **4**, the electron-donating OH is involved in a through-resonance effect with the electron-withdrawing NO_2 fragment. This leads to nitro–*aci*-nitro tautomerization, yielding an additional quinoid form [45].

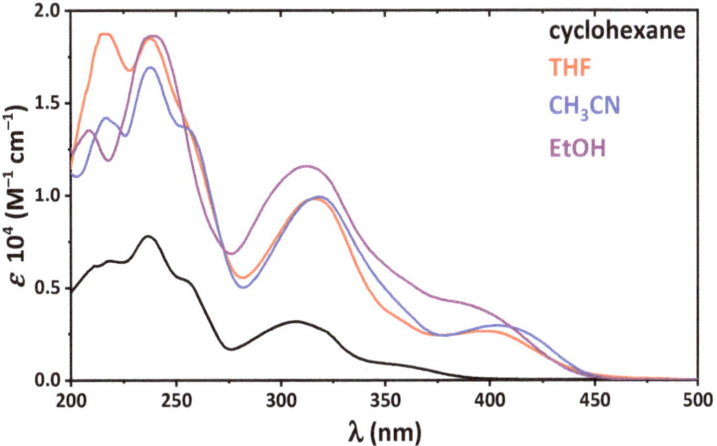

Figure 1. UV-vis spectra of **4** in the applied solvents.

Optical properties of **4** were also studied in a series of alcohols, namely MeOH, EtOH, *n*PrOH, *i*PrOH, and *n*BuOH. Interestingly, in the absorption spectra of **4** in all alcohols, except *n*PrOH, the band of the *cis*-keto-enamine form was observed with the most remarkable one in MeOH (Figure 2), which is the most polar and acidic within the applied alcohols. The most striking observation is the absence of the band of the *cis*-keto-enamine form in the spectrum of **4** in *n*PrOH (Figure 2). This might be explained by possible specific solute–solvent interactions.

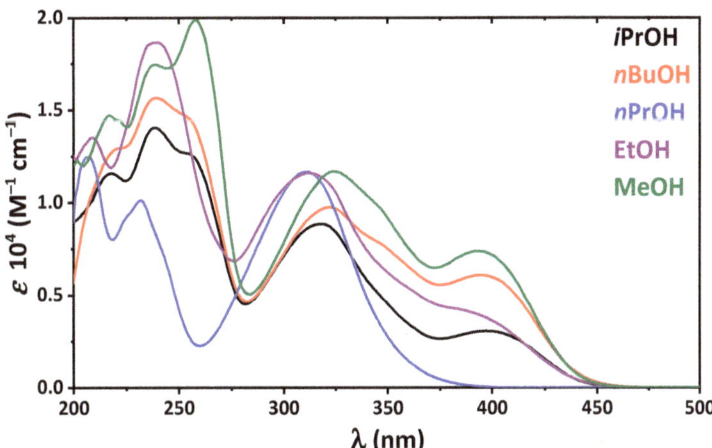

Figure 2. UV-vis spectra of **4** in the applied solvents.

Recently, we established that **2** was emissive in EtOH [17]. The resulting emission of **2** in EtOH was due to two emission bands arising from two conformers of the *cis*-keto-enamine* form. Herein we report the emission properties of **4** in different alcohols. Interestingly, **4** was found to be emissive in *i*PrOH and *n*PrOH, while almost no emission was found in EtOH and MeOH (Figure 3).

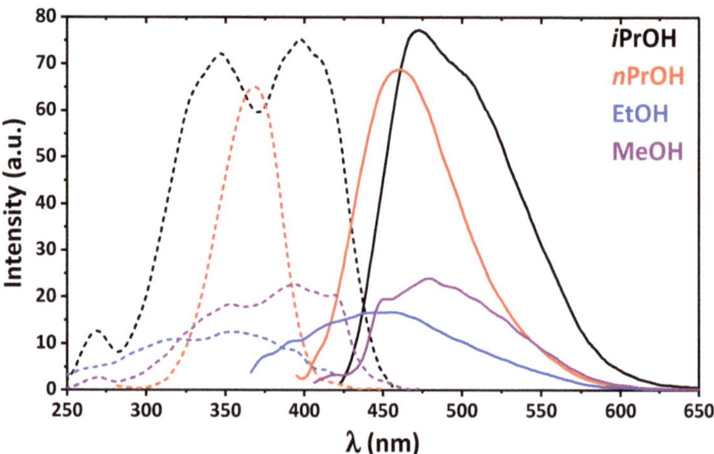

Figure 3. Emission (solid line) and excitation (dashed line) spectra of **4** in MeOH (λ_{exc} = 395 nm, λ_{em} = 480 nm), EtOH (λ_{exc} = 350 nm, λ_{em} = 440 nm), *n*PrOH (λ_{exc} = 360 nm, λ_{em} = 460 nm) and *i*PrOH (λ_{exc} = 400 nm, λ_{em} = 475 nm).

The emission spectra of **4** in *i*PrOH and *n*PrOH exhibit an intense band at 475 and 460 nm, respectively. Notably, the former band is accompanied with an intense low-energy shoulder (Figure 3). The deconvolution process of the spectrum in *i*PrOH has allowed revealing three single bands at 463, 493, and 526 nm, and a ratio of these bands is 26.4, 30.0, and 43.6%, respectively (Figure S8). The same deconvolution of the emission spectrum of **4** in *n*PrOH also revealed three bands, but with the maxima remarkably shifted to higher energies (444, 466, and 496 nm) and with a ratio of 17.2, 46.2, and 36.6%, respectively (Figure S8). Based on the absorption and excitation spectra of **4** in *i*PrOH we were able to conclude that these three emission bands corresponded to two *cis*-keto-enamine* and *trans*-keto-enamine* conformers [8–18]. Contrarily, all three bands in the emission spectrum of **4** in *n*PrOH arose exclusively from the emission of different *cis*-keto-enamine* conformers.

We next studied the solution of **4** upon the gradual addition of two different bases, namely NEt$_3$ and NaOH, to probe the influence of the nature of bases (electronic properties and steric demands). Upon addition of both bases, the absorption band at 313 nm disappeared and two new bands, centered at about 360 and 390 nm, appeared (Figure 4 and Figure S9). Notably, the ratio of intensities of the new bands was in favor of the low-energy band when NaOH was used.

The gradual addition of the methanesulfonic acid into a solution of **4** in EtOH decreased the intensity of the band of the *cis*-keto-enamine form, which was due to protonation of the imine N-atom, while the NO_2 group remained unchanged, as seen from the high-energy band, which remained constant during the titration process (Figure 5). Protonation of the imine N-atom prevented the formation of the intramolecular O–H···N bonding.

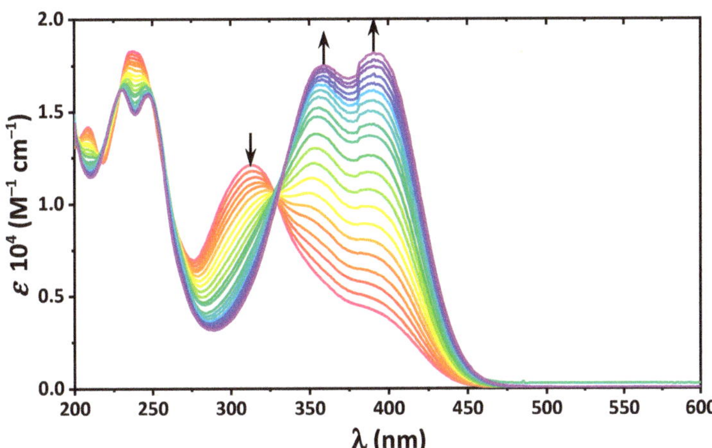

Figure 4. UV-vis spectra of **4** in EtOH upon gradual addition of NaOH up to 2 eqv., with a step of 0.1 eqv.

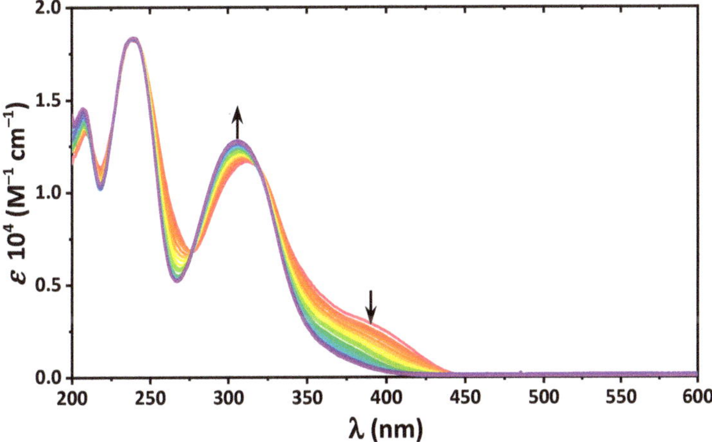

Figure 5. UV-vis spectra of **4** in EtOH upon gradual addition of CH_3SO_3H up to 2 eqv., with a step of 0.1 eqv.

Since compound **3** also tends to clearly exhibit the *cis*-keto-enamine form in EtOH (Figure S7), although less pronounced than **4**, but more pronounced that **1** and **2**, its optical properties were further studied upon gradual addition of NaOH and CH_3SO_3H. In general, the behavior of a solution of **3** in EtOH in the presence of NaOH and CH_3SO_3H was similar to a solution of **4** but much less pronounced; however, emission spectra of **3** differed significantly from those of **4**.

After the addition of NaOH, the absorption band at 330 nm disappeared with the simultaneous increase of a low-intense band in the visible region (Figure 6). However, the addition of CH_3SO_3H to a solution of **3** in EtOH vanished the same band, arising from the *cis*-keto-enamine tautomer (Figure 7).

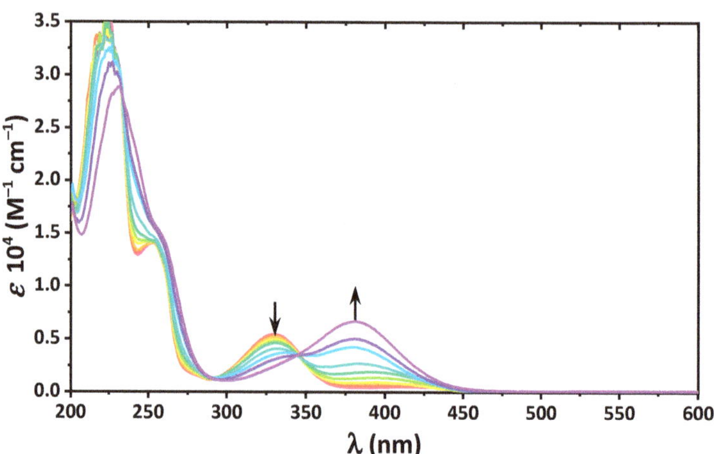

Figure 6. UV-vis spectra of **3** in EtOH upon gradual addition of NaOH up to 5 eqv., with a step of 0.5 eqv.

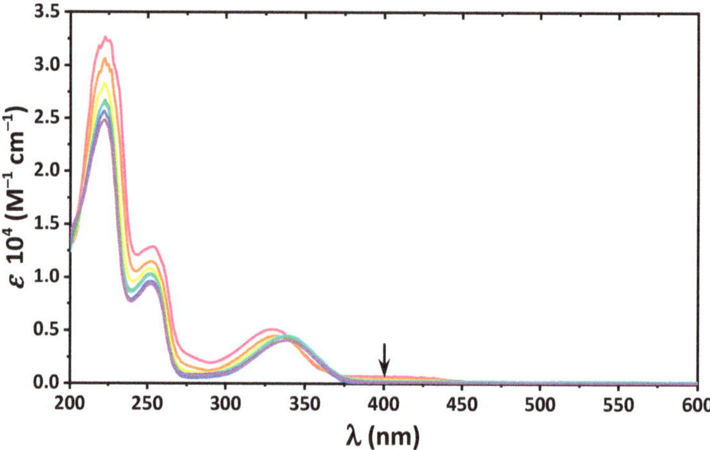

Figure 7. UV-vis spectra of **3** in EtOH upon gradual addition of CH_3SO_3H up to 3.5 eqv., with a step of 0.5 eqv.

Upon the addition of NaOH to a solution of **3** in EtOH, an emission band appeared (Figure 8), which deconvolution revealed two bands at 480 and 505 nm and with an area ratio of 30.1 and 69.9%, respectively (Figure S10). Based on the comparison of the absorption (Figure 6) and excitation (Figure 8) spectra, the emission band was assigned to two conformers of the deprotonated form of **3**. Solutions of **3** in EtOH were found to be non-emissive regardless of the added amount of CH_3SO_3H.

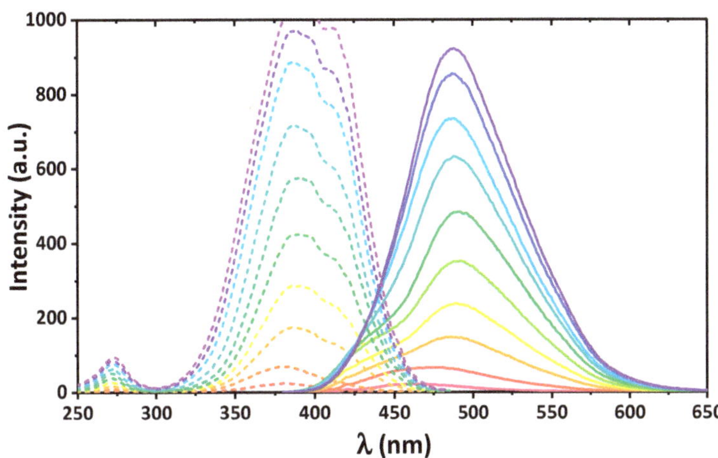

Figure 8. Emission (solid line) and excitation (dashed line) spectra of **3** in EtOH gradual addition of NaOH up to 5 eqv., with a step of 0.5 eqv. (λ_{exc} = 375 nm, λ_{em} = 480 nm).

We have further applied the density functional theory (DFT) calculations to examine the fine features of **1–4**. It was established that the calculated values of bond lengths, bond angles, and dihedral angles (Table 2) were in good agreement with the values recently obtained from single-crystal X-ray diffraction [17]. The observed differences between the calculated and experimental geometrical parameters are obviously explained by the fact that the DFT computations were performed in the gas phase.

Table 2. Selected bond lengths (Å) and angles (°) in the structures of **1–4**, obtained by using the B3LYP/6-311++G(d,p) method [1].

	1	**2**	**3**	**4**
	Bond lengths			
C=N	1.2816	1.2821	1.2800	1.2787
C–O(H)	1.3410	1.3448	1.3390	1.3298
C=O	1.2034	1.2036	1.2030	1.2031
C–O(Et)	1.3462	1.3461	1.8529	1.3440
	Bond angles			
C=N–C	119.04	118.98	119.17	119.43
	Dihedral angles [2]			
N–C–C=O	29.12	−31.15	24.53	21.83
O=C–O–C	1.54	−1.44	1.53	1.46
C–O–C–C	−179.86	179.95	−0.21	179.87
	Hydrogen bonds			
O–H	0.9927	0.9913	0.9933	0.9985
H···N	1.7472	1.7546	1.7464	1.7200
O···N	5.6362	2.6395	2.6342	2.6160
∠(O–H···N)	147.05	146.66	146.79	147.22

[1] The computational results have been compared according to the crystallographic data [17]. [2] The corresponding dihedral angles must be compared by their magnitudes.

According to the DFT calculations, the dipole moments of the fully optimized ground state geometry of the enol-imine forms were 3.279813 Debye for **1**, 4.290475 Debye for **2**, 4.070018 Debye for **3**, and 6.97922 Debye for **4**. Notably, the transformation from the enol-imine to the *trans*-keto-enamine through the *cis*-keto-enamine tautomers was followed by a remarkable increase in the dipole moments for all the structures (Table 3). Significantly higher dipole moments of different tautomers of **4** in comparison to the corresponding values of tautomers of **1–3** were obviously explained by the presence of the highly polar NO_2 substituent. The energies of the frontier molecular orbitals for the highest occupied molecular orbital (HOMO) and lowest-lying unoccupied molecular orbital (LUMO) are shown in Table 3. Both orbitals were mainly delocalized over the 2-OH(5-X)C_6H_3CH=N–CH_2C(=O) fragment for all the structures except for the LUMO of **4**, where the orbital was mainly spread over the 2-OH(5-X)C_6H_3 fragment (Figure 9).

Table 3. Total energy, dipole moment, frontier molecular HOMO and LUMO orbitals, gap value, and descriptors for **1–4** in gas phase, obtained by using the B3LYP/6-311++G(d,p) method.

	1	2	3	4
enol-imine tautomer				
Total energy (eV)	−19,255.75967	−22,372.91079	−89,285.41965	−24,822.27922
Dipole moment (Debye)	3.279813	4.290475	4.070018	6.97922
E_{HOMO} (eV)	−0.23040	−0.21087	−0.23196	−0.26088
E_{LUMO} (eV)	−0.06421	−0.06388	−0.07426	−0.09961
$\Delta E_{LUMO-HOMO} = E_{LUMO} - E_{HOMO}$ (eV)	0.16619	0.14699	0.15770	0.16127
Ionization energy, $I = -E_{HOMO}$ (eV)	0.23040	0.21087	0.23196	0.26088
Electron affinity, $A = -E_{LUMO}$ (eV)	0.06421	0.06388	0.07426	0.09961
Electronegativity, $\chi = (I + A)/2$ (eV)	0.14731	0.13738	0.15311	0.18025
Chemical potential, $\mu = -\chi$ (eV)	−0.14731	−0.13738	−0.15311	−0.18025
Global chemical hardness, $\eta = (I - A)/2$ (eV)	0.08310	0.07350	0.07885	0.08064
Global chemical softness, $S = 1/(2\eta)$ (eV^{-1})	6.01721	6.80318	6.34115	6.20078
Global electrophilicity index, $\omega = \mu^2/(2\eta)$ (eV)	0.13057	0.12839	0.14865	0.20145
Global nucleophilicity index, $E = \mu \times \eta$ (eV^2)	−0.01224	−0.01010	−0.01207	−0.01453
Maximum additional electric charge, $\Delta N_{max} = -\mu/\eta$	1.77273	1.86917	1.94179	2.23532
cis-keto-enamine tautomer				
Total energy (eV)	−19,255.55920	−22,372.72632	−89,285.24115	−24,822.17304
Dipole moment (Debye)	5.623354	6.171161	6.678792	10.089039
E_{HOMO} (eV)	−0.20510	−0.19161	−0.20980	−0.23553
E_{LUMO} (eV)	−0.07356	−0.07108	−0.08356	−0.10179
$\Delta E_{LUMO-HOMO} = E_{LUMO} - E_{HOMO}$ (eV)	0.13154	0.12053	0.12624	0.13374
Ionization energy, $I = -E_{HOMO}$ (eV)	0.20510	0.19161	0.20980	0.23553
Electron affinity, $A = -E_{LUMO}$ (eV)	0.07356	0.07108	0.08356	0.10179
Electronegativity, $\chi = (I + A)/2$ (eV)	0.13933	0.13135	0.14668	0.16866
Chemical potential, $\mu = -\chi$ (eV)	−0.13933	−0.13135	−0.14668	−0.16866
Global chemical hardness, $\eta = (I - A)/2$ (eV)	0.06577	0.06027	0.06312	0.06687
Global chemical softness, $S = 1/(2\eta)$ (eV^{-1})	7.60225	8.29669	7.92142	7.47719
Global electrophilicity index, $\omega = \mu^2/(2\eta)$ (eV)	0.14758	0.14313	0.17043	0.21270
Global nucleophilicity index, $E = \mu \times \eta$ (eV^2)	−0.00916	−0.00792	−0.00926	−0.01128
Maximum additional electric charge, $\Delta N_{max} = -\mu/\eta$	2.11844	2.17946	2.32383	2.52221
trans-keto-enamine tautomer				
Total energy (eV)	−19,255.22025	−22,372.44811	−89,284.91121	−24,821.82359
Dipole moment (Debye)	7.396322	7.781535	7.644000	10.221661
E_{HOMO} (eV)	−0.20158	−0.19035	−0.20722	−0.23080
E_{LUMO} (eV)	−0.07988	−0.07497	−0.08963	−0.11216
$\Delta E_{LUMO-HOMO} = E_{LUMO} - E_{HOMO}$ (eV)	0.12170	0.11538	0.11759	0.11864
Ionization energy, $I = -E_{HOMO}$ (eV)	0.20158	0.19035	0.20722	0.23080
Electron affinity, $A = -E_{LUMO}$ (eV)	0.07988	0.07497	0.08963	0.11216
Electronegativity, $\chi = (I + A)/2$ (eV)	0.14073	0.13266	0.14843	0.17148
Chemical potential, $\mu = -\chi$ (eV)	−0.14073	−0.13266	−0.14843	−0.17148
Global chemical hardness, $\eta = (I - A)/2$ (eV)	0.06085	0.05769	0.05880	0.05932
Global chemical softness, $S = 1/(2\eta)$ (eV^{-1})	8.21693	8.66701	8.50412	8.42886
Global electrophilicity index, $\omega = \mu^2/(2\eta)$ (eV)	0.16274	0.15253	0.18735	0.24785
Global nucleophilicity index, $E = \mu \times \eta$ (eV^2)	−0.00856	−0.00765	−0.00873	−0.01017
Maximum additional electric charge, $\Delta N_{max} = -\mu/\eta$	2.31274	2.29953	2.52445	2.89076

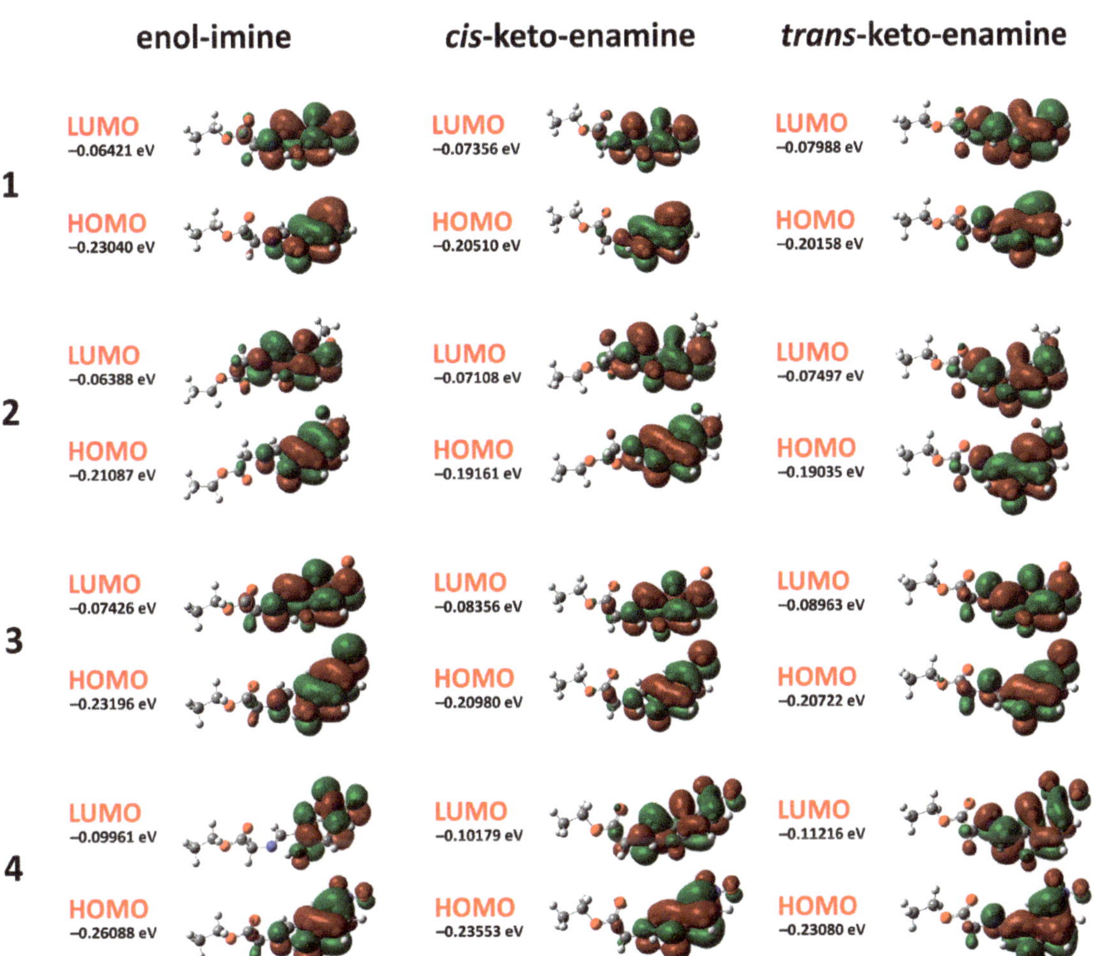

Figure 9. Energy levels and front views on the electronic isosurfaces of the high occupied and low unoccupied molecular orbitals of the ground state of **1–4**, obtained by using the B3LYP/6-311++G(d,p) method.

The so-called ionization potential (I) and the electron affinity (A) value of the molecules were established as follows: $I = -E_{HOMO}$ and $A = -E_{LUMO}$ (Table 3) [46], which determined the electron-donating ability and the ability to accept an electron, respectively. As such, as lower values of I as better donation of an electron, while as higher values of A as better ability to accept electrons. Both the I and A values for all the tautomers of **1–4** are remarkably lower than unity (Table 3), indicating that the reported dyes each exhibit high electron-donating and low electron-accepting properties. Notably, the corresponding *cis*-keto-enamine and *trans*-keto-enamine tautomers of **1–4** are slightly better electron donors and electron acceptors in comparison to their enol-imine derivatives (Table 3).

To estimate the relative reactivity of molecules of **1–4**, we have further established values of the so-called global chemical reactivity descriptors derived from the HOMO–LUMO energy gap (Table 3). The values of chemical potential (μ) for **1–4** were in the range from −0.13135 eV to −0.18025 eV for all tautomers, indicating the poor electron-accepting ability and the strong donating ability, which was further supported with low values of electronegativity, χ (Table 3). Notably, the values of electronegativity for all the *cis*-keto-enamine tautomers were lower than those for the enol-imine and *trans*-keto-enamine forms of **1–4** (Table 3). Chemical hardness (η) describes the resistance towards deformation/polarization of the electron cloud of the molecule upon a chemical reaction, while softness (S) is a reverse of chemical hardness [46]. Compounds **1–4** are characterized by low values of η and high values of S, respectively, indicating a remarkable tendency to exchange their electron clouds with the surrounding environment for all the structures (Table 3). It should be noted that the *trans*-keto-enamine tautomers are more pronounced to exchange their electron clouds with the surrounding environment in comparison to the corresponding *cis*-keto-enamine tautomers, and even much more pronounced in comparison to the enol-imine tautomers (Table 3). The electrophilicity index (ω) describes the energy of stabilization to accept electrons [46]. The ω values for all forms of **1–4** were found in the range from about 0.12 eV to 0.25 eV (Table 3). These values are low, indicating the strong nucleophilic nature of **1–4**. Finally, compounds **1–4** can accept about 1.77–2.24 electrons for the enol-imine forms, 2.12–2.52 electrons for the *cis*-keto-enamine forms, and 2.30–2.89 electrons for the *trans*-keto-enamine forms, respectively, as evidenced from the ΔN_{max} values, of which the highest values correspond to **4** (Table 3).

The electrophilic and nucleophilic sites in **1–4** were examined using the molecular electrostatic potential (MEP) analysis. The red and blue colors of the MEP surface correspond to electron-rich (nucleophilic) and electron-deficient (electrophilic) regions, respectively. On the MEP surfaces of the enol-imine tautomers of **1–4** the most pronounced nucleophilic centers are located on the carbonyl and hydroxyl oxygen atoms, while the other negative electrostatic potential sites in the enol-imine form of **4** are located on the oxygen atoms of the NO_2 group (Figure 10). In the *cis*-keto-enamine and *trans*-keto-enamine tautomers of **1–4** the carbonyl oxygen atom attached to the aromatic ring is the most remarkable nucleophilic center, alongside with both oxygen atoms of the NO_2 group in **4**, while the carbonyl oxygen atom of the carboxyl group becomes a less pronounced nucleophilic center (Figure 10). As the most electrophilic region the $CH=N–CH_2$ fragment for the enol-imine tautomers and the $CH–NH–CH_2$ fragment for the *cis*-keto-enamine and *trans*-keto-enamine tautomers can be highlighted for the structures of **1–4** (Figure 10).

The calculated absorption spectra of the fully optimized ground state geometry of all the three tautomers of **1–4** (Figures S11–S13) are in agreement with experimental spectra. In particular, the experimental UV-vis spectra for the enol-imine tautomers exhibited bands at 215–260 and 331–353 nm, and the calculated spectra for the same tautomers contained bands at 225–250 and 302–351 nm (Table S1). The *cis*-keto-enamine tautomers are shown as a band centered at 382–418 nm in both the experimental and calculated absorption spectra (Table S1). The calculated UV-vis spectra for the *trans*-keto-enamine tautomers of **1–4** each exhibited a low-energy band at 407–428 nm, while no similar bands were observed in the corresponding experimental spectra (Table S1), thus, testifying to the absence of the *trans*-keto-enamine tautomers of **1–4** in the applied solvents.

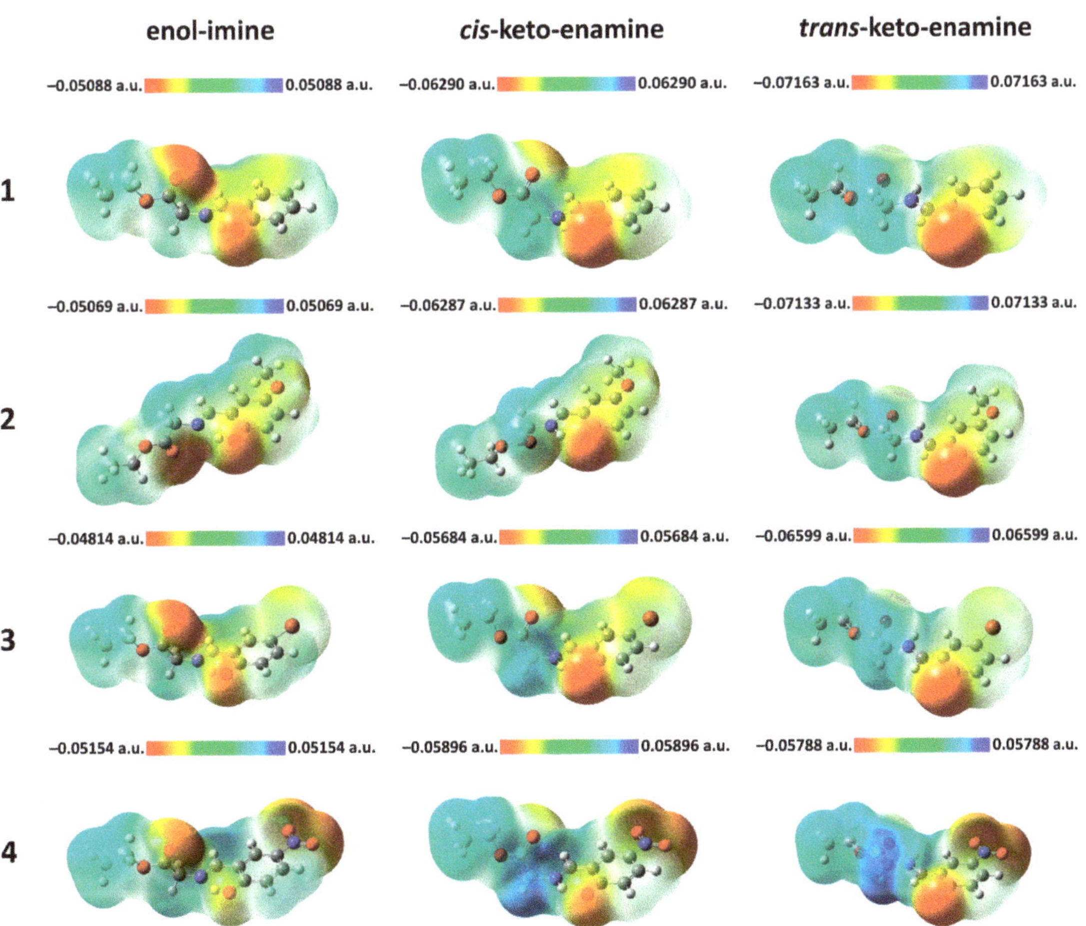

Figure 10. Views of the molecular electrostatic potential surface of **1–4**, obtained by using the B3LYP/6-311++G(d,p) method.

The calculated UV-vis spectra of the enol-imine forms of **1–4** exhibit absorption bands with three (for **1**) or two (for **2–4**) maxima exclusively in the UV region centered at about 220–255 nm and 300–350 nm (Figure S11). These bands mostly corresponded to the HOMO–1 → LUMO, HOMO → LUMO, and HOMO → LUMO+3 transitions (Table 4, Figure S11). The absorption bands in the spectra of **3** and **4** were additionally supported by the HOMO → LUMO+1 transition and further described by the HOMO–4 → LUMO transition in the spectrum of **3**, and by the HOMO–3 → LUMO+1 and HOMO–1 → LUMO+1 transitions in the spectrum of **4**, respectively (Table 4, Figure S11). As for the calculated UV-vis spectra of the *cis*-keto-enamine and *trans*-keto-enamine tautomers of **1–4**, absorption bands are observed in both the UV and visible regions up to about 500–550 nm (Figure S12) and 600 nm (Figure S13), respectively. These bands are characterized by two maxima centered at about 250–290 nm and 380–430 nm (Figures S12 and S13). Notably, the calculated UV-vis spectrum of the *cis*-keto-enamine form of **4** contained an additional maxima at 320 nm (Figure S12). The corresponding transitions, responsible for the observed bands in the calculated UV-vis spectra of the *cis*-keto-enamine and *trans*-keto-enamine tautomers of **1–4**, are shown in Figures S14–S17 and collected in Tables S2 and S3.

Table 4. Values for the calculated UV-vis spectra of the ground state of the enol-imine tautomers of **1–4**, obtained by using the TD-DFT/B3LYP/6-311++G(d,p) method.

	1			2	
λ_{max} (nm)	Oscillator strength	Transitions	λ_{max} (nm)	Oscillator strength	Transitions
220.6	0.1419	HOMO–3 → LUMO (14.0%)	239.1	0.2994	HOMO–3 → LUMO (9.0%)
		HOMO–1 → LUMO (8.1%)			HOMO–1 → LUMO (26.8%)
		HOMO–1 → LUMO+3 (2.5%)			HOMO → LUMO+3 (50.2%)
		HOMO–1 → LUMO+9 (2.0%)			HOMO → LUMO+10 (3.4%)
		HOMO → LUMO+1 (2.7%)			HOMO → LUMO+13 (2.3%)
		HOMO → LUMO+2 (10.7%)			
		HOMO → LUMO+3 (44.7%)			
		HOMO → LUMO+9 (7.2%)			
222.6	0.0311	HOMO → LUMO+1 (5.6%)	244.3	0.0188	HOMO–1 → LUMO (2.8%)
		HOMO → LUMO+2 (45.0%)			HOMO → LUMO+1 (3.7%)
		HOMO → LUMO+3 (24.2%)			HOMO → LUMO+2 (85.9%)
		HOMO → LUMO+4 (8.6%)			HOMO → LUMO+5 (3.9%)
		HOMO → LUMO+5 (7.5%)			
228.5	0.0315	HOMO–3 → LUMO (70.8%)	251.2	0.0720	HOMO–1 → LUMO (21.3%)
		HOMO–3 → LUMO+1 (4.0%)			HOMO → LUMO+1 (7.1%)
		HOMO–3 → LUMO+3 (2.4%)			HOMO → LUMO+3 (45.5%)
		HOMO → LUMO+2 (12.3%)			HOMO → LUMO+4 (23.2%)
		HOMO → LUMO+3 (3.5%)			
252.8	0.2896	HOMO–1 → LUMO (76.5%)	256.6	0.0742	HOMO–1 → LUMO (33.4%)
		HOMO → LUMO (3.2%)			HOMO → LUMO+3 (49.0%)
		HOMO → LUMO+1 (4.7%)			HOMO → LUMO+4 (10.9%)
		HOMO → LUMO+2 (9.8%)			
312.1	0.1055	HOMO–1 → LUMO (2.3%)	350.5	0.1086	HOMO → LUMO (96.9%)
		HOMO → LUMO (93.9%)			

	3			4	
λ_{max} (nm)	Oscillator strength	Transitions	λ_{max} (nm)	Oscillator strength	Transitions
224.7	0.0487	HOMO–5 → LUMO (10.6%)	240.4	0.0556	HOMO–3 → LUMO (85.4%)
		HOMO → LUMO+4 (57.1%)			HOMO–1 → LUMO+1 (9.6%)
		HOMO → LUMO+5 (23.2%)			
228.4	0.2498	HOMO–5 → LUMO (3.1%)	244.8	0.2633	HOMO–3 → LUMO (8.1%)
		HOMO–4 → LUMO (32.2%)			HOMO–3 → LUMO+1 (29.7%)
		HOMO–1 → LUMO (16.9%)			HOMO–1 → LUMO+1 (55.7%)
		HOMO → LUMO+1 (17.2%)			
		HOMO → LUMO+2 (6.0%)			
		HOMO → LUMO+3 (7.5%)			

Table 4. *Cont.*

237.8	0.2157	HOMO → LUMO+5 (2.3%) HOMO → LUMO+8 (3.9%) HOMO–4 → LUMO (28.8%) HOMO–3 → LUMO (2.9%) HOMO–1 → LUMO (16.3%) HOMO → LUMO+1 (32.7%) HOMO → LUMO+2 (2.6%) HOMO → LUMO+3 (9.3%) HOMO → LUMO+4 (2.1%)	273.4	0.1054	HOMO–1 → LUMO (77.0%) HOMO → LUMO+1 (18.3%)
252.7	0.1308	HOMO–1 → LUMO (57.9%) HOMO → LUMO+1 (35.1%)	299.4	0.0744	HOMO–1 → LUMO (13.6%) HOMO → LUMO (34.4%) HOMO → LUMO+1 (48.9%)
329.3	0.0858	HOMO → LUMO (95.7%)	308.7	0.2162	HOMO–1 → LUMO (5.1%) HOMO–1 → LUMO+1 (3.1%) HOMO → LUMO (61.8%) HOMO → LUMO+1 (28.9%)

3. Materials and Methods

3.1. Materials

All reagents and solvents were commercially available and used without further purification. Dyes **1–4** were obtained according to the described synthetic procedure [18].

3.2. Physical Measurements

Absorption and fluorescent spectra from the freshly prepared solutions (10^{-4} M) in freshly distilled solvents were recorded on an Agilent 8453 instrument.

3.3. Computational Details

The ground state geometry of **1–4** was fully optimized without symmetry restrictions. The calculations were performed by means of the GaussView 6.0 molecular visualization program [47] and Gaussian 09, Revision D.01 program package [48] using the density functional theory (DFT) method with Becke three-parameter Lee-Yang-Parr (B3LYP) hybrid functional [49,50] and 6-311++G(d,p) [49,51] basis set. The crystal structure geometry was used as a starting model for structural optimization. The vibration frequencies were calculated for the optimized structures in the gas phase and no imaginary frequencies were obtained. The Cartesian atomic coordinates for the optimized structures are gathered in Tables S4–S7. The electronic isosurfaces of the HOMO and LUMO orbitals and MEP surfaces of all tautomers of **1–4** were generated from the fully optimized ground state geometry obtained by using the B3LYP/6-311++G(d,p) method. The absorption spectra of the fully optimized ground state geometry of all the three tautomers of **1–4** were simulated at the TD-DFT/B3LYP/6-311++G(d,p) level.

4. Conclusions

In summary, herein we discuss structural studies using the Hirshfeld molecular surface approach, as well as photophysical properties of four dyes: ethyl *N*-salicylideneglycinate (**1**), ethyl *N*-(5-methoxysalicylidene)glycinate (**2**), ethyl *N*-(5-bromosalicylidene)glycinate (**3**), and ethyl *N*-(5-nitrosalicylidene)glycinate (**4**).

The intermolecular H···H, H···C, and H···O contacts are the dominant contributors to the molecular surface of all the reported compounds. 3 is further described by a remarkable proportion of the intermolecular H···Br contacts. Furthermore, a distinct proportion of the C···C contacts was found in the molecular surface of 3 and 4, which is explained by intermolecular π···π stacking interactions between the phenylene rings.

The absorption spectra of **4** in THF and CH$_3$CN exhibit a band for the *cis*-keto-enamine form, while only the enol-imine tautomer was found in the absorption spectrum of **4** in cyclohexane and in the absorption spectra of **1–3** in cyclohexane, THF, and CH$_3$CN. Furthermore, in the absorption spectra of **4** in MeOH, EtOH, *i*PrOH, and *n*BuOH the *cis*-keto-enamine form is clearly observed, with the most remarkable one in MeOH, while the same band is absent in the spectrum of **4** in *n*PrOH, that might be explained by possible specific solute-solvent interactions. Dye **4** is emissive in *i*PrOH and *n*PrOH, arising from two conformers of the *cis*-keto-enamine* form and from the *trans*-keto-enamine* form, while the origin of emission in *n*PrOH is exclusively from different *cis*-keto-enamine* conformers. Titration of the solutions of **3** and **4** in EtOH by NaOH leads to the gradual increasing of the band in the visible region of the corresponding UV-vis spectra. The same band gradually decreases and finally vanishes upon titration of the solutions of **3** and **4** in EtOH by CH$_3$SO$_3$H due to the protonation of the imine N-atom. Notably, upon addition of NaOH to the solution of **3** in EtOH, an emission band appeared and increased with increasing NaOH concentration.

We have also demonstrated that photophysical properties of a reported series of closely related compounds **1–4** can efficiently be tuned not only by changing the corresponding substituent in the phenolic ring, thus changing electronic properties, but also by the nature of a solvent (non-polar vs. polar aprotic vs. polar protic) and pH (NEt$_3$ or NaOH, and CH$_3$SO$_3$H). All these factors allow to fine influence to the keto-enamine–enol-imine tautomerization both in the ground and excited states, yielding two or even three emission bands with a certain ratio and, as a result, different resulting emission color. No doubt, all these findings are of potential interest for molecular optics. Furthermore, the presence of the ethyl glycinate fragment in **1–4** might play a pivotal role for the application of these compounds, e.g., as luminescent sensors, in biological systems.

According to the DFT calculation results, it was established that all the tautomers of **1–4** each exhibit high electron-donating and low electron-accepting properties. The most pronounced nucleophilic centers of the enol-imine tautomers are located on the carbonyl and hydroxyl oxygen atoms, while the other negative electrostatic potential sites in the enol-imine form of **4** are located on the oxygen atoms of the NO$_2$ group. In the *cis*-keto-enamine and *trans*-keto-enamine tautomers of **1–4**, the carbonyl oxygen atom attached to the aromatic ring is the most remarkable nucleophilic center, alongside both oxygen atoms of the NO$_2$ group in **4**, while the carbonyl oxygen atom of the carboxyl group becomes a less pronounced nucleophilic center. As the most electrophilic region, the CH=N–CH$_2$ fragment for the enol-imine tautomers and the CH–NH–CH$_2$ fragment for the *cis*-keto-enamine and *trans*-keto-enamine tautomers are highlighted for the structures of **1–4**. The calculated absorption spectra of the fully optimized ground state geometry of all the three tautomers of **1–4** are in good agreement with experimental spectra.

Supplementary Materials: The following are available online, Figures S1–S4: 2D and decomposed 2D fingerprint plots of observed contacts for the crystal structure of **1–4**, Figures S5–S7: UV-vis spectra of **1–3** in the applied solvents, Figure S8: Emission and excitation spectra of **4** in *n*PrOH and *i*PrOH, Figure S9: UV-vis spectra of **4** in EtOH upon gradual addition of NEt$_3$, Figure S10: Emission and excitation spectra of **3** in EtOH after addition of 5 eqv. of NaOH, Figure S11: The calculated UV-vis spectra of the ground states of the enol-imine tautomers of **1–4**, Figure S12: The calculated UV-vis spectra of the ground states of the *cis*-keto-enamine tautomers of **1–4**, Figure S13: The calculated UV-vis spectra of the ground states of the *trans*-keto-enamine tautomers of **1–3**, Figures S14–S17: Energy levels and views on the electronic isosurfaces of the selected molecular orbitals of the ground state of **1–4**, Table S1: Values for the main maxima in the experimental UV-vis spectra of **1–4** in different solvents, and in the calculated UV-vis spectra for different tautomers of **1–4**, Table S2: Values for the calculated UV-vis spectra of the ground state for the *cis*-keto-enamine tautomers of **1–4**, Table S3: Values for the calculated UV-vis spectra of the ground state for the *trans*-keto-enamine tautomers of **1–4**, Tables S4–S7: Cartesian atomic coordinates for optimized structures of **1–4**, obtained by using the DFT/B3LYP/6–311++G(d,p) method.

Author Contributions: Conceptualization, M.G.B. and D.A.S.; methodology, D.A.S.; formal analysis, L.E.A., M.G.B., and D.A.S.; investigation, L.E.A., M.G.B., and D.A.S.; data curation, L.E.A., M.G.B., and D.A.S.; writing—original draft preparation, L.E.A. and M.G.B.; writing—review and editing, L.E.A. and D.A.S.; supervision, D.A.S.; project administration, D.A.S. All authors have read and agreed to the published version of the manuscript.

Funding: This research received no external funding.

Institutional Review Board Statement: Not applicable.

Informed Consent Statement: Not applicable.

Data Availability Statement: Not available.

Conflicts of Interest: The authors declare no conflict of interest.

Sample Availability: Samples of the compounds ethyl *N*-salicylideneglycinate (**1**), ethyl *N*-(5-methoxysalicylidene)glycinate (**2**), ethyl *N*-(5-bromosalicylidene)glycinate (**3**), and ethyl *N*-(5-nitrosalicylidene)glycinate (**4**) are available from the authors.

References

1. Sacconi, L. Tetrahedral complexes of nickel (II) and copper (II) with schiff bases. *Coord. Chem. Rev.* **1996**, *1*, 126–132. [CrossRef]
2. Yamada, S. Recent aspects of the stereochemistry of schiff-base-metal complexes. *Coord. Chem. Rev.* **1966**, *1*, 415–437. [CrossRef]
3. Nath, M.; Goyal, S. Organotin (Iv) complexes of schiff bases: A review. *Main Group Met. Chem.* **1996**, *19*, 75–102. [CrossRef]
4. Aggarwal, N.; Kumar, R.; Dureja, P.; Diwan, S.; Rawat, D.S. Schiff bases as potential fungicides and nitrification inhibitors. *J. Agric. Food Chem.* **2009**, *57*, 8520–8525. [CrossRef] [PubMed]
5. Weng, Q.; Yi, J.; Chen, X.; Luo, D.; Wang, Y.; Sun, W.; Kang, J.; Han, Z. Controllable Synthesis and Biological Application of Schiff Bases from d-Glucosamine and Terephthalaldehyde. *ACS Omega* **2020**, *5*, 24864–24870. [CrossRef] [PubMed]
6. Ferraz de Paiva, R.E.; Marçal Neto, A.; Santos, I.A.; Jardim, A.C.G.; Corbi, P.P.; Bergamini, F.R.G. What is holding back the development of antiviral metallodrugs? A literature overview and implications for SARS-CoV-2 therapeutics and future viral outbreaks. *Dalton Trans.* **2020**, *49*, 16004–16033. [CrossRef]
7. Wang, P.H.; Keck, G.J.; Lien, E.J.; Lai, M.M.C. Design, Synthesis, Testing, and Quantitative Structure-Activity Relationship Analysis of Substituted Salicylaldehyde Schiff Bases of 1-Amino-3-Hydroxyguani-dine Tosylate as New Antiviral Agents against Coronavirus. *J. Med. Chem.* **1990**, *33*, 608–614. [CrossRef]
8. Safin, D.A.; Robeyns, K.; Garcia, Y. Crown ether-containing *N*-salicylidene aniline derivatives: Synthesis, characterization and optical properties. *CrystEngComm* **2012**, *14*, 5523–5529. [CrossRef]
9. Safin, D.A.; Robeyns, K.; Garcia, Y. Solid-state thermo- and photochromism in *N,N'*-bis(5-X-salicylidene)diamines (X = H, Br). *RSC Adv.* **2012**, *2*, 11379–11388. [CrossRef]
10. Safin, D.A.; Garcia, Y. First evidence of thermo- and two-step photochromism of tris-anils. *RSC Adv.* **2013**, *3*, 6466–6471. [CrossRef]
11. Safin, D.A.; Bolte, M.; Garcia, Y. Photoreversible solid state negative photochromism of *N*-(3,5-dichlorosalicylidene)-1-aminopyrene. *CrystEngComm* **2014**, *16*, 5524–5526. [CrossRef]
12. Safin, D.A.; Babashkina, M.G.; Robeyns, K.; Bolte, M.; Garcia, Y. *N*-Salicylidene aniline derivatives based on the *N'*-thiophosphorylated thiourea scaffold. *CrystEngComm* **2014**, *16*, 7053–7061. [CrossRef]
13. Safin, D.A.; Bolte, M.; Garcia, Y. Solid-state photochromism and thermochromism of *N*-salicylidene pyrene derivatives. *CrystEngComm* **2014**, *16*, 8786–8793. [CrossRef]
14. Safin, D.A.; Babashkina, M.G.; Robeyns, K.; Garcia, Y. C–H···Br–C vs. C–Br···Br–C vs. C–Br···N bonding in molecular self-assembly of pyridine-containing dyes. *RSC Adv.* **2016**, *6*, 53669–53678. [CrossRef]
15. Safin, D.A.; Robeyns, K.; Babashkina, M.G.; Filinchuk, Y.; Rotaru, A.; Jureschi, C.; Mitoraj, M.P.; Hooper, J.; Brela, M.; Garcia, Y. Polymorphism driven optical properties of an anil dye. *CrystEngComm* **2016**, *18*, 7249–7259. [CrossRef]
16. Safin, D.A.; Robeyns, K.; Garcia, Y. 1,2,4-Triazole-based molecular switches: Crystal structures, Hirshfeld surface analysis and optical properties. *CrystEngComm* **2016**, *18*, 7284–7294. [CrossRef]
17. Shapenova, D.S.; Shiryaev, A.A.; Bolte, M.; Kukułka, M.; Szczepanik, D.W.; Hooper, J.; Babashkina, M.G.; Mahmoudi, G.; Mitoraj, M.P.; Safin, D.A. Resonance Assisted Hydrogen Bonding Phenomenon Unveiled from Both Experiment and Theory–an Example of New Family of Ethyl *N*-salicylideneglycinate Dyes. *Chem. Eur. J.* **2020**, *26*, 12987–12995. [CrossRef]
18. Shiryaev, A.A.; Burkhanova, T.M.; Mahmoudi, G.; Babashkina, M.G.; Safin, D.A. Photophysical properties of ethyl *N*-(5-bromosalicylidene)glycinate and ethyl *N*-(5-nitrosalicylidene)glycinate in CH$_2$Cl$_2$. *J. Lumin.* **2020**, *226*, 117454. [CrossRef]

19. Hadjoudis, E.; Vittorakis, M.; Moustakali-Mavridis, I. Photochromism and thermochromism of schiff bases in the solid state and in rigid glasses. *Tetrahedron* **1987**, *43*, 1345–1360. [CrossRef]

20. Hadjoudis, E.; Mavridis, I.M. Photochromism and thermochromism of Schiff bases in the solid state: Structural aspects. *Chem. Soc. Rev.* **2004**, *33*, 579–588. [CrossRef]

21. Amimoto, K.; Kawato, T. Photochromism of organic compounds in the crystal state. *J. Photochem. Photobiol. C* **2005**, *6*, 207–226 [CrossRef]

22. Haneda, T.; Kawano, M.; Kojima, T.; Fujita, M. Thermo-to-photo-switching of the chromic behavior of salicylideneanilines by inclusion in a porous coordination network. *Angew. Chem. Int. Ed.* **2007**, *46*, 6643–6645. [CrossRef]

23. Filarowski, A.; Koll, A.; Sobczyk, L. Intramolecular Hydrogen Bonding in o-hydroxy Aryl Schiff Bases. *Curr. Org. Chem.* **2009**, *13*, 172–193. [CrossRef]

24. Bertolasi, V.; Gilli, P.; Gilli, G. Crystal Chemistry and Prototropic Tautomerism in 2-(1-Iminoalkyl)- phenols (or naphthols) and 2-Diazenyl-phenols (or naphthols). *Curr. Org. Chem.* **2009**, *13*, 250–268. [CrossRef]

25. Hadjoudis, E.; Chatziefthimiou, S.D.; Mavridis, I.M. Anils: Photochromism by H-transfer. *Curr. Org. Chem.* **2009**, *13*, 269–286. [CrossRef]

26. Minkin, V.I.; Tsukanov, A.V.; Dubonosov, A.D.; Bren, V.A. Tautomeric Schiff bases: Iono-, solvato-, thermo-and photochromism. *J. Mol. Struct.* **2011**, *998*, 179–191. [CrossRef]

27. Inokuma, Y.; Kawano, M.; Fujita, M. Crystalline molecular flasks. *Nat. Chem.* **2011**, *3*, 349–358. [CrossRef]

28. Mahmudov, K.T.; Pombeiro, A.J.L. Resonance-Assisted Hydrogen Bonding as a Driving Force in Synthesis and a Synthon in the Design of Materials. *Chem. Eur. J.* **2016**, *22*, 16356–16398. [CrossRef] [PubMed]

29. Jana, S.; Dalapati, S.; Guchhait, N. Proton Transfer Assisted Charge Transfer Phenomena in Photochromic Schiff Bases and Effect of–NEt2 Groups to the Anil Schiff Bases. *J. Phys. Chem. A* **2012**, *116*, 10948–10958. [CrossRef]

30. Germino, J.C.; Barboza, C.A.; Quites, F.J.; Vazquez, P.A.M.; Atvars, T.D.Z. Dual Emissions of Salicylidene-5-chloroaminepyridine Due to Excited State Intramolecular Proton Transfer: Dynamic Photophysical and Theoretical Studies. *J. Phys. Chem. C* **2015**, *119*, 27666–27675. [CrossRef]

31. Zhao, L.; Sui, D.; Chai, J.; Wang, Y.; Jiang, S. Digital Logic Circuit Based on a Single Molecular System of Salicylidene Schiff Base. *J. Phys. Chem. B* **2006**, *110*, 24299–24304. [CrossRef] [PubMed]

32. Kleij, A.W. New Templating Strategies with Salen Scaffolds (Salen = *N,N′*-Bis(salicylidene)ethylenediamine Dianion). *Chem. Eur. J.* **2008**, *14*, 10520–10529. [CrossRef] [PubMed]

33. Cheng, J.; Ma, X.; Zhang, Y.; Liu, J.; Zhou, X.; Xiang, H. Optical Chemosensors Based on Transmetalation of Salen-Based Schiff Base Complexes. *Inorg. Chem.* **2014**, *53*, 3210–3219. [CrossRef] [PubMed]

34. Martín, C.; Fiorani, G.; Kleij, A.W. Recent Advances in the Catalytic Preparation of Cyclic Organic Carbonates. *ACS Catal.* **2015**, *5*, 1353–1370. [CrossRef]

35. Fiorani, G.; Guo, W.; Kleij, A.W. Sustainable conversion of carbon dioxide: The advent of organocatalysis. *Green Chem.* **2015**, *17*, 1375–1389. [CrossRef]

36. Beens, H.; Grellmann, K.H.; Gurr, M.; Weller, A.H. Effect of solvent and temperature on proton transfer reactions of excited molecules. *Discuss. Faraday Soc.* **1965**, *39*, 183–193. [CrossRef]

37. Zhao, J.; Ji, S.; Chen, Y.; Guo, H.; Yang, P. Excited state intramolecular proton transfer (ESIPT): From principal photophysics to the development of new chromophores and applications in fluorescent molecular probes and luminescent materials. *Phys. Chem. Chem. Phys.* **2012**, *14*, 8803–8817. [CrossRef]

38. Padalkar, V.S.; Seki, S. Excited-state intramolecular proton-transfer (ESIPT)-inspired solid state emitters. *Chem. Soc. Rev.* **2016**, *45*, 169–202. [CrossRef]

39. Spackman, M.A.; Jayatilaka, D. Hirshfeld surface analysis. *CrystEngComm* **2009**, *11*, 19–32. [CrossRef]

40. Spackman, M.A.; McKinnon, J.J. Fingerprinting intermolecular interactions in molecular crystals. *CrystEngComm* **2002**, *4*, 378–392. [CrossRef]

41. Wolff, S.K.; Grimwood, D.J.; McKinnon, J.J.; Turner, M.J.; Jayatilaka, D.; Spackman, M.A. *CrystalExplorer 3.1*; University of Western Australia: Perth, Australia, 2012.

42. Jelsch, C.; Ejsmont, K.; Huder, L. The enrichment ratio of atomic contacts in crystals, an indicator derived from the Hirshfeld surface analysis. *IUCrJ* **2014**, *1*, 119–128. [CrossRef] [PubMed]

43. Alarcón, S.H.; Olivieri, A.C.; Sanz, D.; Claramunt, R.M.; Elguero, J. Substituent and solvent effects on the proton transfer equilibrium in anils and azo derivatives of naphthol. Multinuclear NMR study and theoretical calculations. *J. Mol. Struct.* **2004**, *705*, 1–9. [CrossRef]

44. Nagy, P.I.; Fabian, W.M.F. Theoretical Study of the Enol Imine ↔ Enaminone Tautomeric Equilibrium in Organic Solvents. *J. Phys. Chem. B* **2006**, *110*, 25026–25032. [CrossRef] [PubMed]

45. Krygowski, T.M.; Woźniak, K.; Anulewicz, R.; Pawlak, D.; Kolodziejski, W.; Grech, E.; Szady, A. Through-Resonance Assisted Ionic Hydrogen Bonding in 5-Nitro-N-salicylideneethylamine. *J. Phys. Chem. A* **1997**, *101*, 9399–9404. [CrossRef]

46. Geerlings, P.; De Proft, F.; Langenaeker, W. Conceptual Density Functional Theory. *Chem. Rev.* **2003**, *103*, 1793–1873. [CrossRef]

47. Dennington, R.; Keith, T.A.; Millam, J.M. *GaussView, Version 6.0*; Semichem Inc., Shawnee Mission: Shawnee, Kansas, 2016.

48. Frisch, M.J.; Trucks, G.W.; Schlegel, H.B.; Scuseria, G.E.; Robb, M.A.; Cheeseman, J.R.; Scalmani, G.; Barone, V.; Mennucci, B.; Petersson, G.A.; et al. *Gaussian 09, Revision D.01*; Gaussian, Inc.: Pittsburgh, PA, USA, 2016.

49. Krishnan, R.; Binkley, J.S.; Seeger, R.; Pople, J.A. Self-consistent molecular orbital methods. XX. A basis set for correlated wave functions. *J. Chem. Phys.* **1980**, *72*, 650–654. [CrossRef]
50. Becke, A.D. Density-functional thermochemistry. III. The role of exact exchange. *J. Chem. Phys.* **1993**, *98*, 5648–5652. [CrossRef]
51. Frisch, M.J.; Pople, J.A.; Binkley, J.S. Self-consistent molecular orbital methods 25. Supplementary functions for Gaussian basis sets. *J. Chem. Phys.* **1984**, *8*, 3265–3269. [CrossRef]

 *molecules*

Article

Naphthazarin Derivatives in the Light of Intra- and Intermolecular Forces

Karol Kułacz [†], Michał Pocheć [†], Aneta Jezierska * and Jarosław J. Panek *

Faculty of Chemistry, University of Wrocław, ul. F. Joliot-Curie 14, 50-383 Wrocław, Poland;
karol.kulacz@chem.uni.wroc.pl (K.K.); michal.pochec@chem.uni.wroc.pl (M.P.)
* Correspondence: aneta.jezierska@chem.uni.wroc.pl (A.J.); jaroslaw.panek@chem.uni.wroc.pl (J.J.P.);
  Tel.: +48-71-3757-224 (A.J. & J.J.P.); Fax: +48-71-3282-348 (A.J. & J.J.P.)
† These authors contributed equally to this work.

Abstract: Our long-term investigations have been devoted the characterization of intramolecular hydrogen bonds in cyclic compounds. Our previous work covers naphthazarin, the parent compound of two systems discussed in the current work: 2,3-dimethylnaphthazarin (**1**) and 2,3-dimethoxy-6-methylnaphthazarin (**2**). Intramolecular hydrogen bonds and substituent effects in these compounds were analyzed on the basis of Density Functional Theory (DFT), Møller–Plesset second-order perturbation theory (MP2), Coupled Clusters with Singles and Doubles (CCSD) and Car-Parrinello Molecular Dynamics (CPMD). The simulations were carried out in the gas and crystalline phases. The nuclear quantum effects were incorporated a posteriori using the snapshots taken from ab initio trajectories. Further, they were used to solve a vibrational Schrödinger equation. The proton reaction path was studied using B3LYP, ωB97XD and PBE functionals with a 6-311++G(2d,2p) basis set. Two energy minima (deep and shallow) were found, indicating that the proton transfer phenomena could occur in the electronic ground state. Next, the electronic structure and topology were examined in the molecular and proton transferred (PT) forms. The Atoms In Molecules (AIM) theory was employed for this purpose. It was found that the hydrogen bond is stronger in the proton transferred (PT) forms. In order to estimate the dimers' stabilization and forces responsible for it, the Symmetry-Adapted Perturbation Theory (SAPT) was applied. The energy decomposition revealed that dispersion is the primary factor stabilizing the dimeric forms and crystal structure of both compounds. The CPMD results showed that the proton transfer phenomena occurred in both studied compounds, as well as in both phases. In the case of compound **2**, the proton transfer events are more frequent in the solid state, indicating an influence of the environmental effects on the bridged proton dynamics. Finally, the vibrational signatures were computed for both compounds using the CPMD trajectories. The Fourier transformation of the autocorrelation function of atomic velocity was applied to obtain the power spectra. The IR spectra show very broad absorption regions between 700 cm^{-1}–1700 cm^{-1} and 2300 cm^{-1}–3400 cm^{-1} in the gas phase and 600 cm^{-1}–1800 cm^{-1} and 2200 cm^{-1}–3400 cm^{-1} in the solid state for compound **1**. The absorption regions for compound **2** were found as follows: 700 cm^{-1}–1700 cm^{-1} and 2300 cm^{-1}–3300 cm^{-1} for the gas phase and one broad absorption region in the solid state between 700 cm^{-1} and 3100 cm^{-1}. The obtained spectroscopic features confirmed a strong mobility of the bridged protons. The inclusion of nuclear quantum effects showed a stronger delocalization of the bridged protons.

Keywords: intramolecular hydrogen bonds; gas phase; crystalline phase; DFT; MP2; CCSD; AIM; SAPT; nuclear quantum effects; CPMD

Citation: Kułacz, K.; Pocheć, M.; Jezierska, A.; Panek, J.J. Naphthazarin Derivatives in the Light of Intra- and Intermolecular Forces. *Molecules* **2021**, *26*, 5642. https://doi.org/10.3390/molecules26185642

Academic Editors: Mirosław Jabłoński and Benedito José Costa Cabral

Received: 12 July 2021
Accepted: 14 September 2021
Published: 17 September 2021

Publisher's Note: MDPI stays neutral with regard to jurisdictional claims in published maps and institutional affiliations.

1. Introduction

The nature of hydrogen bonds is complex and still presents open questions. In addition to conventional hydrogen bonds, during recent decades so called unconventional hydrogen bonds have appeared as important scientific topics [1–8]. Hydrogen bonds are located on the interaction strength ladder in the middle position, between covalent, ionic, and van der Waals interactions [9]. They are much weaker than chemical covalent bonds, but their presence is of great significance in nature. The strength of most hydrogen bonds is between 10 kJ/mol and 40 kJ/mol [10]; however, they are ubiquitous and cannot be neglected in the discussion of factors decisive for the structure of bulk materials: liquids and solid states. They are key elements of many processes at the molecular level, as well as influencing molecular and macroscopic properties of various systems [11–15]. They were found to be, e.g., present in enzymatic reactions [16,17], responsible for structure stabilization [18–22], arrangement of molecules in crystals [23,24] and supporting molecular engineering [25,26]. Therefore, it is evident that hydrogen bonds are non-covalent interactions relevant in various branches of contemporary science [27].

Hydrogen bonds are generally divided into intra- and intermolecular ones. While the latter provide the possibility of supramolecular assembly and molecular recognition, the former are relevant to the molecular structure, stabilizing the hydrogen-bonded conformations and giving rise to the tautomeric forms. The intramolecular hydrogen bonds are more open to types of enhancement such as resonance-assisted phenomena [28,29], but on the other hand the charge-assisted bonds can be easily formed between two separate molecules [30–32]. Another aspect related to hydrogen bonds is the proton transfer phenomenon and associated changes in geometric and electronic structure parameters [33–35]. This phenomenon has been studied using experimental and theoretical approaches, because its role cannot be neglected in biomolecules and other compounds where tautomeric forms occur, e.g., see refs. [36–42].

The current study is a continuation of our long-term research effort to shed light on the hydrogen bridge dynamics in molecular crystals with diverse chemical compositions— cyclic compounds in particular. In order to make our study comprehensive, we focus not only on intramolecular interactions, but also on intermolecular forces. Our studies covered, e.g., description of molecular properties in monocyclic aromatic o-hydroxy Schiff and Mannich bases [43,44], bicyclic N-oxides [45–47] and proton sponges [41,48]. Following the line of our research devoted to intramolecular hydrogen bonds investigations in compounds possessing fused rings, we have focused our attention on naphthazarin and its derivatives. The 5,6-dihydroxy-1,4-naphtoquinone (commonly referred to as naphthazarin) and its derivatives are a part of the wider naphthoquinone group. Members of this class of compounds are widely distributed in natural sources [49] and have been proven to possess many interesting biological properties. Recent years have brought reports on their antibacterial, antifungal and biostatic activities [50,51]. Naphthazarin, alone or in conjunction with other compounds, can be used as a potent biopesticide or insecticide [52]. Medicine also has many potential ways of utilizing naphthazarin and its derivatives. They have been proven to possess high anti-inflammatory potential and the positive effect on the healing of wounds [53]. More elaborate derivatives can also be applied in oncological treatment—they have reported to have an inhibitory effect on DNA Topoisomerase-I [54], heat-shock factor and glutathione status in the aftermath of hypoxia [55]. All of the aforementioned properties show that there is still need to study both the possible ways of synthesis and design of derivatives with desired properties [56,57]. One of the most pronounced characteristics of naphthazarin is the presence of two hydrogen bonds between the hydroxyl groups and their neighboring carbonyl oxygen atoms. This not only forms two distinct quasi-rings in the structure, but also allows the opportunity to study the effects of substitution in the fused rings and double hydrogen bonding properties [58]. In case of the naphthazarin and its selected derivatives, investigations into physico-chemical properties are reported by [59–61]. The computational studies have been also used to assess compatibility of experimental and theoretical data in the IR and Raman spectra of naphthazarin, which in turn allowed precise assignment of bands [62].

Here, we present our theoretical results obtained for two naphthazarin derivatives: 2,3-Dimethyl-5,8-dihydroxy-1,4-naphthoquinone (**1**) and 5,8-Dihydroxy-2,3-dimethoxy-6-methylnaphtho-1,4-quinone (**2**), presented in Figures 1 and S1. The motivation for the current work and the choice of the compounds was the comparison of symmetric and asymmetric substitution with diverse, but not very strong, substituent properties represented by the classical physico-chemical parameters, Hammett constants. The –Me and –OMe substituents possess different properties regarding their electro- and nucleophilicity. The methyl groups are relatively mild on the Hammett scale (their classical Hammett constants are only −0.07 and −0.17 for σ_m and σ_p, respectively [63]. The –OMe substituents are reported to reach the values of +0.12 and −0.27 for σ_m and σ_p [63]. Such values indicate the possibility of an interesting interplay between local properties of the substituents and an environmental influence in the solid state (with the presence of other molecules). The physico-chemical properties of both compounds were studied on the basis of X-ray methods as well as, e.g., NMR spectroscopy for **1** by Rodríguez et al. [64] and for **2** by Cannon et al. [65]. Concerning compound **1** the crystal structure is built of molecules, which are stacked up the c axis and the molecules overlap forming the charge-transfer complex [64]. Compound **2** contains two methoxyl groups, which are slightly different with regard to geometry: one of the groups lies in the plane of the ring, but the methyl group of the methoxyl deviates from the ring plane by 1.08 Å [65]. This experimentally observed difference in geometry, which may result from packing forces, would not necessarily be detected in solution. According to the authors, electronic effects were responsible for observed non-equivalence of the methoxyl resonances in the NMR spectrum in solution. The electron-donating properties of the methyl group are well known and characterized [63]. Inductive effects usually cover short distances; however, in the case of compound **2**, they could influence the methoxy groups as well [65]. Therefore, the main aim of the study was further examination of internal and external forces responsible for the molecular features of the compounds.

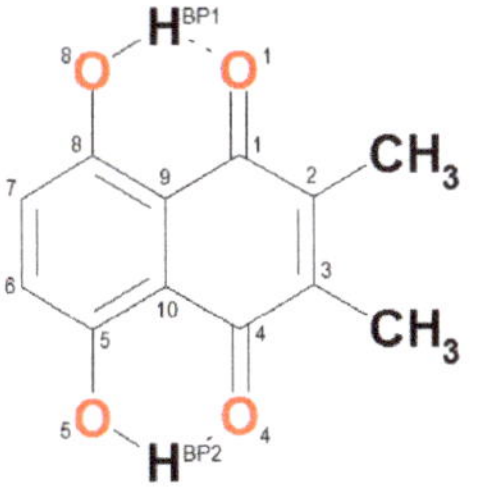

1 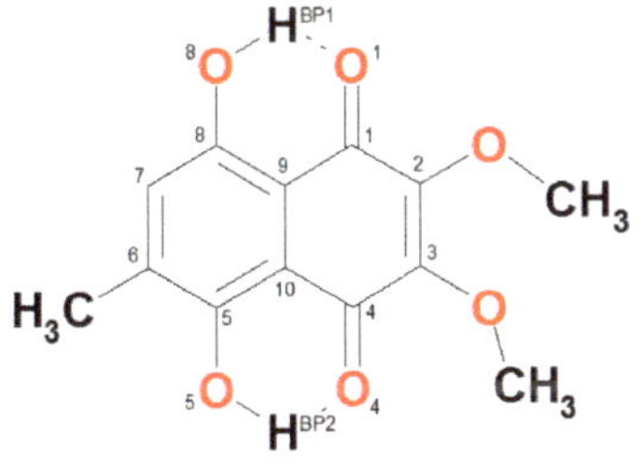   **2**

Figure 1. Molecular structures of the studied 2,3-dimethylnaphthazarin (**1**) and 2,3-dimethoxy-6-methylnaphthazarin (**2**) with the atom numbering scheme applied in the current study.

In order to achieve the goals delineated in the previous paragraphs, the fundamental issues of comparisons of intra- and intermolecular phenomena, influence of substituents, and correlation of bridge proton motions, diverse theoretical approaches were considered and employed. Static and dynamical models were developed on the basis of Density Functional Theory (DFT) [66,67] and Car–Parrinello molecular dynamics [68]. The simulations were carried out in the gas phase and in the solid state. Our particular attention was focused on: (i) The proton reaction path and related energy changes in the monomeric forms of the studied compounds. We put emphasis on the substituent effects on the hydrogen bond properties; (ii) The electronic structure and topology changes—the comparison of molecular and proton transferred (PT) forms on the basis of the Atoms in Molecules (AIM) theory [69]; (iii) The energy partitioning in the dimers extracted from the X-ray data as well as obtained theoretically on the basis of Symmetry-Adapted Perturbation Theory (SAPT) [70]; (iv) The hydrogen bridges dynamics in the gas and crystalline phases, which enabled us to detect differences derived from environmental effects—gas phase vs. solid state comparison and the structural impact of the nuclear quantum effects (NQE) for the bridge protons; (v) Vibrational signatures present in the studied compounds, but particularly associated with the intramolecular hydrogen bond—gas phase vs. solid state comparison. To the best of our knowledge this is the first study that examines these particular naphthazarin derivatives considering intra- and intermolecular forces.

2. Results and Discussion

2.1. Geometric and Electronic Structure Description of Naphthazarin Derivatives Monomers with Special Emphasis on Intramolecular Hydrogen Bonds

This section contains results from diverse theoretical approaches, ranging from gradient-corrected DFT to post-Hartree–Fock schemes. However, the common trait is that the computational models represent the gas phase molecules using atom-centered Gaussian basis sets providing wavefunctions, Kohn–Sham orbitals and electron density. The chosen DFT functionals belong to the most widely used approaches: the PBE functional is of the Generalized Gradient Approximation (GGA) type and does not use the exact (Hartree–Fock) exchange. B3LYP is a hybrid functional with Hartree–Fock admixture. The ωB97XD is also a hybrid functional, additionally including empirical dispersion correction. The chosen post-Hartree–Fock schemes are Møller–Plesset second order perturbative calculus, MP2, and Coupled Cluster theory with single and double excitations, CCSD. The Car–Parrinello scheme, relevant to the further sections, is based on the delocalized plane-wave basis sets with inherent periodicity. This makes CPMD calculations technically easier for solids and liquids than for the gas phase, where periodicity has to be artificially removed. Moreover, exact (Hartree–Fock) exchange is not easily implementable on the plane-wave basis. Therefore, CPMD calculations utilize almost solely the GGA functionals, such as PBE, due to their efficiency and well tested performance.

The molecular structures of the monomeric forms of studied 2,3-dimethylnaphthazarin (**1**) and 2,3-dimethoxy-6-methylnaphthazarin (**2**) are presented in Figure 1. The intramolecular non-covalent interactions present in the molecules are classified as Resonance-Assisted Hydrogen Bonds (RAHBs) [28]. The geometry of the studied compounds was modified to analyze the energy of various isomers (see Table S1 for details; this table includes not only the electronic energy, but also the energy values corrected for the vibrational zero point energy (ZPE) contribution). The lowest energy was found for the molecular forms. The double-PT forms are slightly higher in energy (we will use the convention: electronic/ZPE-corrected value in kcal/mol): for **1**, the difference is 1.94/1.91 kcal/mol (B3LYP), 1.61/1.38 kcal/mol (PBE) and 2.02/1.54 kcal/mol (ωB97XD). For **2**, the corresponding increases in energy of the double-PT form with respect to the molecular form are: 2.94/2.74 kcal/mol (B3LYP), 2.78/2.29 kcal/mol (PBE) and 2.91/2.58 kcal/mol (ωB97XD). This shows that from a thermodynamic point of view the molecular forms are preferred, but not strongly, over the PT forms. The trans forms through which the double-proton transfer proceeds [61] are even higher in energy, e.g., for **1** the differences with respect to the molecular forms are: 5.30/4.65 kcal/mol (B3LYP), 2.82/1.26 kcal/mol (PBE) and 6.81/6.23 kcal/mol (ωB97XD). However, these values are all well within the range thermally accessible to the extended molecules at room temperature, especially when the effects of donor–acceptor distance modulation are accounted for (see the Car–Parrinello study below).

The geometric details of the intramolecular hydrogen bonds (comparison of experimental and computational data) are summarized in Table S2. It is shown that the DFT method was able to reproduce the metric parameters related to the hydrogen bonding with a good agreement. The impact of the substituents, as already noted in the Introduction, corresponds to their inductive and resonance properties. The Hammett constants are, respectively: $\sigma_m = -0.07$ and $\sigma_p = -0.17$ for the -Me, while for –OMe they are $\sigma_m = +0.12$ and $\sigma_p = -0.27$ [63]. The bridge protons are located at the non-substituted ring of **1** and methyl-substituted ring of **2**, which shows that substitution affects aromaticity and the modified rings prefer participation in the quinoid-like structure.

The proton potential functions for the proton motion are presented in Figure 2. The proton reaction energy paths were investigated using B3LYP, ωB97XD and PBE functionals with the 6-311++G(2d,2p) basis set. As shown in the figure, two energy minima were obtained in the case of both compounds and both studied bridges. In the case of compound **1**, only one hydrogen bridge was analyzed due to the symmetry exhibited by the compound. Concerning compound **2**, both intramolecular hydrogen bridges were analyzed. The compound is not symmetrical due to the presence of the methyl group. Before discussing the details of the proton potential functions, it is necessary to consider whether single- or double-proton transfer should be pursued. Our DFT and CPMD results for naphthazarin [61] indicate that the simultaneous double-proton transfer is less probable and leads to higher barriers, which is assumed to be the effect of deeper modification of aromaticity than in the case of the single-proton transfer event. However, the single-proton transfer enables the second proton transfer (PT) event to happen very fast but not simultaneously, in the order of several O-H stretching periods.

The deeper energy minimum is localized at ca. 1 Å of the $O8\text{-}H^{BP1}/O5\text{-}H^{BP2}$ covalent bond length in both compounds. The elongation of the bond towards the acceptor atom (O1/O4) provided information of the energy barrier, which was found—depending on the applied functional—to be very similar in both studied compounds. The highest energy barrier was obtained for the ωB97XD functional (10.1 ± 0.05) kcal/mol for Bridge 1 of both compounds and (9.55 ± 0.05) kcal/mol for Bridge 2 of compound **2**. The results obtained with assistance of the B3LYP functional are (7.95 ± 0.05) kcal/mol for Bridge 1 and (7.45 ± 0.05) kcal/mol for Bridge 2 in the case of compound **2**. The lowest energy barrier was noticed as a result of PBE functional application (3.55 ± 0.05) kcal/mol for all cases. The DFT results were validated with the single-point energy calculations at the post-Hartree–Fock MP2 and CCSD level for the PBE geometries. Both of these approaches yielded the same ordering of relative energies than the DFT functionals: the barriers for Bridge 1 are almost equal for **1** and **2**, and the barrier for Bridge 2 of compound **2** is slightly lower. The MP2 perturbative calculus provided a barrier height of 6.72 kcal/mol for **1**, 6.86 kcal/mol for Bridge 1 of **2**, and 6.59 kcal/mol for Bridge 2 of **2**. The corresponding CCSD values of barrier height estimate are: 9.59 kcal/mol for **1**, 9.37 kcal/mol for Bridge 1 of **2**, and 9.13 kcal/mol for Bridge 2 of **2**. Our previous results on naphthazarin [61] have shown that MP2 and CCSD methods provide PT barrier heights correspondingly lower and higher from the accurate CCSD(T) barrier height, and we expect similar performances from these methods in the current study. This indicates that the barrier height estimates from the post-Hartree–Fock methods and DFT functionals are in agreement.

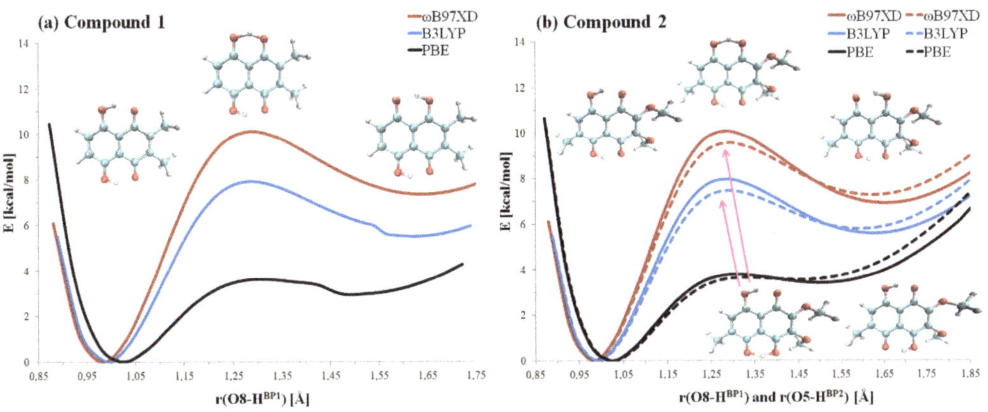

Figure 2. The potential energy profiles for the proton motion in the hydrogen bridges of compounds **1** (**a**) and **2** (**b**), respectively. The hydrogen bridges denoted as $O8\text{-}H^{BP1}...O1$ (Bridge 1) for compound **1** and $O8\text{-}H^{BP1}...O1$ (Bridge 1) and $O5\text{-}H^{BP2}...O4$ (Bridge 2) for compound **2** are presented. In the case of **2**, the solid line denotes Bridge 1 while the dotted line denotes Bridge 2.

Returning to the discussion of structure–energy relations, we note that the second energy minimum is shallow; therefore, we could expect that the bridged proton is mostly localized on the donor (O8/O5) atom. However, the presence of the second energy minimum indicates that the bridge protons are labile and they could approach the proton-acceptor atom domain. Almost identical energy barriers showed that the substituent effects as well as the lack of the symmetry (in the case of compound **2**) have not significantly influenced the proton transfer reaction path in the investigated napththazarin derivatives.

The electronic structure analysis was carried out based on AIM theory. The selected results of the analysis are presented in Tables 1 and 2. The partial atomic charges are reported for atoms forming quasi-rings in the studied compounds (see Table 1). We have analyzed molecular and tautomeric (proto transferred (PT)) forms of both compounds. The electron density of the donor atom (O8/O5) is smaller when the bridged proton is attached to it. A decrease in the electron density at the acceptor atom (O1/O4) is observed for the tautomeric (PT) form. It can be seen that the hydrogen atoms (H^{BP1} and H^{BP2}) are more positively charged when they are transferred to the acceptor atom side. Next, the sum of partial atomic charges in the quasi-rings was computed. It was found that for compound **1**, Bridge 1, the sum decreased from -0.1532 [e] in the molecular form to -0.1591 [e] in the proton transferred form. A similar observation was made for compound **2**—the sum of the quasi-ring (Bridge 1) atomic charges decreased from -0.1042 [e] to -0.1089 [e]. Concerning the hydrogen bridge denoted as Bridge 2 (see Table 1), in compound **1**, there was a decrease in the sum of the partial atomic charges in the quasi-ring from -0.1554 [e] to -0.1583. However, an opposite situation was found in the case of compound **2**— there was an increase in the values of the sum of the atomic charges from -0.1707 to -0.1663 [e]. This could be associated with the presence of the methyl group in the vicinity of the quasi-ring and asymmetry introduced by it to compound **2**. It is also known from the crystal structure of the compound [65] that the methoxy groups are sterically not equivalent; moreover, these groups are not chemically equivalent due to their having different relative positions with regard to the methyl group. There was also an interaction between the methoxy group and O4 proton-acceptor atom (for details, see the text below). The interatomic O8...O1 and O5...O4 distances determined experimentally are equal to 2.551 Å and 2.589 Å, respectively. This could also be the reason why an opposite tendency concerning the electron density distribution was observed for compound **2** (Bridge 2). The values of electron density and its Laplacian at Bond Critical Points (BCPs) of intramolecular hydrogen bonds for both compounds are shown in Table 2. The electron density values at the hydrogen bridge BCPs are consistent with our previous calculations performed for 2,3-dichloronaphthazarin [71]. The covalent O8-H^{BP1}/O5-H^{BP2} bonds are stronger than those formed after proton transfer (O1-H^{BP1}/O4-H^{BP2}). This observation was made after the electron density and its Laplacian examination at BCPs. The electron density values at BCPs are higher for the OH covalent bonds in the molecular forms of compounds **1** and **2**. However, the intramolecular hydrogen bonds are stronger (higher electron density values at BCPs) for the proton transfer (PT) forms. The values at BCPs obtained based on AIM theory are rather similar for molecular and PT forms. They do not much differ, even comparing compound **1** with compound **2**. This could suggest that the proton transferred form is best described not as $O^{-}...^{+}H$-O, but as simply O...H-O, in parallel with the molecular form O-H...O. The topology maps of electron density are presented in Figure 3. They contain molecular properties common for the AIM description of the electronic structure: critical points (BCPs and RCPs), which are stationary points of the electron density field (i.e., the density gradient is zero at the critical point). In the graphical presentation of Figure 3, these critical points are recognizable as maxima (nuclear positions), saddle points, and minima. Due to the presence of intramolecular hydrogen bonds, the typical quasi-rings were formed and recognized by the BCPs of covalent bonds and the indicated bond paths of the hydrogen bridges. In addition, in the case of compound **2**, some intramolecular contacts were detected between the methoxy groups as well as between the hydrogen of the methoxy group with the O4 atom from the second hydrogen bridge. The presence of an intramolecular hydrogen bonds stabilizes the conformation of the molecules. However, the topology maps showed (in the case of compound **2**), that the electron density distribution in the hydrogen bridge (Bridge 2) could be affected by the competitive interactions introduced by the methoxy substituents. As is shown, two additional quasi-rings were found, indicating that the C-H...O intramolecular hydrogen bonds were formed. They are characterized by the presence of the BCPs and RCPs. However, the presence of such intramolecular interactions was not identified experimentally in the crystal structure [65]. Therefore, the presence of the interactions could be driven by steric effects and degrees of freedom introduced to the isolated molecule model.

Table 1. Atoms In Molecules (AIM) atomic charges calculated for selected atoms of the studied compounds, compounds **1** and **2**, and their proton transferred (PT) forms at the B3LYP/6-311++G(2d,2p) level of theory.

Atomic Charge [e]	Compound 1		Compound 2	
	Molecular Form	PT Form	Molecular Form	PT Form
	Hydrogen Bridge 1			
O8	−1.134	−1.094	−1.134	−1.097
H^{BP1}	0.641	0.643	0.641	0.642
O1	−1.099	−1.139	−1.081	−1.119
C8	0.597	0.874	0.597	0.874
C9	−0.027	−0.030	−0.029	−0.027
C1	0.869	0.587	0.902	0.618
	Hydrogen Bridge 2			
O5	−1.134	−1.094	−1.139	−1.104
H^{BP2}	0.641	0.642	0.642	0.645
O4	−1.098	−1.137	−1.102	−1.136
C5	0.596	0.875	0.582	0.857
C10	−0.028	−0.030	−0.027	−0.029
C4	0.868	0.585	0.873	0.601

Table 2. Atoms In Molecules (AIM) Bond Critical Point properties calculated for selected bonds of the studied compounds, compounds **1** and **2**, and their proton transferred forms (PT) at the B3LYP/6-311++G(2d,2p) level of theory. Electron density ρ_{BCP} is given in $e \cdot a_0^{-3}$ atomic units, and its Laplacian $\nabla^2\rho_{BCP}$ is given in $e \cdot a_0^{-5}$ units.

BCP	Compound 1		Compound 2	
	ρ_{BCP}	$\nabla^2\rho_{BCP}$	ρ_{BCP}	$\nabla^2\rho_{BCP}$
	Molecular Form			
O8-H^{BP1}	0.339	−2.536	0.340	−2.540
H^{BP1}-O1	0.051	0.138	0.050	0.136
O5-H^{BP2}	0.339	−2.533	0.337	−2.511
H^{BP2}-O4	0.051	0.137	0.053	0.140
	Proton-Transferred Form (PT)			
O8-H^{BP1}	0.054	0.139	0.053	0.137
H^{BP1}-O1	0.335	−2.487	0.335	−2.488
O5-H^{BP2}	0.052	0.137	0.057	0.141
H^{BP2}-O4	0.336	−2.505	0.330	−2.441

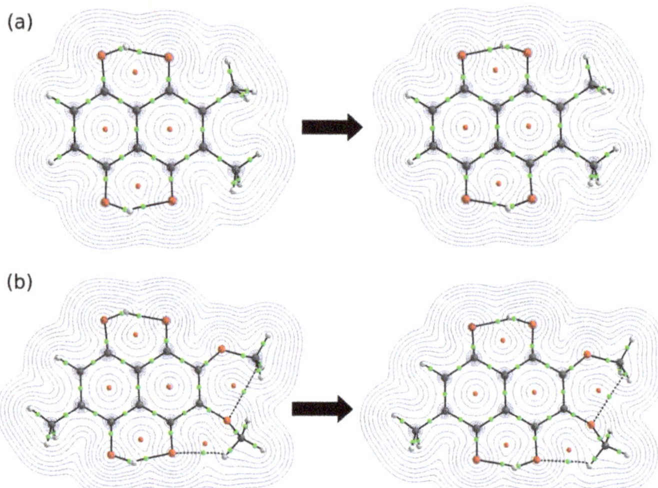

Figure 3. Topology maps of electron density obtained on the basis of AIM theory at the B3LYP/6-311++G(2d,2p) level of theory for compounds **1** (**a**) and **2** (**b**). The molecular (left) and proton transferred (right) forms are presented. The black solid and dashed lines indicate the intramolecular interaction paths. The green and red dots mark the presence of BCPs and RCPs, respectively.

2.2. Intermolecular Forces in Naphthazarin Derivatives Dimers Based on Symmetry-Adapted Perturbation Theory (SAPT)

The presence of two distinct types of stacked dimers in the crystals of **1** or **2** (anti-parallel vs. parallel arrangement of molecules, respectively; see Figure 4) indicates that even the relatively mild substitution can influence crystal packing forces. It is therefore necessary to investigate the molecules of **1** and **2** on the basis of interaction energy partitioning schemes. Symmetry-Adapted Perturbation Theory (SAPT, see Ref. [70]) has become a de facto standard for such investigations, although many other approaches exist, some of them capable of tackling covalent bonding, for example, a DFT-based energy decomposition analysis [72] or localized orbital energy decomposition, LMOEDA, useful for non-covalent forces such as beryllium bonds [73]. While analyzing the results presented in this section, it is necessary to remember that SAPT is a perturbative approach, in which intra- and intermonomer correlations are treated separately. SAPT0 omits the intramonomer correlation, while SAPT2 includes this effect up to the second order of perturbation. Both levels account for the intermonomer electron correlation, which is the source of polarization and dispersion contributions. The SAPT partitioning divides the interaction energy into "static" contributions (electrostatic interaction of frozen electron densities, and Pauli exchange repulsion) and "correlated" terms (induction–mutual polarization of monomers—and dispersion).

The crystal structures of both compounds contain the following basic types of dimeric structures, depicted in Figure 5: d1—the head-to-head planar arrangements, d2—molecular planes tilted at a shallow angle, d3—stacking, d4—planar arrangement with a side-to-side skew (present only for **2**). Direct use of the experimental structures in the SAPT calculations leads to the results gathered in Table 3, while the DFT-optimized structures are described in Table 4. It must be stressed that the DFT optimization leads to the collapse of the d2-type dimers into the stacking arrangement, underlining the role of the confinement of molecules leading to the formation of diverse structural motifs.

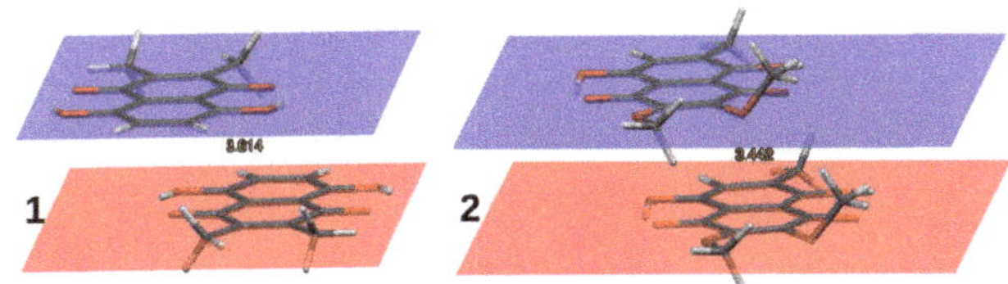

Figure 4. Two distinct types of stacking in the crystal structures of **1**, 2,3-dimethylnaphthazarin (anti-parallel stacking), and **2**, 2,3-dimethoxy-6-methylnaphthazarin (parallel stacking). The interplanar distances are 3.614 Å and 3.442 Å, respectively.

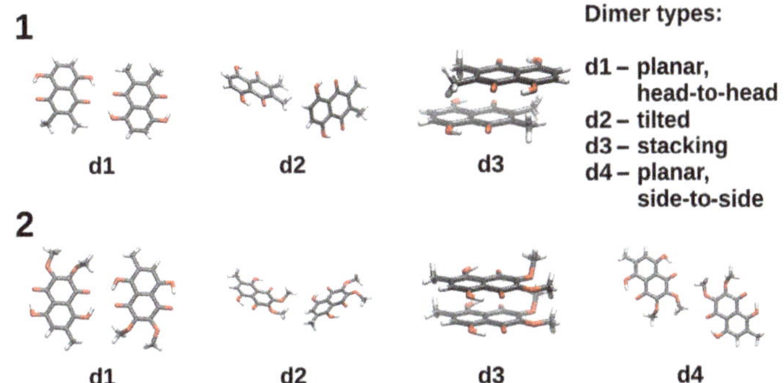

Figure 5. Dimers extracted from the crystal structures of compounds **1** (upper part) and **2** (lower part), used in the SAPT study.

Table 3. SAPT2/jun-cc-pVDZ results of energy partitioning for the dimers of compounds **1** and **2** (see Figure 5) with structures taken directly from the diffraction experiments. All energy terms in kcal/mol: Elst—electrostatics; Exch—exchange (Pauli) repulsion; Ind—induction (polarization); Disp—dispersion; SAPT0 and SAPT2 are defined according to Ref. [74].

Compound	Dimer	Elst	Exch	Ind	Disp	SAPT0	SAPT2
1	d1	−3.617	4.754	−0.676	−3.739	−3.342	−3.278
1	d2	−0.710	1.716	−0.181	−2.104	−1.328	−1.279
1	d3	−3.210	7.190	−0.884	−11.991	−9.233	−8.894
2	d1	−4.426	3.857	−0.760	−4.178	−6.305	−5.507
2	d2	−0.750	1.598	−0.349	−2.084	−1.601	−1.585
2	d3	−6.162	15.081	−2.063	−21.160	−14.795	−14.304
2	d4	−5.099	3.273	−1.069	−2.838	−6.849	−5.733

Table 4. SAPT2/jun-cc-pVDZ results of energy partitioning for the dimers of compounds **1** and **2** (see Figure 5) with structures taken from the DFT structural optimization. All energy terms in kcal/mol: Elst—electrostatics; Exch—exchange (Pauli) repulsion; Ind—induction (polarization); Disp—dispersion; SAPT0 and SAPT2 are defined according to Ref. [74].

Compound	Dimer	Elst	Exch	Ind	Disp	SAPT0	SAPT2
1	d1	−6.116	7.711	−1.112	−4.766	−5.410	−4.283
1	d3	−9.129	20.285	−2.485	−21.867	−13.681	−13.196
2	d1	−5.830	8.537	−1.187	−5.958	−5.443	−4.438
2	d3	−14.669	29.178	−3.754	−29.914	−19.906	−19.158
2	d4	−7.830	8.427	−2.254	−4.258	−7.290	−5.916

The results gathered in Tables 3 and 4 show that the energetically most important structural motif (stacking dimer d3) is formed with the dominant role of dispersion. The role of dispersion is visible especially when the d3 dimers of experimental solid state structure are compared with their DFT-optimized analogues. Surprisingly, the latter are more strongly bound. This is an outcome of two competing factors. On the one hand, the crystal electrostatic and steric field tends to squeeze the molecules together, so that no empty voids remain in the structure. This promotes smaller intermolecular separations and stronger stacking forces. On the other hand, the presence of neighbouring molecules means that the capacity of the molecule to interact with its neighbours must split between much more interactions than in the dimer. The latter factor prevails, and the DFT-optimized stacking dimers are bound stronger by ca. 4–5 kcal/mol than their crystal structure equivalents. It is interesting to note that the DFT-optimized d3 structures exhibit not only stronger dispersion, but also electrostatic and induction contributions.

It seems paradoxical that the d1, d2 and d4 dimers, relying mostly on electrostatic forces including hydrogen bonds, present more equalized distribution of the interaction energy terms than the stacked d3 dimers (both in the gas phase and in the arrangement from the crystal structure). However, there is another factor which is closely related to the type of force dominating the interactions. Two levels of theory, SAPT0 and SAPT2, are provided in Tables 3 and 4 to explain this factor. We note that the total SAPT0 and SAPT2 interaction energies are very close to each other when the studied molecules do not engage in hydrogen bonding, highlighting the role of intramonomer electron correlation in the formation of hydrogen bonds. For example, the d3 dimer of **1** has the interaction energy of -13.681 kcal/mol at the SAPT0 level and -13.196 kcal/mol at the SAPT2 level. This means that the hydrogen bonds and electrostatic forces, displaying larger differences between the SAPT0 and SAPT2 energies, contain significant contributions of higher-order corrections connected with electron correlation, not present at the SAPT0 level. On the other hand, the presence of hydrogen bonds in the dimers is associated in this case with a relatively weak interaction (ca. 4–5 kcal/mol). The weakest dimers (d2 type) are rather just multipolar, electrostatic contacts and their particular shape is governed by steric hindrance of the substituents (especially for **2**).

Summarizing the SAPT study, we stress that the stacked arrangement is the principal structural motif of the crystal from the geometric point of view. This fact agrees with the role of dispersion forces as the most important factor from the energetic point of view. However, the details of the solid state structure are modified by the substituents and the polar nature of the compounds introduced not only by the intramolecular hydrogen bonds, but also by the substituents, even relatively mild on the Hammett scale (the methyl groups in **1**, with classical Hammett constants of only -0.07 and -0.17 for σ_m and σ_p, respectively [63]).

2.3. First-Principle Molecular Dynamics (FPMD) in the Gas and Crystalline Phases

The applied Car–Parrinello molecular dynamics (CPMD) enables the investigation of molecular and spectroscopic features of the naphthazarin derivatives based on ab initio Potential Energy Surface (PES), which is of great importance when we are expecting to register proton transfer phenomena events. The time-evolution study provides an insight into the dynamical nature of the hydrogen bonding present in the studied systems. Therefore, special attention was paid to the intramolecular hydrogen bridges present in both compounds. The CPMD simulations were performed in the gas phase and in the solid state. The two phase study enabled detection of differences related to the environmental effects' influence on the hydrogen bond dynamics, e.g., the crystal field and the presence of neighbouring molecules. The details of the hydrogen bonds' average metric parameters are presented in Table S2. The reported values are in good agreement with the experimental data available [64,65], as well as the static DFT results.

Figures 6 and 7, showing the time evolution of the distances related to the hydrogen bridges, use the same color coding to aid data interpretation. The black line corresponds to the O...O donor–acceptor distance, and it simply oscillates around an equilibrium value throughout the simulation time. The red line is the donor-proton bond length, of rather small amplitude, while the green line is the proton-acceptor distance, oscillating in a wide range. The objects of our interest, proton transfer events, are accompanied by a sudden increase in the donor-proton bond length, accompanied by the lowering of the proton-acceptor separation (the red and green lines cross over).

In Figure 6, the hydrogen bridge dynamics is presented for compound **1**. The upper part shows data obtained in the gas phase. There are many proton-sharing events during the CPMD simulation run. The bridged protons exhibit strong mobility, which results in proton transfer phenomena registered during the 20 ps of the CPMD run. The protons moved to the acceptor-atom side, stayed there for a short time and kept moving again to the proton-donor side. In the solid state (lower part of the Figure 6), the proton transfer events were noticed as well. However, there were less proton-sharing events comparing to the gas phase results (see also the discussion of proton possession statistics two paragraphs below). This could be explained by the presence of intermolecular hydrogen bonds and molecular overlapping present in the crystal structure [64]. The presence of neighbouring molecules forming intermolecular hydrogen bonds (O-H...O), where the O-H group is involved in the intramolecular hydrogen bond, as well as interacts with an oxygen atom (proton-acceptor from another molecule) and introduces competition in the interactions. This shows a significant difference between the isolated molecule and the crystalline phase dynamics, where many factors are included during the CPMD simulations. There is a visible correlation in the bridged protons dynamics in both phases. Compound **1** exhibits symmetry; therefore, one could expect that the dynamical nature of the bridged protons will be similar, but it will depend on the phase discussed—the crystal packing lowers the effective symmetry perceived by the analyzed molecule.

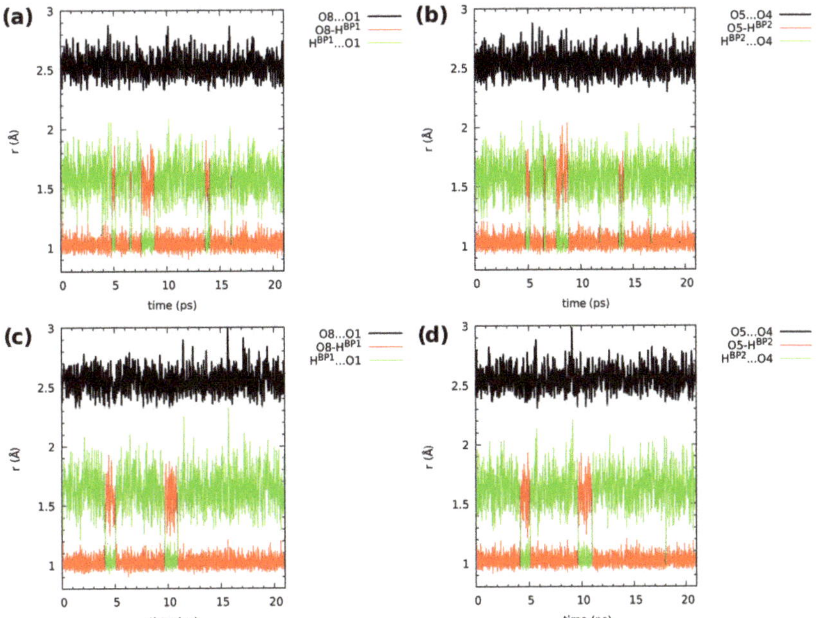

Figure 6. Hydrogen bridge structural parameters during the CPMD simulation of 2,3-dimethylnaphthazarin (**1**). The graphs show gas phase results for (**a**) Bridge 1 and (**b**) Bridge 2, and solid state results for (**c**) Bridge 1 and (**d**) Bridge 2. For atom numbering scheme, see Figure 1.

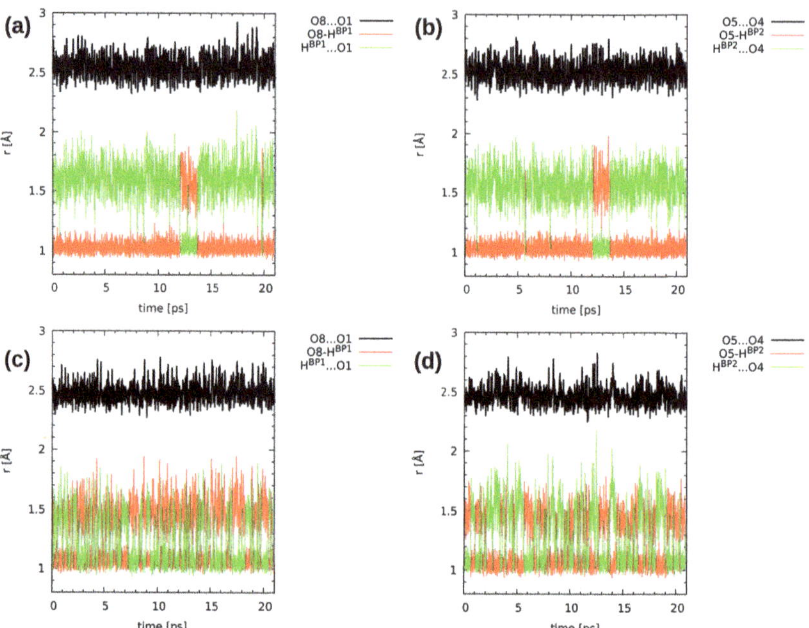

Figure 7. Hydrogen bridge structural parameters during the CPMD simulation of 2,3-dimethoxy-6-methylnaphthazarin (**2**). The graphs show gas phase results for (**a**) Bridge 1 and (**b**) Bridge 2, and solid state results for (**c**) Bridge 1 and (**d**) Bridge 2. For atom numbering scheme, see Figure 1.

The CPMD results concerning the intramolecular hydrogen bond dynamics of compound **2** are presented in Figure 7. The bridged protons exhibit strong mobility in both studied phases. In the gas phase (upper part of Figure 7), there are frequent proton-sharing events and proton transfer phenomena were noticed as well. During the 20 ps run, there were 3 ps long proton transfer events, and after this time, the bridged protons moved back again to the proton-donor atom. There is also a correlation in the bridged protons dynamics—the 3 ps long PT events happened at the same time for both bridges. A solid state study provided a different picture of the proton mobility in the hydrogen bridges (lower part of Figure 7). The bridged protons were strongly delocalized between the donor and acceptor atoms in both hydrogen bonds. The compound did not exhibit symmetry due to the presence of the methyl group in the sixth position as well as methoxy groups, which are not equivalent [65]. There were also intermolecular hydrogen bonds, involving (similarly to compound **1**) OH groups from the intramolecular hydrogen bonds and proton-acceptor atoms from the neighbouring molecules. There was also molecular overlapping according to the X-ray measurements [65]. Comparing gas phase results with the solid state of compound **2**, it is visible from the structural data analysis that external forces influence the bridged protons dynamics. In the case of compound **2**, we can draw the conclusion that the presence of methoxy groups and the lack of symmetry introduced inductive and steric effects, which provided us with a different dynamical nature of the intramolecular hydrogen bonds present in compound **2** with respect to **1**.

The diverse properties of hydrogen bonds were further analyzed using statistics-based approaches. First, we calculated the proton possession statistics, i.e., percentages of the time spent by the given bridge proton at the donor or acceptor site. The proper association of the proton with its site at a given time was determined by the Voronoi geometric criterion—donor-proton vs. proton-acceptor distance comparison. The results, gathered in Table 5, indicate that in case of compound **1**, gas phase and solid state statistics are very similar. The degree of convergence of the dynamics trajectory can be estimated by comparison of the two equivalent bridges, O8...O1 vs. O5...O4—the differences are no more than 0.5%. The differences between the gas phase and solid state results are 0.2% for the O8...O1 bridge and 1.1% for the O5...O4 bridge, which is close to the difference between the two bridges. This means that the solid state environment does not seem to change the overall partitioning of the proton residence time between the donor and the acceptor sites, and it promotes slower dynamics (fewer proton sharing events). The results for the gas phase CPMD simulation of **2** are also similar to the case of **1**: the two bridges, which are not equivalent, are still similar enough to provide the same statistics of proton possession. The solid state case is more interesting: the packing forces (the presence of neighbours and their electrostatic field) lead to almost equally shared protons; however, the H^{BP1} tends to reside more at the acceptor site than the H^{BP2} proton.

Table 5. Proton possession statistics for the CPMD runs. Percentages of the time spent by the bridge proton at the donor or acceptor site, determined by Voronoi geometric criterion–distance comparison.

$O8\text{-}H^{BP1}...O1$		$O5\text{-}H^{BP2}...O4$	
1, gas phase			
O8 donor	O1 acceptor	O5 donor	O4 acceptor
89.7%	10.3%	90.2%	9.8%
1, solid state			
O8 donor	O1 acceptor	O5 donor	O4 acceptor
89.5%	10.5%	89.1%	10.9%
2, gas phase			
O8 donor	O1 acceptor	O5 donor	O4 acceptor
91.6%	8.4%	91.8%	8.2%
2, solid state			
O8 donor	O1 acceptor	O5 donor	O4 acceptor
41.6%	58.4%	53.1%	46.9%

The second part of the CPMD trajectory statistical analysis is provided by the histograms for the donor-proton positions in the two hydrogen bridges—see Figure 8. The histograms (probability density plots) show how the proton positions are correlated in the sense of averaging over the CPMD run. It is visible that for compound **1**, regardless of the simulation conditions—the gas phase or the solid state—the two protons H^{BP1} and H^{BP2} are strongly correlated and located mostly at the donor site. This is also true for the gas phase trajectory of **2**. These results are in agreement with the data for naphthazarin [61]: it was shown that when an asynchronous proton jump occurs, it is very probable that a second proton transfer will follow within a few O-H oscillation periods. From this point of view, it is interesting to note that the solid state simulation of **2**, where the protons are more delocalized, also exhibits important motion correlations. The histogram shown in panel (d) of Figure 8 consists of four more populated regions forming a square shape. These regions correspond to the molecular form, the PT form, and the two less stable forms with single-proton transfer. There are no indications of a synchronous double-proton transfer, which would result in formation of a populated region in the center of the square shape.

While the current study was carried out within the Newtonian classical nuclear dynamics, corresponding to the Born–Oppenheimer picture, it is recognized that in some instances the nuclear quantum effects (NQE) are important for qualitative and quantitative agreement with experiments. An excess proton in water migrates due to a complicated mechanism in which quantum fluctuations, rather than tunneling, play a crucial role [75,76]. Quantum disorder in the hydrogen bonds is required to explain the X-ray absorption spectra of water and ice [77]. Intramolecular hydrogen bonds can be diversely affected by nuclear quantization. Picolinic acid N-oxide with a very strong O-H...O bond requires anharmonic, quantum treatment of the proton motion to rationalize enormous red shifts of the ν_{OH} mode [78]. Weaker hydrogen bonds, such as those in o-hydroxy Mannich bases, exhibit a single-well potential with the minimum clearly at the donor side [79], while in the N-oxides of Mannich bases the potentials are very flat and broad, allowing the proton to move almost freely within the bridge [45]. The current study contains an assessment of the importance of nuclear quantum effects for the H^{BP1} proton. The results, shown in Figure 9, are obtained with the snapshot-based a posteriori approach [79,80] involving numerical solution of a vibrational Schrödinger equation [81]. Our attention was focused on the impact of the NQE phenomena on the $O8\text{-}H^{BP1}$ distance. It is visible in Figure 9 that the NQE tend to increase the proton delocalization between the donor and acceptor sites, making the H^{BP1} atom shift towards the center of the O8...O1 bridge (the red crosses indicating the NQE-corrected positions, are located closer to the half of the actual O8...O1 distance than are the green circles—classical positions). For each of the four investigated cases, one of the snapshots presents the PT structure, where the $O8\text{-}H^{BP1}$ distance is larger than 1.5 Å. In such cases, the NQE shift the proton position towards lower $O8\text{-}H^{BP1}$ values. The impact of NQE is not decisive for the proton localization in the studied compounds **1** and **2**, with a very interesting exception of the crystalline phase of **2**. The proton at the O8...O1 distances above 2.5 Å behaves in a way similar to the other cases, but at 2.48 Å the impact of NQE is particularly large. The same distance for **1** and gas phase **2** does not lead to such large NQE; therefore, it seems that this is the precise region of bridge length at which the combination of the molecular structure of **2** and the crystal environment make the NQE (including tunneling) very effective. However, when the bridge is compressed even further—to 2.37 Å—the impact of NQE is again very small. The explanation is as follows: at such a short bridge length, the proton potential is already of the flat single-well type, making this bridge temporarily a "low-barrier hydrogen bond" for which the tunneling effects are negligible [75]. As a final remark to the study of NQE, we note that the classical CPMD trajectory is able to sample this region of the molecular phase space, as seen in Figure 8. This fact indicates that the NQE should not have a qualitative impact on the properties of the investigated systems.

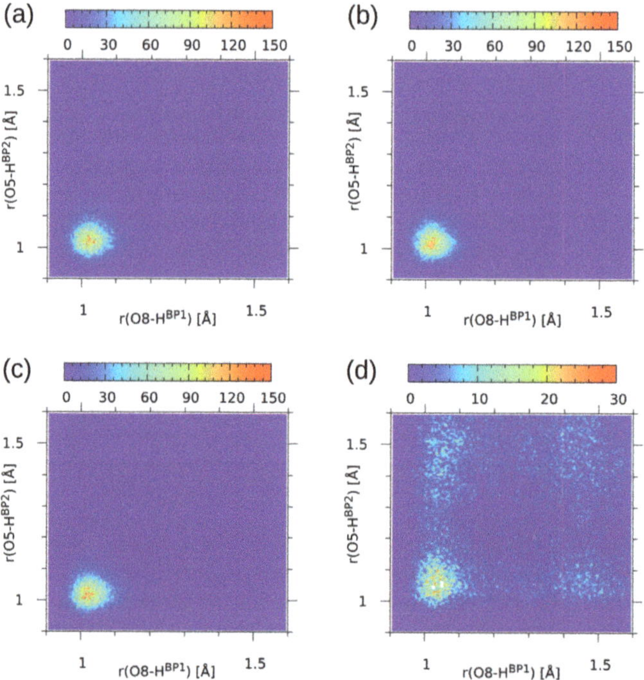

Figure 8. Histograms for the donor-proton distances in the two hydrogen bridges of the studied compounds—results of the CPMD simulation for (**a**) **1** in the gas phase, (**b**) **1** in the solid state, (**c**) **2** in the gas phase, (**d**) **2** in the solid state. Color scale represents probability density in Å^{-2}.

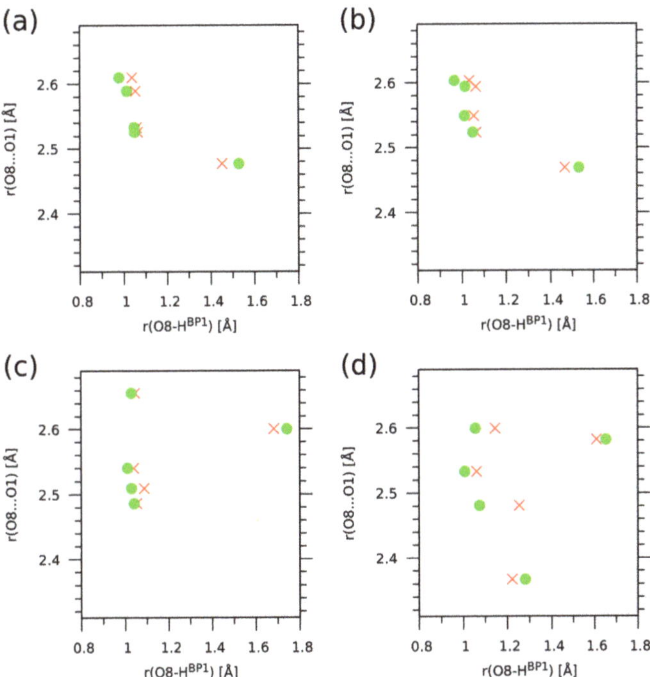

Figure 9. Impact of nuclear quantum effects for the H^{BP1} bridge proton on the $O8\text{-}H^{BP1}$ distance. Green circles—classical value of the distance; red crosses—quantum expectation value of the $O8\text{-}H^{BP1}$ distance operator. Results of a posteriori quantum treatment of CPMD trajectory for (**a**) **1** in the gas phase, (**b**) **1** in the solid state, (**c**) **2** in the gas phase, (**d**) **2** in the solid state.

Vibrational signatures of the bridged protons, corresponding to the ν_{OH} at the high-wavenumber region, are presented in Figure 10. Since the IR spectra of these compounds are not available in the literature, the most natural source of comparison is the parent compound, naphthazarin. Investigation of the bridge proton features has two main goals. First, it is possible to trace the presence of strong interactions in the crystal. On the other hand, the bridges in **2** are not symmetrical, and their asymmetry can lead to slightly diverse positions of the normal modes. This is an interesting issue in relation to the parent naphthazarin itself, where the skeleton, devoid of the substituents, does not prefer any of the proton positions. The broad absorptions of the νOH/OD and γOH/OD stretching modes were experimentally identified in naphthazarin at 3060/2200 cm^{-1} and 793/560 cm^{-1}, respectively; the upper wavenumber region is the most relevant for the fast proton dynamics corresponding to the stretching mode [62].

The first goal, detection of strong interactions, can be accomplished by comparison of the ν_{OH} band positions. Compound **1** exhibits similar positions of this band in the gas phase (from 2300 to 3400 cm^{-1}) and crystal (from 2200 to 3400 cm^{-1}). The band center at ca. 2800 cm^{-1}–2900 cm^{-1} is at a slightly lower wavenumber than the experimental value of 3060 cm^{-1} for naphthazarin [62]. The lower wavenumber absorptions, 700–1700 cm^{-1} in the gas phase and 600–1800 cm^{-1} in the solid state, should be attributed to the mechanical influence of the heavy-atom motions. These values indicate on the one hand a middle-strong O-H...O hydrogen bonding, and on the other a relatively small impact of the crystal packing effects on the vibrational features of **1**. These facts agree well with the not too frequent proton transfer events in this compound (see Figure 6, which also confirms that the PT occurrence in the gas phase and in the crystal is very similar). We have noted already that the PT events are not strictly synchronous, but they are strongly correlated. This makes the vibrational signatures of H^{BP1} and H^{BP2} virtually identical. This is not strictly true for compound **2**, where the chemical nature of the substituents is different in the vicinity of H^{BP1} than in the vicinity of H^{BP2}. The difference is almost not visible in the results of the gas phase simulation of **1**—the ν_{OH} vibrational features of both protons fall into the 2300 cm^{-1} to 3300 cm^{-1} range (the 700 cm^{-1}–1700 cm^{-1} region is associated with heavy atom motions, as already noted for compound **1**), and the signature of H^{BP1} is centered at ca. 2800 cm^{-1}, while the signature of the H^{BP2} proton peaks at ca. 100 cm^{-1} has a lower wavenumber. The difference is small, and it is also in agreement with the time evolution of the distance parameters (see Figure 7). The lowering of the band center position with respect to naphthazarin (3060 cm^{-1} in the experimental spectrum) is also not large. Quite unexpectedly (if one has not yet appreciated the solid state distance parameters shown in Figure 7), the crystal field makes the bridge protons very strongly delocalized. The resulting vibrational signature is extremely broad and forms a continuous background feature from ca. 700 to 3100 cm^{-1}. This feature does not differentiate the two bridge protons. The reason for such a profound change in the bridge proton dynamics should be sought after for a particular arrangement of molecules in crystal; thus, the competition between inter- and intramolecular contacts turns out to be cooperation in the case of the solid state compound, compound **2**.

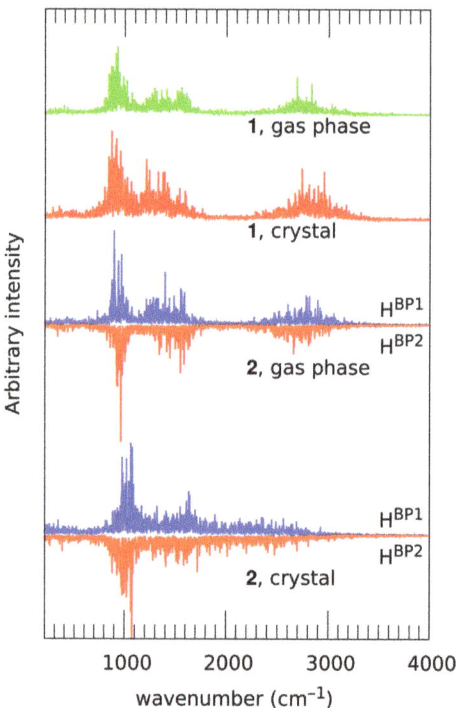

Figure 10. Vibrational signatures (atomic velocity power spectra) of the bridge protons calculated from the CPMD simulation of **1** and **2**. In the case of **2**, signatures of the non-equivalent bridge protons are presented as separate curves placed back to back. For atom numbering scheme, see Figure 1.

3. Computational Methodology

3.1. Static Models on the Basis of Density Functional Theory (DFT)

The models of monomers and dimers were constructed on the basis of X-ray structures of 2,3-dimethylnaphthazarin (**1**) (CCDC deposition number—1125030) and 2,3-dimethoxy-

6-methylnaphthazarin (**2**) (CCDC deposition number—1161869) [64,65,82]. The geometry optimization for the molecular forms of monomers was performed using Density Functional Theory (DFT) [66,67] and three functionals: B3LYP [83], PBE [84,85] and ωB97XD [86] with valence-split triple-zeta Pople's style basis set denoted as 6-311++G(2d,2p) [87,88]. The choice of functionals was devised to represent the current spectrum of the most widely used approaches: the PBE functional is of the Generalized Gradient Approximation (GGA) type used frequently in the context of plane-wave calculations (including Car–Parrinello MD), and does not use the exact exchange. On the other hand, B3LYP is a hybrid functional, and so is the ωB97XD, but the latter includes empirical dispersion correction. Following the geometry optimization, harmonic frequencies were computed to confirm that the obtained structures correspond with the minimum on the Potential Energy Surface (PES). Additionally, models with diverse proton positions were constructed and optimized as well using the DFT method (for details, see Table S1 of the Supplementary Information). In the next step, the single-point simulations at the MP2 [89] and CCSD [90,91] levels with def2-TZVP basis set [92] were carried out for the structures of the minima and transition state on the PT pathway. Next, the structures with OH groups on the proton-donor side were taken to investigate the proton potential paths using the scan method with geometry optimization (the O-H increment was set to 0.05 Å, the O8H^{BP1}O1 and O5H^{BP2}O4 valence angles were frozen while the remaining parts of the molecules were optimized). The results of the scans formed a discrete set of points, from which a proton potential function was derived. Thus, the barrier height is determined with accuracy depending on the discrete steps of energy in the vicinity of the transition state; the error estimate is the internal property of the procedure based on the discrete series of points, not the absolute uncertainty of a particular DFT functional. The zero-point vibrational correction is not included in the reported values. Finally, the wavefunctions for the Atoms In Molecules (AIM) theory [69] analysis were prepared with assistance of the B3LYP functional and 6-311++G(2d,2p) basis set for molecular and proton transferred forms of monomers. The theory was applied for the electronic structure as well as molecular topology investigations. Special attention was paid to the electron density and its Laplacian values at Bond and Ring Critical Points (BCPs and RCPs) related to the intramolecular hydrogen bonding. Next, for the dimeric structures extracted from the crystal data of compounds **1** and **2** [64,65], the energy minimization was performed using the ωB97XD functional [86] and 6-311++G(2d,2p) basis set. The simulations were carried out in the gas phase with the Gaussian 09 rev. D.01 [93] and Gaussian 16 rev. C.01. suite of programs [94]. The single-point MP2 and CCSD calculations were conducted with the Turbomole 6.5 program [95]. The AIM analysis was performed using the AIMAll program [96]. In addition, sets of coordinates are provided in the Supplementary Information for the current study.

3.2. An Application of Symmetry-Adapted Perturbation Theory (SAPT) to Dimers

The Symmetry-Adapted Perturbation Theory (SAPT) [70] enables energy decomposition between interacting molecules, in our case dimers. The method divides an exact Hamiltonian into Hartree–Fock contribution of monomers, $\hat{F}_A$ and $\hat{F}_B$, correlation components interacting inside the monomers, $\hat{W}_A$ and $\hat{W}_B$, and the contribution covering interaction between monomers, $\hat{V}$:

$$\hat{H} = \hat{F}_A + \hat{F}_B + \hat{W}_A + \hat{W}_B + \hat{V} \tag{1}$$

An important advantage of the SAPT scheme is the fact the individual components could be grouped into four principal groups with precisely defined physical interpretation: (i) electrostatic (E_{elst})—approximate Coulombic interactions of electron density decomposition of isolated monomers (without the effect of polarization by the neighboring molecule); (ii) exchange (E_{exch}—which is the short-range Pauli repulsion; (iii) Induction (E_{ind}) and exchange-induction (E_{ex-ind}—which is based on mutual polarization of the monomers; (iv) dispersion (E_{disp})—consideration of short-lived instantaneous multipoles. Depending on the considered energy components, the SAPT hierarchy of interactions is obtained. The

SAPT levels most commonly used are SAPT0 (in agreement with Hartree–Fock method) and SAPT2 (with accuracy approximate to the MP2 method):

$$E_{SAPT0} = E_{elst}^{10} + E_{exch}^{10} + E_{ind,r}^{20} + E_{ex-ind,r}^{20} + \delta E^{HF} + E_{disp}^{20} + E_{ex-disp}^{20} \tag{2}$$

$$E_{SAPT2} = E_{SAPT0} + E_{elst,r}^{12} + E_{exch}^{11} + E_{exch}^{12} +{}^{t} E_{ind}^{22} +{}^{t} E_{ex-ind}^{22} \tag{3}$$

These equations show the fundamental difference between the SAPT0 and SAPT2 approximations: the SAPT0 components never use intramonomer electron correlation, so—generally speaking—the resulting components of interaction energy are based on the non-correlated Hartree–Fock wavefunctions of the monomers. SAPT2, on the other hand, includes intramonomer correlation up to the second perturbative order, which is especially important for very weak interactions. In our experience with hydrogen-bonded systems, SAPT0 results are overestimated in comparison to the more accurate SAPT2 approach, but the general trends are reproduced with quite a high degree of correlation between the methods. Regarding the computational efficiency and memory requirements, SAPT2 can be prohibitively demanding for systems of ca. 60 atoms. However, due to the electron density expansion on specially fitted basis functions (density fitting technique), the SAPT0 computational cost is comparable to the MP2 method.

The energy decomposition of the naphthazarin derivative dimers (see Figure 5) was performed for: (i) data extracted from the X-ray structures of the investigated compounds [64,65] in order to reproduce the intermolecular forces in the crystal structure responsible for the crystal unit cell arrangement; (ii) the data obtained as a result of gas phase DFT simulations at the ωB97XD/6-311++G(2d,2p) level of theory. The interaction energy was calculated at the SAPT2/jun-cc-pVDZ level of theory (truncation of the diffuse functions in the jun-cc-pVDZ basis is derived in [97]). The basis set superposition error (BSSE) correction [98] was included in the simulations of the dimers (the studied dimers were divided into "monomers" in order to fulfil the requirements of the Boys–Bernardi method). The SAPT calculations were carried out using the Psi4 1.2.1 [99] program.

3.3. Car-Parrinello Molecular Dynamics in the Gas Phase and Solid State

The dynamical nature of the studied naphthazarin derivatives (compounds denoted as **1** and **2**, see Figures 1 and S2) [64,65] were examined in the light of First-Principle Molecular Dynamics (FPMD) method. The simulations were performed for the isolated molecules as well as for the molecular crystals. The gas phase simulations results were further used for the comparative study of differences introduced by the interatomic forces present in the solid state. Our attention was placed on the intramolecular hydrogen bonds' dynamics and properties. We have analyzed the hydrogen bridges dynamics as a function of simulation time. For this purpose, detailed analysis of metric parameters was performed for O1...O8/O5...O4 interatomic distance, O1-H^{BP1}/O2-H^{BP2} covalent bonds and H^{BP1}...O8/H^{BP2}...O4 intramolecular hydrogen bonds in compound **1**. Compound **1** is symmetric; therefore, we could expect that the bridged proton dynamics will be similar. However, we placed emphasis on a detailed view of protons motion in the hydrogen bridges. Compound **2** has a broken symmetry due to the presence of the CH$_3$ substituent. Both hydrogen bridges were taken into consideration in the analysis of metric parameters. We were looking for any correlations in the hydrogen bridge dynamics. Another aspect related to the data analyses were vibrational signatures provided by the OH groups. The Fourier transformation of the autocorrelation function of atomic velocity was employed to develop power spectra. The models used for Car–Parrinello molecular dynamics (CPMD) in the gas phase are presented in Figure 1. The initial geometries for the isolated molecules were extracted from the X-ray data [64,65] and placed in cubic boxes with a = 15 Å for compound **1** and a= 16 Å for compound **2**. The models for CPMD in the solid state were prepared on the basis of crystallographic unit cells [64,65]. The unit cell dimensions for compound **1** are as follows: a = 16.429 Å, b = 6.524 Å, c = 9.136 Å and β = 90.19° with Z = 4, while for compound **2**, a = 3.873 Å, b = 20.21 Å, c = 15.00 Å and β = 96.05° with Z = 4. The computational setup for the simulations in both studied phases was prepared bearing in mind the fact that intramolecular hydrogen bond dynamics were being studied. The simulations were divided into geometry optimization of the studied compounds, **1** and **2**, and subsequent CPMD runs in the gas phase and solid state. The exchange correlation functional by Perdew, Burke and Ernzerhof (PBE) [84,85] and Troullier–Martins [100] pseudopotentials were applied. The fictitious electron mass (EMASS) was equal to 400 a.u. and the time-step was set to 3 a.u. The kinetic energy cutoff for the plane-wave basis set was 80 Ry. The CPMD calculations were performed at 295 K, controlled by Nosé–Hoover thermostat chain assigned to ions [101,102]; the electronic system was thermostatted at the orbital kinetic energy values determined in separate short non-thermostatted runs for each system. Hockney's scheme [103] was applied to remove interactions with periodic images of the cubic cell during the gas phase dynamics. The translational and rotational movements were removed from the CPMD data collection as well. The crystalline phase CPMD was carried out with Γ point approximation [104] and Periodic Boundary Conditions (PBCs). The real-space electrostatic summation was set to TESR = 8 nearest neighbours in each direction. The CPMD simulations were divided into two parts: (i) equilibration (the initial part of the trajectory—ca. 10,000 steps—was removed from further analyses); (ii) production run, which lasted for 21 ps.

The CPMD simulations were performed using the CPMD 3.17.1 program [105]. The post-processing was carried out using home-made scripts and the VMD 1.9.3 [106] program. The graphical presentation of the obtained results in the current study was conducted with assistance of the VMD 1.9.3 [106] and Gnuplot [107] programs.

3.4. Estimation of the Nuclear Quantum Effects on the Structural Properties in the Gas Phase and Solid State

The nuclear quantum effects for the bridge proton motion were studied using an a posteriori approach based on the CPMD trajectory [79,80]. In short, the method consists of selecting several snapshots from the CPMD trajectory, calculating proton potential functions for each snapshots, and then, finally, solving the vibrational Schrödinger equation (see, e.g., [79,81]). The particular details for the current study are as follows. Four cases were considered: compounds **1** and **2** in the gas phase and solid state. For each case, five snapshots were extracted from the CPMD trajectory with constant time intervals. For each snapshot, a set of 16 to 20 bridge proton positions (depending on the donor–acceptor distance) was generated for the scan using the donor, proton and acceptor coordinates to define a fragment of an arc. The generated proton positions were then used to calculate single-point energies for the studied systems using the corresponding computational setup of the CPMD code—see the section above. Then, each of the generated proton potential profiles was fitted with a 9th degree polynomial, and a one-dimensional vibrational Schrödinger equation was solved using a grid basis set of 400 points spanning the O8-H^{BP1} region from 0.7 Å to 2.0 Å. Finally, the expectation value of the O8-H^{BP1} distance operator at 295 K was calculated taking into account the three lowest-lying vibrational levels. The electronic structure calculations were carried out with the CPMD 3.17.1 program [105], while the quantum vibrational effects were studied with the software developed by Stare and Mavri [81].

4. Conclusions

We have presented computational results of two naphthazarin derivatives substituted with methyl and methoxy groups in diverse manner. We have examined various factors influencing the molecular features exhibited by the aforementioned derivatives in relation to the properties of the substituents and symmetry breaking by their introduction. The presence of the substituents and changes in the chemical composition have led to changes in the bridged proton dynamics and intermolecular interactions in comparison to the parent compound, naphthazarin. The computations were performed in the electronic ground state, both in the gas phase and solid state. In order to shed light on the intermolecular interactions, the dimers of compounds **1** and **2** were investigated. Our computational findings were compared with the experimental data available (structural and spectroscopic). The application of the DFT method with three different functionals, each using a 6-311++G(2d,2p) basis set, complemented with the single-point MP2 and CCSD calculations with the def2-TZVP basis set, provided information of the proton reaction path and the energy barrier for the proton transfer. The highest DFT energy barrier equals ca. 10 kcal/mol, while MP2 and CCSD provided the barrier heights of ca. 6.8 and 9.6 kcal/mol, respectively. Moreover, two energy minima were located in both molecules and in both examined hydrogen bridges. The application of the AIM theory gave a quantitative picture of the electron density distribution in the molecular and proton transferred forms of the studied compounds. The topological analysis confirmed the presence of the intramolecular hydrogen bonds (in agreement with experimental X-ray findings in the literature). Additionally, it was shown, on the basis of electron density and its Laplacian values, that the hydrogen bonds are stronger in the tautomeric PT forms. The SAPT analysis gave an insight into energy partitioning and provided information on the primary factors responsible for dimer stabilization. It was found that the primary factors are the dispersive forces. Using the SAPT method, we could identify and describe quantitatively external forces influencing the molecular features of compounds **1** and **2**. The CPMD results showed that protons in the hydrogen bridges are very labile. Proton transfer phenomena were observed in the gas phase as well as in the solid state. In compound **2**, there is a clearly visible influence of environmental factors on the hydrogen bridge dynamics. The vibrational analysis confirmed, by the broad absorption regions observed in the computed power spectra, a strong anharmonicity of the studied hydrogen bonds as well as their dynamics. It is especially visible in compound **2**, where in the solid state only one very broad absorption (700 cm^{-1}–3100 cm^{-1}) region was found. The incorporation of nuclear quantum effects to the hydrogen bridges showed a stronger delocalization of the bridged protons, especially at shorter, but not the shortest, distances between the donor and acceptor heavy atoms.

Supplementary Materials: The following are available online. Figure S1. The structures of the investigated naphthazarin derivatives: 2,3-dimethylnaphthazarin (**1**) and 2,3-dimethoxy-6-methylnaphthazarin (**2**), with atom numbering scheme for hydrogen bridges. Coloring scheme: oxygen atom—red, carbon atom—grey and hydrogen atom—white. Figure S2. The models for gas phase and solid state CPMD simulations. Left—the isolated molecule model of 2,3-dimethylnaphthazarin (**1**); right—the model used for solid state simulations of 2,3-dimethoxy-6-methylnaphthazarin (**2**). Table S1. Energy for compounds **1** and **2** with different proton positions in the hydrogen bridges computed using DFT method. Electronic as well as vibrational zero point-corrected values are given. Table S2. Selected geometric parameters related to the intramolecular hydrogen bonds of 2,3-dimethylnaphthazarin (**1**) and 2,3-dimethoxy-6-methylnaphthazarin (**2**). Comparison of experimental and computed data. Metric parameters are given in Å and degrees. CPMD results are presented as average ± standard deviation. Sets of coordinates for the minima and transition state estimates from the DFT scans (XYZ format).

Author Contributions: Conceptualization, A.J.; methodology, A.J. and J.J.P.; computations, K.K., M.P., A.J. and J.J.P.; data analysis, K.K., M.P., A.J. and J.J.P.; writing—original draft preparation, K.K., M.P., A.J. and J.J.P.; writing—review and editing, K.K., M.P., A.J. and J.J.P.; supervision, A.J. All authors have read and agreed to the published version of the manuscript.

Funding: This research was supported by the National Science Centre grant number UMO-2015/17/B/ST4/03568.

Institutional Review Board Statement: Not applicable.

Informed Consent Statement: Not applicable.

Data Availability Statement: All the relevant processed data (energy values, AIM data, SAPT energy terms, structural information, time evolution of distances in the hydrogen bridges, vibrational signatures) are reported within the manuscript.

Acknowledgments: The authors thank the Wrocław Center for Networking and Supercomputing (WCSS) in Wrocław and Academic Computer Centre Cyfronet AGH (PL-Grid infrastructure Prometheus) in Kraków for generous allocation of computing time and technical support. In addition, A.J. and J.J.P. thank the National Science Centre (Poland) for supporting this work under the grant no. UMO-2015/17/B/ST4/03568.

Conflicts of Interest: The authors declare no conflict of interest.

Abbreviations

The following abbreviations are used in this manuscript:

DFT	Density Functional Theory
CPMD	Car–Parrinello Molecular Dynamics
PT	Proton Transfer
AIM	Atoms In Molecules
SAPT	Symmetry-Adapted Perturbation Theory
NQE	Nuclear Quantum Effects
RAHB	Resonance-Assisted Hydrogen Bond
PES	Potential Energy Surface
MP2	Møller–Plesset second-order perturbation theory
CCSD	Coupled Clusters with Singles and Doubles
BCP	Bond Critical Point
RCP	Ring Critical Point
BSSE	Basis Set Superposition Error
FPMD	First-Principle Molecular Dynamics
PBCs	Periodic Boundary Conditions

References and Note

1. Belkova, N.V.; Shubina, E.S.; Epstein, L.M. Diverse World of Unconventional Hydrogen Bonds. *Acc. Chem. Res.* **2005**, *38*, 624–631. [CrossRef] [PubMed]
2. Desiraju, G.R.; Steiner, T. Other weak and non-conventional hydrogen bonds. In *The Weak Hydrogen Bond*; Oxford University Press: Oxford, UK, 2001; pp. 122–292. [CrossRef]
3. Honacker, C.; Kappelt, B.; Jabłoński, M.; Hepp, A.; Layh, M.; Rogel, F.; Uhl, W. Aluminium Functionalized Germanes: Intramolecular Activation of Ge-H Bonds, Formation of a Dihydrogen Bond and Facile Hydrogermylation of Unsaturated Substrates. *Eur. J. Inorg. Chem.* **2019**, *2019*, 3287–3300. [CrossRef]
4. Müller-Dethlefs, K.; Hobza, P. Noncovalent Interactions: A Challenge for Experiment and Theory. *Chem. Rev.* **2000**, *100*, 143–167. [CrossRef]
5. Hobza, P.; Zahradník, R.; Müller-Dethlefs, K. The World of Non-Covalent Interactions: 2006. *Collect. Czechoslov. Chem. Commun.* **2006**, *71*, 443–531. [CrossRef]
6. Hobza, P.; Řezáč, J. Introduction: Noncovalent Interactions. *Chem. Rev.* **2016**, *116*, 4911–4912. [CrossRef]
7. Puzzarini, C.; Spada, L.; Alessandrini, S.; Barone, V. The challenge of non-covalent interactions: Theory meets experiment for reconciling accuracy and interpretation. *J. Phys. Condens. Matter* **2020**, *32*, 343002. [CrossRef]
8. Zierkiewicz, W.; Michalczyk, M.; Scheiner, S. Competition between Intra and Intermolecular Triel Bonds. Complexes between Naphthalene Derivatives and Neutral or Anionic Lewis Bases. *Molecules* **2020**, *25*, 635. [CrossRef] [PubMed]
9. Desiraju, G.R. Hydrogen Bridges in Crystal Engineering: Interactions without Borders. *Acc. Chem. Res.* **2002**, *35*, 565–573. [CrossRef] [PubMed]
10. Israelachvili, J.N. *Intermolecular and Surface Forces*; Academic Press: London, UK, 1992.
11. Jeffrey, G.; Saenger, W. *Hydrogen Bonding in Biological Structures*; Springer: Berlin, Germany, 1991.
12. Scheiner, S. *Hydrogen Bonding: A Theoretical Perspective*; Oxford University Press: New York, NY, USA, 1997.
13. Jeffrey, G. *An Introduction to Hydrogen Bonding*; Oxford University Press: New York, NY, USA, 1997.
14. Grabowski, S.J. What Is the Covalency of Hydrogen Bonding? *Chem. Rev.* **2011**, *111*, 2597–2625. [CrossRef]
15. Herschlag, D.; Pinney, M.M. Hydrogen Bonds: Simple after All? *Biochemistry* **2018**, *57*, 3338–3352. [CrossRef]
16. Trylska, J. The role of hydrogen bonding in the enzymatic reaction catalyzed by HIV-1 protease. *Protein Sci.* **2004**, *13*, 513–528. [CrossRef]
17. Simón, L.; Goodman, J.M. Enzyme Catalysis by Hydrogen Bonds: The Balance between Transition State Binding and Substrate Binding in Oxyanion Holes. *J. Org. Chem.* **2009**, *75*, 1831–1840. [CrossRef] [PubMed]
18. Sobczyk, L.; Grabowski, S.J.; Krygowski, T.M. Interrelation between H-Bond and Pi-Electron Delocalization. *Chem. Rev.* **2005**, *105*, 3513–3560. [CrossRef] [PubMed]

19. Filarowski, A.; Koll, A.; Sobczyk, L. Intramolecular Hydrogen Bonding in o-hydroxy Aryl Schiff Bases. *Curr. Org. Chem.* **2009**, *13*, 172–193. [CrossRef]

20. Kwocz, A.; Panek, J.J.; Jezierska, A.; Hetmańczyk, Ł.; Pawlukojć, A.; Kochel, A.; Lipkowski, P.; Filarowski, A. A molecular roundabout: Triple cyclically arranged hydrogen bonds in light of experiment and theory. *New J. Chem.* **2018**, *42*, 19467–19477 [CrossRef]

21. Li, J.; Wang, S.; Liao, L.; Ma, Q.; Zhang, Z.; Fan, G. Stabilization of an intramolecular hydrogen-bond block in an s-triazine insensitive high-energy material. *New J. Chem.* **2019**, *43*, 10675–10679. [CrossRef]

22. Ozeryanskii, V.A.; Marchenko, A.V.; Pozharskii, A.F.; Filarowski, A.; Spiridonova, D.V. Combination of "Buttressing" and "Clothespin" Effects for Reaching the Shortest NHN Hydrogen Bond in Proton Sponge Cations. *J. Org. Chem.* **2021**, *86*, 3637–3647 [CrossRef] [PubMed]

23. Abe, Y.; Harata, K.; Fujiwara, M.; Ohbu, K. Molecular Arrangement and Intermolecular Hydrogen Bonding in Crystals of Methyl 6-O-Acyl-d-glycopyranosides. *Langmuir* **1996**, *12*, 636–640. [CrossRef]

24. Vella-Zarb, L.; Braga, D.; Orpen, A.G.; Baisch, U. The influence of hydrogen bonding on the planar arrangement of melamine in crystal structures of its solvates, cocrystals and salts. *CrystEngComm* **2014**, *16*, 8147. [CrossRef]

25. Mes, T.; Smulders, M.M.J.; Palmans, A.R.A.; Meijer, E.W. Hydrogen-Bond Engineering in Supramolecular Polymers: Polarity Influence on the Self-Assembly of Benzene-1,3,5-tricarboxamides. *Macromolecules* **2010**, *43*, 1981–1991. [CrossRef]

26. Sikder, A.; Ghosh, S. Hydrogen-bonding regulated assembly of molecular and macromolecular amphiphiles. *Mater. Chem. Front.* **2019**, *3*, 2602–2616. [CrossRef]

27. Han, K.L.; Zhao, G.J. (Eds.) *Hydrogen Bonding and Transfer in the Excited State*; John Wiley & Sons, Ltd.: Chichester, UK, 2010. [CrossRef]

28. Gilli, G.; Bellucci, F.; Ferretti, V.; Bertolasi, V. Evidence for resonance-assisted hydrogen bonding from crystal-structure correlations on the enol form of the .beta.-diketone fragment. *J. Am. Chem. Soc.* **1989**, *111*, 1023–1028. [CrossRef]

29. Gilli, P.; Bertolasi, V.; Ferretti, V.; Gilli, G. Evidence for resonance-assisted hydrogen bonding. 4. Covalent nature of the strong homonuclear hydrogen bond. Study of the O-H-O system by crystal structure correlation methods. *J. Am. Chem. Soc.* **1994**, *116*, 909–915. [CrossRef]

30. Mautner, M.M.N. The Ionic Hydrogen Bond. *Chem. Rev.* **2005**, *105*, 213–284. [CrossRef]

31. Gilli, G.; Gilli, P. Towards an unified hydrogen-bond theory. *J. Mol. Struct.* **2000**, *552*, 1–15. [CrossRef]

32. Jesus, A.J.L.; Redinha, J.S. Charge-Assisted Intramolecular Hydrogen Bonds in Disubstituted Cyclohexane Derivatives. *J. Phys. Chem. A* **2011**, *115*, 14069–14077. [CrossRef] [PubMed]

33. Hammes-Schiffer, S.; Stuchebrukhov, A.A. Theory of Coupled Electron and Proton Transfer Reactions. *Chem. Rev.* **2010**, *110*, 6939–6960. [CrossRef] [PubMed]

34. Jacquemin, D.; Zúñiga, J.; Requena, A.; Céron-Carrasco, J.P. Assessing the Importance of Proton Transfer Reactions in DNA. *Acc. Chem. Res.* **2014**, *47*, 2467–2474. [CrossRef]

35. Ishikita, H.; Saito, K. Proton transfer reactions and hydrogen-bond networks in protein environments. *J. R. Soc. Interface* **2014**, *11*, 20130518. [CrossRef]

36. Jana, S.; Dalapati, S.; Guchhait, N. Proton Transfer Assisted Charge Transfer Phenomena in Photochromic Schiff Bases and Effect of –NEt2 Groups to the Anil Schiff Bases. *J. Phys. Chem. A* **2012**, *116*, 10948–10958. [CrossRef]

37. Panek, J.J.; Filarowski, A.; Jezierska-Mazzarello, A. Impact of proton transfer phenomena on the electronic structure of model Schiff bases: An AIM/NBO/ELF study. *J. Chem. Phys.* **2013**, *139*, 154312. [CrossRef]

38. Joshi, H.C.; Antonov, L. Excited-State Intramolecular Proton Transfer: A Short Introductory Review. *Molecules* **2021**, *26*, 1475. [CrossRef] [PubMed]

39. Stuchebrukhov, A.A. Mechanisms of proton transfer in proteins: Localized charge transfer versus delocalized soliton transfer. *Phys. Rev. E* **2009**, *79*. [CrossRef] [PubMed]

40. Narayan, S.; Wyatt, D.L.; Crumrine, D.S.; Cukierman, S. Proton Transfer in Water Wires in Proteins: Modulation by Local Constraint and Polarity in Gramicidin A Channels. *Biophys. J.* **2007**, *93*, 1571–1579. [CrossRef] [PubMed]

41. Jezierska, A.; Panek, J.J. Theoretical study of intramolecular hydrogen bond in selected symmetric "proton sponges" on the basis of DFT and CPMD methods. *J. Mol. Model.* **2020**, *26*, 37. [CrossRef]

42. Król-Starzomska, I.; Filarowski, A.; Rospenk, M.; Koll, A.; Melikova, S. Proton Transfer Equilibria in Schiff Bases with Steric Repulsion. *J. Phys. Chem. A* **2004**, *108*, 2131–2138. [CrossRef]

43. Jezierska, A.; Panek, J.J. First-Principle Molecular Dynamics Study of Selected Schiff and Mannich Bases: Application of Two-Dimensional Potential of Mean Force to Systems with Strong Intramolecular Hydrogen Bonds. *J. Chem. Theory Comput.* **2008**, *4*, 375–384. [CrossRef]

44. Jezierska-Mazzarello, A.; Panek, J.J.; Vuilleumier, R.; Koll, A.; Ciccotti, G. Direct observation of the substitution effects on the hydrogen bridge dynamics in selected Schiff bases—A comparative molecular dynamics study. *J. Chem. Phys.* **2011**, *134*, 034308. [CrossRef]

45. Jezierska, A.; Panek, J.J.; Koll, A. Spectroscopic Properties of a Strongly Anharmonic Mannich Base N-oxide. *ChemPhysChem* **2008**, *9*, 839–846. [CrossRef] [PubMed]

46. Panek, J.J.; Błaziak, K.; Jezierska, A. Hydrogen bonds in quinoline N-oxide derivatives: First-principle molecular dynamics and metadynamics ground state study. *Struct. Chem.* **2016**, *27*, 65–75. [CrossRef]

47. Panek, J.J.; Jezierska, A. N-oxide Derivatives: Car–Parrinello Molecular Dynamics and Electron Localization Function Study on Intramolecular Hydrogen Bonds. *J. Phys. Chem. A* **2018**, *122*, 6605–6614. [CrossRef] [PubMed]
48. Jezierska, A.; Panek, J.J. "Zwitterionic Proton Sponge" Hydrogen Bonding Investigations on the Basis of Car–Parrinello Molecular Dynamics. *J. Chem. Inf. Model.* **2015**, *55*, 1148–1157. [CrossRef] [PubMed]
49. Pekin, G.; Ganzera, M.; Senol, S.; Bedir, E.; Korkmaz, K.; Stuppner, H. Determination of Naphthazarin Derivatives in Endemic Turkish Alkanna Species by Reversed Phase High Performance Liquid Chromatography. *Planta Med.* **2007**, *73*, 267–272. [CrossRef]
50. Sánchez-Calvo, J.M.; Barbero, G.R.; Guerrero-Vásquez, G.; Durán, A.G.; Macías, M.; Rodríguez-Iglesias, M.A.; Molinillo, J.M.G.; Macías, F.A. Synthesis, antibacterial and antifungal activities of naphthoquinone derivatives: A structure–activity relationship study. *Med. Chem. Res.* **2016**, *25*, 1274–1285. [CrossRef]
51. Masuda, K.; Funayama, S.; Komiyama, K.; Unezawa, I.; Ito, K. Constituents of Tritonia crocosmaeflora, I. Tricrozarin A, a Novel Antimicrobial Naphthazarin Derivative. *J. Nat. Prod.* **1987**, *50*, 418–421. [CrossRef]
52. Rudnicka, M.; Ludynia, M.; Karcz, W. The Effect of Naphthazarin on the Growth, Electrogenicity, Oxidative Stress, and Microtubule Array in Z. mays Coleoptile Cells Treated With IAA. *Front. Plant Sci.* **2019**, *9*, 1940. [CrossRef] [PubMed]
53. Papageorgiou, V.P.; Assimopoulou, A.N.; Couladouros, E.A.; Hepworth, D.; Nicolaou, K.C. The Chemistry and Biology of Alkannin, Shikonin, and Related Naphthazarin Natural Products. *Angew. Chem. Int. Ed.* **1999**, *38*, 270–301. [CrossRef]
54. Song, G.; Kim, Y.; You, Y.; Cho, H.; Kim, S.; Sok, D.E.; Ahn, B. Naphthazarin derivatives (VI): Synthesis, inhibitory effect on DNA topoisomerase-I and antiproliferative activity of 2- or 6-(1-oxyiminoalkyl)- 5,8-dimethoxy-1,4-naphthoquinones. *Arch. Pharm.* **2000**, *333*, 87–92. [CrossRef]
55. Park, E.M.; Lee, I.J.; Kim, S.H.; Song, G.Y.; Park, Y.M. Inhibitory effect of a naphthazarin derivative, S64, on heat shock factor (Hsf) activation and glutathione status following hypoxia. *Cell Biol. Toxicol.* **2003**, *19*, 273–284. [CrossRef]
56. Ma, L.; Zhang, Q.; Yang, C.; Zhu, Y.; Zhang, L.; Wang, L.; Liu, Z.; Zhang, G.; Zhang, C. Assembly Line and Post-PKS Modifications in the Biosynthesis of Marine Polyketide Natural Products. In *Comprehensive Natural Products III*; Elsevier: Amsterdam, The Netherlands, 2020; pp. 139–197. [CrossRef]
57. Tchizhova, A.Y.; Anufriev, V.P.; Glazunov, V.P.; Denisenko, V.A. The chemistry of naphthazarin derivatives. *Russ. Chem. Bull.* **2000**, *49*, 466–471. [CrossRef]
58. Herbstein, F.H.; Kapon, M.; Reisner, G.M.; Lehman, M.S.; Kress, R.B.; Wilson, R.B.; Shiau, W.i.; Duesler, E.N.; Paul, I.C.; Curtin, D.Y.; et al. Polymorphism of naphthazarin and its relation to solid-state proton transfer. Neutron and X-ray diffraction studies on naphthazarin C. *Proc. R. Soc. Lond. A. Math. Phys. Sci.* **1985**, *399*, 295–319. [CrossRef]
59. Rusinska-Roszak, D. Energy of Intramolecular Hydrogen Bonding in ortho-Hydroxybenzaldehydes, Phenones and Quinones. Transfer of Aromaticity from ipso-Benzene Ring to the Enol System(s). *Molecules* **2017**, *22*, 481. [CrossRef] [PubMed]
60. Jezierska, A.; Kizior, B.; Szyja, B.M.; Panek, J.J. On the nature of inter- and intramolecular interactions involving benzo[h]quinoline and 10-hydroxybenzo[h]quinoline: Electronic ground state vs excited state study. *J. Mol. Struct.* **2021**, *1234*, 130126. [CrossRef]
61. Jezierska, A.; Błaziak, K.; Klahm, S.; Lüchow, A.; Panek, J.J. Non-Covalent Forces in Naphthazarin—Cooperativity or Competition in the Light of Theoretical Approaches. *Int. J. Mol. Sci.* **2021**, *22*, 8033. [CrossRef] [PubMed]
62. Tabrizi, M.Z.; Tayyari, S.F.; Tayyari, F.; Behforouz, M. Fourier transform infrared and Raman spectra, vibrational assignment and density functional theory calculations of naphthazarin. *Spectrochim. Acta A Mol. Biomol. Spectrosc.* **2004**, *60*, 111–120. [CrossRef]
63. Hansch, C.; Leo, A.; Taft, R.W. A survey of Hammett substituent constants and resonance and field parameters. *Chem. Rev.* **1991**, *91*, 165–195. [CrossRef]
64. Rodríguez, J.; Smith-Verdier, P.; Florencio, F.; García-Blanco, S. Crystal and molecular structure of 2, 3-dimethylnaph-thazarin: A charge-transfer complex. *J. Mol. Struct.* **1984**, *112*, 101–109. [CrossRef]
65. Cannon, J.; Matsuki, Y.; Patrick, V.; White, A. Selective O-Demethylation of 5, 8-Dihydroxy-2, 3-Dimethoxy-6-Methylnaphtho-1, 4-Quinone: The Crystal Structures of 5, 8-Dihydroxy-2, 3-Dimethoxy-6-Methylnaphtho-1, 4-Quinone and 2, 5, 8-Triacetoxy-3-Methoxy-6-Methylnaphtho-1, 4-Quinone. *Aust. J. Chem.* **1987**, *40*, 1191. [CrossRef]
66. Hohenberg, P.; Kohn, W. Inhomogeneous Electron Gas. *Phys. Rev.* **1964**, *136*, B864–B871. [CrossRef]
67. Kohn, W.; Sham, L.J. Self-Consistent Equations Including Exchange and Correlation Effects. *Phys. Rev.* **1965**, *140*, A1133–A1138. [CrossRef]
68. Car, R.; Parrinello, M. Unified Approach for Molecular Dynamics and Density-Functional Theory. *Phys. Rev. Lett.* **1985**, *55*, 2471–2474. [CrossRef]
69. Bader, R. *Atoms in Molecules: A Quantum Theory*; International Ser. of Monogr. on Chem.; Clarendon Press: Oxford, UK, 1994.
70. Jeziorski, B.; Moszynski, R.; Szalewicz, K. Perturbation Theory Approach to Intermolecular Potential Energy Surfaces of van der Waals Complexes. *Chem. Rev.* **1994**, *94*, 1887–1930. [CrossRef]
71. Kizior, B.; Panek, J.J.; Szyja, B.M.; Jezierska, A. Structure-Property Relationship in Selected Naphtho- and Anthra-Quinone Derivatives on the Basis of Density Functional Theory and Car–Parrinello Molecular Dynamics. *Symmetry* **2021**, *13*, 564. [CrossRef]
72. Bickelhaupt, F.M.; Baerends, E.J. Kohn-Sham Density Functional Theory: Predicting and Understanding Chemistry. In *Reviews in Computational Chemistry*; John Wiley and Sons, Ltd.: Hoboken, NJ, USA, 2000; pp. 1–86. [CrossRef]
73. Alkorta, I.; Elguero, J.; Oliva-Enrich, J.M.; Yáñez, M.; Mó, O.; Montero-Campillo, M.M. The Importance of Strain (Preorganization) in Beryllium Bonds. *Molecules* **2020**, *25*, 5876. [CrossRef] [PubMed]
74. Hohenstein, E.G.; Sherrill, C.D. Density fitting of intramonomer correlation effects in symmetry-adapted perturbation theory. *J. Chem. Phys.* **2010**, *133*, 014101. [CrossRef]

75. Marx, D.; Tuckerman, M.E.; Hutter, J.; Parrinello, M. The nature of the hydrated excess proton in water. *Nature* **1999**, *397*, 601–604. [CrossRef]
76. Marx, D.; Tuckerman, M.E.; Parrinello, M. Solvated excess protons in water: Quantum effects on the hydration structure. *J. Phys. Condens. Matter* **2000**, *12*, A153–A159. [CrossRef]
77. Kong, L.; Wu, X.; Car, R. Roles of quantum nuclei and inhomogeneous screening in the X-ray absorption spectra of water and ice. *Phys. Rev. B* **2012**, *86*, 134203. [CrossRef]
78. Stare, J.; Panek, J.; Eckert, J.; Grdadolnik, J.; Mavri, J.; Hadži, D. Proton Dynamics in the Strong Chelate Hydrogen Bond of Crystalline Picolinic Acid N-Oxide. A New Computational Approach and Infrared, Raman and INS Study. *J. Phys. Chem. A* **2008**, *112*, 1576–1586. [CrossRef]
79. Jezierska, A.; Panek, J.J.; Koll, A.; Mavri, J. Car-Parrinello simulation of an O–H stretching envelope and potential of mean force of an intramolecular hydrogen bonded system: Application to a Mannich base in solid state and in vacuum. *J. Chem. Phys.* **2007**, *126*, 205101. [CrossRef]
80. Denisov, G.S.; Mavri, J.; Sobczyk, L. Potential energy shape for the proton motion in hydrogen bonds reflected in infrared and NMR spectra. In *Hydrogen Bonding—New Insights*, 1st ed.; Challenges and Advances in Computational Chemistry and Physics, 3; Grabowski, S.J., Ed.; Springer: Dordrecht, The Netherlands, 2006; pp. 377–416.
81. Stare, J.; Mavri, J. Numerical solving of the vibrational time-independent Schrödinger equation in one and two dimensions using the variational method. *Comput. Phys. Commun.* **2002**, *143*, 222–240. [CrossRef]
82. CCDC Structural Database. Available online: https://www.ccdc.cam.ac.uk/ (accessed on 7 May 2021).
83. Becke, A.D. A multicenter numerical integration scheme for polyatomic molecules. *J. Chem. Phys.* **1988**, *88*, 2547–2553. [CrossRef]
84. Perdew, J.P.; Ernzerhof, M.; Burke, K. Rationale for mixing exact exchange with density functional approximations. *J. Chem. Phys.* **1996**, *105*, 9982–9985. [CrossRef]
85. Perdew, J.P.; Burke, K.; Ernzerhof, M. Generalized Gradient Approximation Made Simple. *Phys. Rev. Lett.* **1997**, *78*, 1396. [CrossRef]
86. Chai, J.-D.; Head-Gordon, M. Long-range corrected hybrid density functionals with damped atom–atom dispersion corrections. *Phys. Chem. Chem. Phys.* **2008**, *10*, 6615. [CrossRef] [PubMed]
87. McLean, A.D.; Chandler, G.S. Contracted Gaussian basis sets for molecular calculations. I. Second row atoms, Z = 11–18. *J. Chem. Phys.* **1980**, *72*, 5639–5648. [CrossRef]
88. Krishnan, R.; Binkley, J.S.; Seeger, R.; Pople, J.A. Self-consistent molecular orbital methods. XX. A basis set for correlated wave functions. *J. Chem. Phys.* **1980**, *72*, 650–654. [CrossRef]
89. Møller, C.; Plesset, M.S. Note on an Approximation Treatment for Many-Electron Systems. *Phys. Rev.* **1934**, *46*, 618–622. [CrossRef]
90. Čížek, J. On the Use of the Cluster Expansion and the Technique of Diagrams in Calculations of Correlation Effects in Atoms and Molecules. In *Advances in Chemical Physics*; John Wiley and Sons, Ltd.: London, UK, 1969; pp. 35–89. [CrossRef]
91. Purvis, G.D.; Bartlett, R.J. A full coupled-cluster singles and doubles model: The inclusion of disconnected triples. *J. Chem. Phys.* **1982**, *76*, 1910–1918. [CrossRef]
92. Weigend, F.; Ahlrichs, R. Balanced basis sets of split valence, triple zeta valence and quadruple zeta valence quality for H to Rn: Design and assessment of accuracy. *Phys. Chem. Chem. Phys.* **2005**, *7*, 3297–3305. [CrossRef]
93. Frisch, M.J.; Trucks, G.W.; Schlegel, H.B.; Scuseria, G.E.; Robb, M.A.; Cheeseman, J.R.; Scalmani, G.; Barone, V.; Mennucci, B.; Petersson, G.A.; et al. *Gaussian 09 Revision E.01*; Gaussian Inc.: Wallingford, CT, USA, 2009.
94. Frisch, M.J.; Trucks, G.W.; Schlegel, H.B.; Scuseria, G.E.; Robb, M.A.; Cheeseman, J.R.; Scalmani, G.; Barone, V.; Petersson, G.A.; Nakatsuji, H.; et al. *Gaussian~16 Revision A.03*; Gaussian Inc.: Wallingford, CT, USA, 2016.
95. TURBOMOLE V6.5, a Development of University of Karlsruhe and Forschungszentrum Karlsruhe GmbH, 1989–2007. TURBOMOLE GmbH, Since 2007. Available online: http://www.turbomole.com (accessed on 1 June 2021).
96. Keith, T.A. *AIMAll*, Version 19.10.12; TK Gristmill Software: Overland Park, KS, USA, 2019.
97. Papajak, E.; Zheng, J.; Xu, X.; Leverentz, H.R.; Truhlar, D.G. Perspectives on Basis Sets Beautiful: Seasonal Plantings of Diffuse Basis Functions. *J. Chem. Theory Comput.* **2011**, *7*, 3027–3034. [CrossRef]
98. Boys, S.; Bernardi, F. The calculation of small molecular interactions by the differences of separate total energies. Some procedures with reduced errors. *Mol. Phys.* **1970**, *19*, 553–566. [CrossRef]
99. Parrish, R.M.; Burns, L.A.; Smith, D.G.A.; Simmonett, A.C.; DePrince, A.E.; Hohenstein, E.G.; Bozkaya, U.; Sokolov, A.Y.; Remigio, R.D.; Richard, R.M.; et al. Psi4 1.1: An Open-Source Electronic Structure Program Emphasizing Automation, Advanced Libraries, and Interoperability. *J. Chem. Theory Comput.* **2017**, *13*, 3185–3197. [CrossRef] [PubMed]
100. Troullier, N.; Martins, J.L. Efficient pseudopotentials for plane-wave calculations. *Phys. Rev. B* **1991**, *43*, 1993–2006. [CrossRef] [PubMed]
101. Nosé, S. A unified formulation of the constant temperature molecular dynamics methods. *J. Chem. Phys.* **1984**, *81*, 511–519. [CrossRef]
102. Hoover, W.G. Canonical dynamics: Equilibrium phase-space distributions. *Phys. Rev. A* **1985**, *31*, 1695–1697. [CrossRef] [PubMed]
103. Hockney, R.W. Potential Calculation and Some Applications. *Methods Comput. Phys.* **1970**, *9*, 135–211.
104. Kittel, C. *Introduction To Solid State Physics*, 8th ed.; Wiley and Sons: Hoboken, NJ, USA, 2019.
105. *CPMD 3.17.1*; Copyright IBM Corp. (1990–2004) Copyright MPI für Festkoerperforschung Stuttgart (1997–2001).

106. Humphrey, W.; Dalke, A.; Schulten, K. VMD–Visual Molecular Dynamics. *J. Mol. Graph.* **1996**, *14*, 33–38. [CrossRef]
107. Williams, T.; Kelley, C. Gnuplot 4.4: An Interactive Plotting Program. 2010. Available online: http://gnuplot.sourceforge.net/ (accessed on 12 May 2021).

molecules

MDPI

Article

Perturbing the O–H ⋯ O Hydrogen Bond in 1-oxo-3-hydroxy-2-propene

Ibon Alkorta [1,*], José Elguero [1] and Janet E. Del Bene [2,*]

[1] Instituto de Química Médica, CSIC, Juan de la Cierva, 3, E-28006 Madrid, Spain; iqmbe17@iqm.csic.es
[2] Department of Chemistry, Youngstown State University, Youngstown, OH 44555, USA
* Correspondence: ibon@iqm.csic.es (I.A.); jedelbene@ysu.edu (J.E.D.B.)

Abstract: Ab initio MP2/aug'-cc-pVTZ calculations have been carried out to identify and characterize equilibrium structures and transition structures on the 1-oxo-3-hydroxy-2-propene: Lewis acid potential energy surfaces, with the acids LiH, LiF, BeH$_2$, and BeF$_2$. Two equilibrium structures, one with the acid interacting with the C=O group and the other with the interaction occurring at the O–H group, exist on all surfaces. These structures are separated by transition structures that present the barriers to the interconversion of the two equilibrium structures. The structures with the acid interacting at the C=O group have the greater binding energies. Since the barriers to convert the structures with interaction occurring at the O–H group are small, only the isomers with interaction occurring at the C=O group could be experimentally observed, even at low temperatures. Charge-transfer energies were computed for equilibrium structures, and EOM-CCSD spin–spin coupling constants $^{2h}J(O–O)$, $^{1h}J(H–O)$, and $^{1}J(O–H)$ were computed for equilibrium and transition structures. These coupling constants exhibit a second-order dependence on the corresponding distances, with very high correlation coefficients.

Keywords: intramolecular hydrogen bonds; structures and binding energies; charge-transfer interactions; spin–spin coupling constants

Citation: Alkorta, I.; Elguero, J.; Del Bene, J.E. Perturbing the O–H ⋯ O Hydrogen Bond in 1-oxo-3-hydroxy-2-propene. *Molecules* **2021**, *26*, 3086. https://doi.org/10.3390/molecules26113086

Academic Editor: Mirosław Jabłoński

Received: 6 April 2021
Accepted: 12 May 2021
Published: 21 May 2021

Publisher's Note: MDPI stays neutral with regard to jurisdictional claims in published maps and institutional affiliations.

1. Introduction

1-oxo-3-hydroxy-2-propene, the enol of malonaldehyde in the conformation that presents an intramolecular hydrogen bond [1], has played an important role in the fields of hydrogen bonding, proton transfer [2–8], resonance assisted hydrogen bonds [6,9–12], quasi-aromaticity [13], and, in general, in non-covalent interactions. The importance of this molecule is due to the fact that it is the simplest of all 1,3-dicarbonyl compounds that were the first systems in which intramolecular hydrogen bonds had been studied [14–19].

Various aspects of 1-oxo-3-hydroxy-2-propene have been investigated, including substituent effects [20,21], solvent effects in which water molecules interact with this molecule [22], excited state properties [23], and the important problem of proton tunneling [23–27]. Both vibrational [28] and rotational [29–31] spectroscopic studies have been carried out on this molecule. In addition, interactions of 1-oxo-3-hydroxy-2-propene with Lewis acids such as methanol [32] and Li$^+$, Na$^+$ and FH [33], and BeX$_2$ [34] have been reported.

A useful property for obtaining structural information about complexes linked by non-covalent interactions and, in particular by hydrogen bonds, is spin–spin coupling constants (SSCC). SSCC are related to the electronic structure of molecules and complexes through geometry, bond order, polarization, and electron densities. As Cremer and Gräfenstein wrote [35], "The analysis of NMR spin–spin coupling leads to a unique insight into the electronic structure of closed-shell molecules". This was known for molecules from the beginning of the use of NMR spectroscopy [36–38] but was extended to complexes by Limbach [39,40] and Del Bene [41,42]. Through relationships between SSCC and geometry, the problem of the localization of the hydrogen-bonded proton could be solved [43].

175

In the present paper, we report the results of an investigation of 1-oxo-3-hydroxy-2-propene in a series of binary complexes with the acids LiH, LiF, BeH$_2$, and BeF$_2$. These complexes contain intramolecular O–H···O hydrogen bonds and lithium (alkali) and beryllium (alkaline earth) intermolecular bonds [44]. Specifically, we have determined the structures and binding energies of these complexes; the proton transfer barriers; the complex stabilization by charge-transfer interactions; and the spin–spin coupling constants ^{2h}J(O–O), ^{1h}J(H–O), and ^{1}J(O–H) across the O–H···O hydrogen bond. It is the purpose of this paper to present and discuss the results of this study.

2. Results and Discussion

In order to simplify the discussion of the equilibrium and the transition structures on the 1-oxo-3-hydroxy-2-propene:acid potential energy surfaces, we refer to the hydrogen-bonded molecule 1-oxo-3-hydroxy-2-propene as **1** and name the complexes **1**:LiH(OH), **1**:LiH(ts), and **1**:LiH(CO), where **1**:LiH(OH) indicates that the acid LiH interacts with the hydroxyl oxygen, **1**:LiH(CO) indicates that the interaction with the acid occurs at the carbonyl oxygen, and **1**:LiH(ts) identifies the transition structure. These complexes are illustrated in Scheme 1.

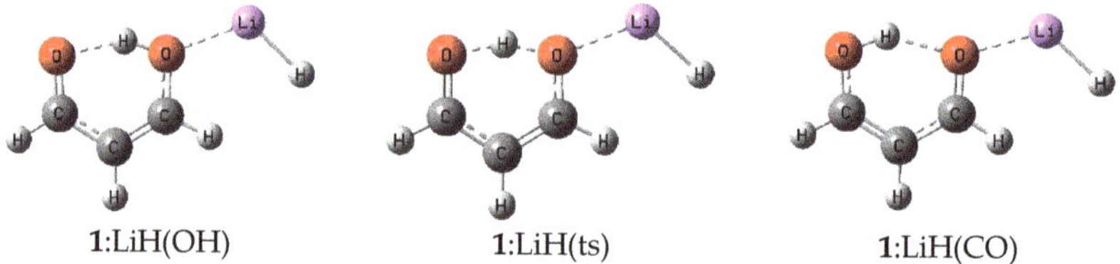

Scheme 1. Some representative complexes.

2.1. Ground State Structures and Binding Energies

Table S1 of the Supporting Information provides the structures, total energies, and molecular graphs of the complexes of 1-oxo-3-hydroxy-2-propene with the Lewis acids LiH, LiF, BeH$_2$, and BeF$_2$. The binding energies, selected distances, and the H–O–O angles in these complexes are reported in Table 1. For the equilibrium complexes, the binding energies range from 65 kJ·mol^{-1} for the complex **1**:LiH(OH) to 100 kJ·mol^{-1} for **1**:BeF$_2$(CO). For each acid, the binding energies decrease in the following order:

Table 1. Binding energies ($-\Delta$E, kJ·mol^{-1}), distances R (Å), and H–O–O angles (<, °) for complexes of C$_3$H$_4$O$_2$ with acids.

Complex	$-\Delta$E	R(O···O)	R(O–H)	R(O···H)	<H–O···O
Isolated monomer C$_3$H$_4$O$_2$	0.0	2.558	1.001	1.648	19.5
ts isolated ^{a}	-11.6 ^{a}	2.365	1.205	1.205	11.1
C$_3$H$_4$O$_2$:LiH at C=O	75.6	2.572	0.995	1.677	20.6
at O–H	64.6	2.492	1.017	1.551	17.3
ts	59.6	2.360	1.171	1.235	11.5
C$_3$H$_4$O$_2$:LiF at C=O	81.9	2.564	0.997	1.664	20.2
at O–H	72.7	2.496	1.015	1.557	17.4
ts	67.1	2.359	1.176	1.229	11.4
C$_3$H$_4$O$_2$:BeH$_2$ at C=O	82.4	2.566	0.992	1.704	23.0
at O–H	67.2	2.473	0.994	1.691	22.5
ts	64.0	2.356	1.156	1.256	12.9
C$_3$H$_4$O$_2$:BeF$_2$ at C=O	99.8	2.573	0.992	1.704	23.0
at O–H	80.9	2.457	1.029	1.514	18.2
ts	78.9	2.358	1.140	1.275	13.3

^{a} The transition structure is 11.6 kJ·mol^{-1} less stable than the equilibrium C$_3$H$_4$O$_2$ structure.

1:acid(CO) > 1:acid (OH) > 1:acid(ts).

When the interaction with the acid occurs at the carbonyl oxygen, the order of decreasing binding energy with respect to the acid is:

BeF_2 > BeH_2 ≈ LiF > LiH

However, when the interaction occurs at the hydroxyl oxygen, the order is:

BeF_2 > LiF > BeH_2 > LiH.

The differences among the binding energies of the equilibrium complexes with the acid at C=O versus O–H range from 9 kJ·mol^{-1} for the complexes with LiF as the acid to 19 kJ·mol^{-1} when BeF_2 is the acid. Figure 1 provides a representation of the binding energies versus the O–O distance for these complexes and transition structures as a function of the acid. It is interesting to note that the binding energies of the transition structures are very similar to those of the complexes with the acid at the O–H group. Moreover, the binding energies of **1:LiF(CO)** and **1:BeH$_2$(CO)** differ by only 0.5 kJ·mol^{-1}.

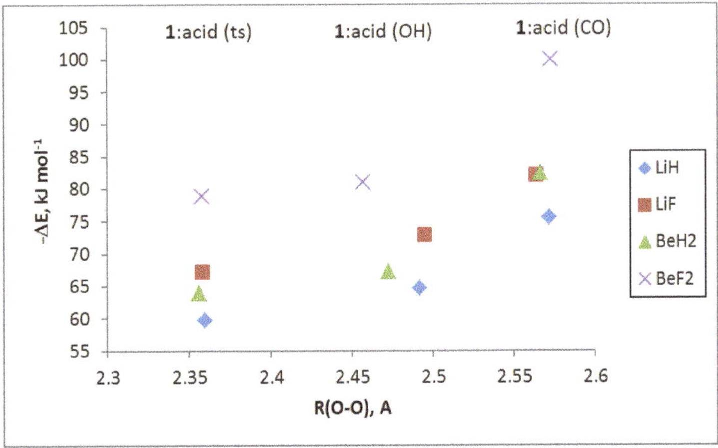

Figure 1. Binding energies versus the O–O distance for 1:acid (OH), 1:acid (ts), and 1:acid (CO) complexes as a function of the acid.

There are many approaches to representing the binding energies of a series of complexes. One of the most interesting and informative can be found in Figure 2, which provides a diagram illustrating the binding energies and the relative binding energies of complexes and transition structures 1:acid(CO), 1:acid(ts), and 1:acid(OH). The transition structures present the barriers that separate the equilibrium structures with the acid at C=O from the structures with the acid at O–H. This barrier is 12 kJ·mol^{-1} for the isolated parent molecule **1**. Interaction of the acid with the C=O group increases the barrier to between 15 and 21 kJ·mol^{-1}, while interaction at the O–H group decreases the barrier to between 2 and 6 kJ·mol^{-1}. These latter barriers and the energy differences indicate that the population of the isomer with the acid at the carbonyl group would be the greater than 98% at room temperature.

The O–O distances across the hydrogen bond in the complexes 1:acid with hydrogen bond formation at the C=O group increase slightly relative to isolated **1**, which has an O–O distance of 2.56 Å. However, when hydrogen bond formation occurs at the O–H group, the O–O distance decreases to between 2.46 to 2.50 Å. As expected, the shortest O–O distances are found in the transition structures for proton transfer, where they decrease to 2.36 Å. An excellent second-order relationship can be obtained when the sum of the O–H distances (R_1 + R_2) in each system is compared to the difference (R_1 − R_2) using the Steiner–Limbach relationship [45,46]. The points with the largest (R_1 + R_2) values in Figure 3 correspond to the 1:acid(OH) complexes, the intermediate ones to the 1:acid(CO) complexes, and the shortest to the 1:acid(TS) complexes. This figure illustrates that the hydrogen-bonded H atom tends to be centered between the two oxygen atoms as they approach each other. The correlation coefficient of the second-order trending in Figure 3 is 0.9996.

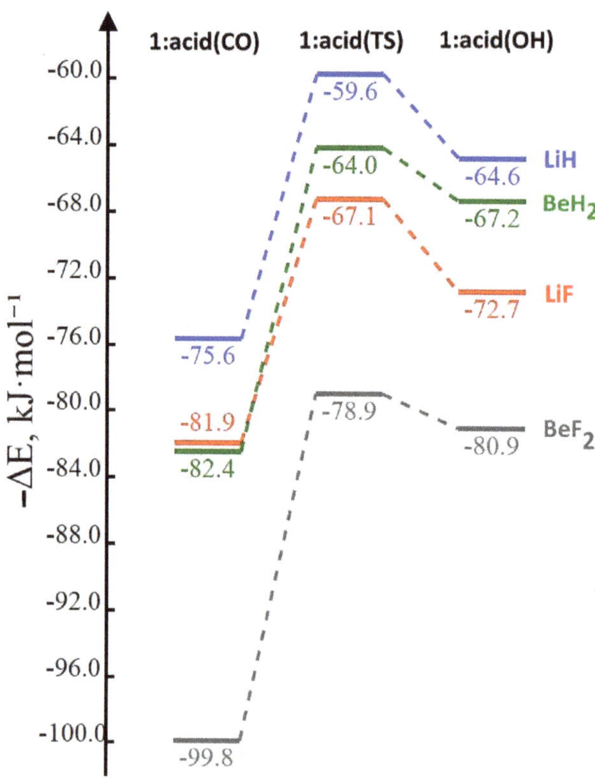

Figure 2. Binding energies of equilibrium and transition structures as a function of the nature of the complex. From these data, the barriers to interconverting the two equilibrium complexes can be readily obtained.

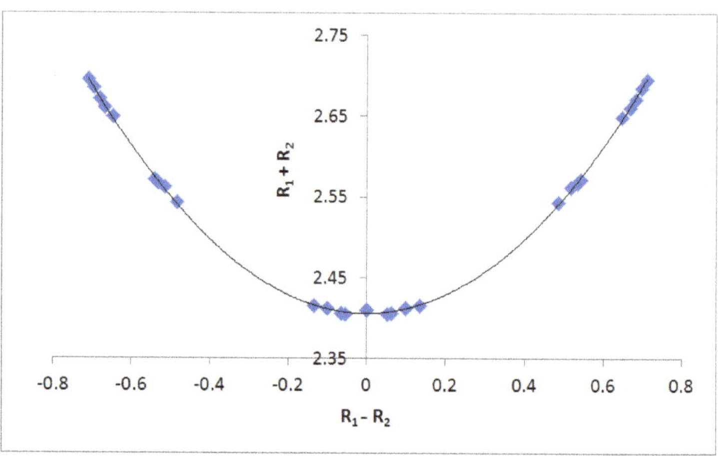

Figure 3. ($R_1 + R_2$) vs. ($R_1 - R_2$) from the Steiner–Limbach relationship.

The hydrogen bonds in all complexes are nonlinear. The deviation from linearity is 20° in isolated **1** and ranges from 17° to 22° in the complexes. The hydrogen bond approaches closer to linearity in the transition structures, where the deviation decreases to between 11° and 13°.

2.2. Orbital Description of the O–H⋯O Hydrogen Bond

There are two canonical lone pair (lp) orbitals associated with the carbonyl oxygen, both in the isolated base (**1**) and in the **1**:acid complexes, and these are illustrated in Figure 4. The orbital lp1 isolated is a lone-pair orbital on O, which has local σ-type symmetry relative to the C=O bond, extending from the carbonyl oxygen in a direction corresponding to a continuation of the O–H bond. Interaction of the O–H group with this orbital leads to a side-wise overlap of a *p*-type orbital on the O–H group with the C=O lp1 orbital. The orbital lp2 is a local π-type orbital on **1**, which is perpendicular to the C=O bond and directed toward the O–H group of **1** with which it interacts. This orbital extends on both sides of the C=O group, where it may also interact with an acid through the lobe of the *p*-type orbital which extends in this direction. This observation is consistent with the greater binding energies of complexes with the base interacting with **1** at the C=O group compared to those with the base interacting at the O–H group.

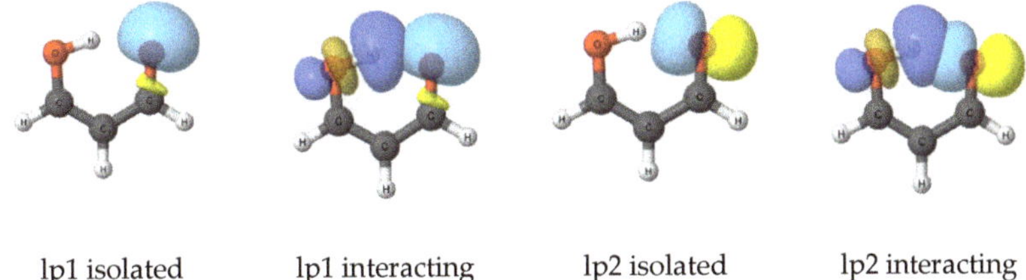

lp1 isolated   lp1 interacting   lp2 isolated   lp2 interacting

Figure 4. Representations of the lone pair (lp) orbitals of **1** isolated and interacting with the O–H group based on the NBO analysis.

2.3. Charge-Transfer Energies

The complexes **1**:acid are stabilized by charge-transfer interactions. The nature of charge transfer and the associated charge-transfer energies are reported in Table 2. Given the nature of the lone-pair orbitals illustrated in Figure 4, it is not surprising that charge transfer from lp2 is the dominant charge-transfer interaction, with energies ranging from 158 kJ·mol^{-1} in **1**:LiF(OH) to 189 kJ·mol^{-1} **1**:BeF$_2$(OH). The charge-transfer energies involving lp1 are much less, with values between 15 and 31 kJ·mol^{-1}. The total charge-transfer energies vary from 100 to 213 kJ·mol^{-1}. It is interesting to note that the strongest complexes occur in **1**:acid(CO) (Table 1 and Figure 2), while the strongest intramolecular hydrogen bond can be seen in **1**:acid(OH). Figure 5 illustrates a linear dependence of these energies on the O–O distance, with a correlation coefficient of 0.965.

Table 2. Intramolecular charge-transfer stabilization energies (kJ·mol^{-1}) for complexes of C$_3$H$_4$O$_2$ with acids.

Complex	O$_{lp1}$→σ*O–H	O$_{lp2}$→σ*O–H	Total
Isolated monomer C$_3$H$_4$O$_2$	16.9	105.5	122.5
C$_3$H$_4$O$_2$:LiH at C=O	15.2	84.9	100.1
at O–H	21.0	162.5	183.6
C$_3$H$_4$O$_2$:LiF at C=O	15.1	91.7	106.8
at O–H	20.7	157.7	178.4
C$_3$H$_4$O$_2$:BeH$_2$ at C=O	26.1	53.6	79.7
at O–H	22.9	168.6	191.5
C$_3$H$_4$O$_2$:BeF$_2$ at C=O	30.5	43.6	74.1
at O–H	24.1	188.7	212.8

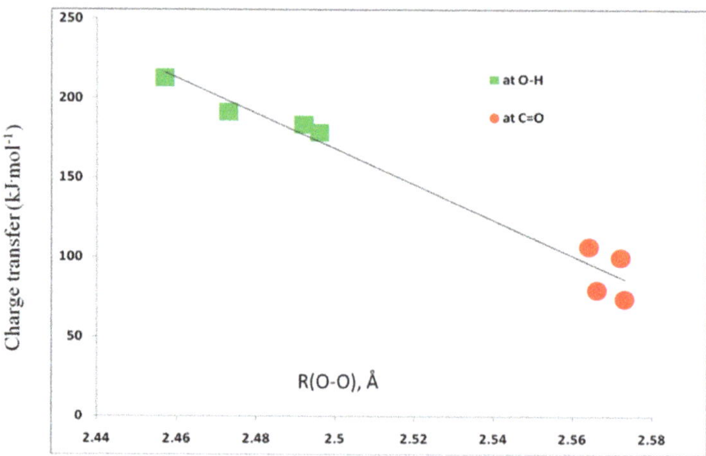

Figure 5. Charge-transfer energies versus the O–O distance.

2.4. Electron Density Analyses

The electron densities of the equilibrium and transition structures were analyzed using the quantum theory of atoms in molecules (QTAIM) methodology. All complexes have two bond critical points. In the equilibrium structures, the first bond critical point is associated with the covalent O–H bond with ρ_{BCP} values of 0.30 au, while the second refers to the hydrogen bond with ρ_{BCP} values around 0.05 au. The transition structures have intermediate values of ρ_{BCP} between 0.15 to 0.20 au. The $\nabla^2\rho_{BCP}$ values are negative for the covalent O–H bonds in the equilibrium and transition structures but positive for the hydrogen bonds. The total energy densities, H_{BCP}, are negative in all structures. Thus, all O–H contacts have some covalent character. The covalency is of medium strength in the equilibrium complexes that have a positive $\nabla^2\rho_{BCP}$ and a negative value of H_{BCP}, while the transition structures have much stronger covalent interactions with a negative value of both $\nabla^2\rho_{BCP}$ and H_{BCP} [47,48]. Excellent exponential correlations are obtained between ρ_{BCP} and H_{BCP} versus the interatomic distance, as illustrated in Figure S1, in agreement with other reports of these parameters as descriptors of intermolecular interactions [49–51].

2.5. Spin–Spin Coupling Constants

The total spin–spin coupling constants [2h]J(O–O), [1h]J(H–O), and [1]J(O–H) are given in Table 3, and the paramagnetic spin–orbit (PSO), diamagnetic spin–orbit (DSO), Fermi contact (FC), and spin–dipole (SD) components are reported in Table S2 of the Supporting Information. For the equilibrium ground states of the complexes, the PSO, FC, and SD components of [2h]J(O–O) are positive. Both the PSO and SD components of [2h]J(O–O) are non-negligible, with the PSO component having values comparable to those of the FC term when interaction with the acid occurs at the carbonyl group. Even though the FC terms for the transition structures have values that are greater and closer to total J, the FC terms are poor approximations to the total coupling constant [2h]J(O–O).

Coupling constants [1h]J(H–O) are also small and positive for the equilibrium structures and are dominated by the FC terms. The PSO terms are smaller and positive for the equilibrium structures, but these are partially canceled by the negative DSO and SD terms. The net result is that the FC terms differ from [1h]J(H–O) by about 1 Hz. The coupling constant [1]J(O–H) is negative and has a significantly greater absolute value than [2h]J(O–O) and [1h]J(H–O) since it refers to a covalent O–H bond. Values of this coupling constant are dominated by the negative FC terms, while the PSO terms make smaller, non-negligible negative contributions to [1]J(O–H) in the equilibrium structures but positive contributions in the transition structures. Thus, all terms should be included in determining [1]J(O–H).

Table 3. Spin–spin coupling constants $^{2h}J(O–O)$, $^{1h}J(H–O)$, and $^{1}J(O–H)$ (Hz) for the O–H$\cdots$O hydrogen bond in complexes $C_3H_4O_2$:Acid.

Complex	$^{2h}J(O–O)$	$^{1h}J(H–O)$	$^{1}J(O–H)$
Isolated monomer $C_3H_4O_2$	11.5	7.9	−77.7
Isolated ts	24.7	−19.7	−19.7
$C_3H_4O_2$:LiH at C=O	10.2	7.9	−79.2
at O–H	13.7	7.4	−81.3
ts	24.0	−14.5	−31.2
$C_3H_4O_2$:LiF at C=O	10.5	8.1	−78.6
at O–H	13.5	7.4	−82.2
ts	24.1	−15.5	−30.0
$C_3H_4O_2$:BeH$_2$ at C=O	10.0	7.8	−81.4
at O–H	14.3	6.4	−89.7
ts	23.8	−12.8	−41.4
$C_3H_4O_2$:BeF$_2$ at C=O	9.6	7.7	−82.1
at O–H	15.4	5.8	−88.7
ts	23.6	−10.3	−47.1

2.5.1. $^{2h}J(O–O)$

Spin–spin coupling constants $^{2h}J(O–O)$ across the O–H$\cdots$O hydrogen bond are reported in Table 3. This coupling constant has a value of 12 Hz in the isolated monomer **1** and then decreases to about 10 Hz in the complexes when the acid interacts with **1** at the C=O bond. When interaction occurs at the O–H bond, $^{2h}J(O–O)$ increases to between 13 and 15 Hz. The value of $^{2h}J(O–O)$ obviously depends on the O–O distance, so it is not surprising that $^{2h}J(O–O)$ increases to about 24 Hz in the transition structures that have the shortest O–O distances. These relationships are seen most easily in Figure 6, which is a plot of $^{2h}J(O–O)$ versus the O–O distance. The second-order trendline has a correlation coefficient of 0.992.

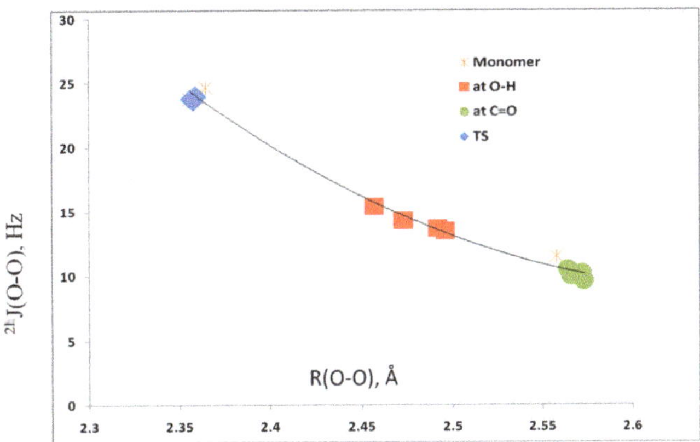

Figure 6. $^{2h}J(O–O)$ versus the O–O distance for complexes **1**:acid.

2.5.2. $^{1h}J(H–O)$

The values of the second coupling constant $^{1h}J(H–O)$ across the hydrogen bond are also reported in Table 3. Its value of 7.9 Hz in **1** changes minimally upon complex formation, ranging from 7.7 to 8.1 Hz when complexation occurs at the O–H group. It decreases to between 5.8 and 7.4 Hz when the acid interacts at the C=O group. Much larger changes are observed in the transition structures as the H–O distance across the hydrogen bond contracts, and $^{1h}J(H–O)$ has a value of −19.7 Hz in **1**. This coupling constant varies from −10.3 Hz in the complex with BeF$_2$ as the acid to −15.5 Hz with LiF as the acid. The dependence of $^{1h}J(H–O)$ on the H–O distance is illustrated in Figure 7, which has a second-order trendline with a correlation coefficient of 0.997. The vertical $^{1h}J(H\text{–}O)$ axis was reversed to account for the negative magnetogyric ratio of ^{17}O. This plot also illustrates the changing nature of the hydrogen bond, from a traditional hydrogen bond in the equilibrium complexes with bond formation at the O–H or C=O groups to a proton-shared hydrogen bond in the transition structures.

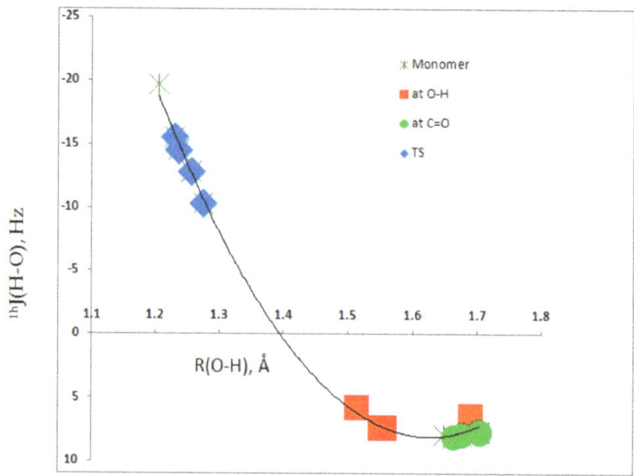

Figure 7. $^{1h}J(O\text{–}H)$ versus the O–H distance for complexes **1**:Acid.

2.5.3. $^{1}J(O\text{–}H)$

The third coupling constant associated with the $O\text{–}H\cdots O$ hydrogen bond is $^{1}J(O\text{–}H)$ for the proton donor O–H group. Its value in **1** of -78 Hz minimally changes in complexes in which the acid interacts with the C=O group, varying between -79 and -82 Hz. However, interaction with the O–H group leads to a significant increase in the absolute value of $^{1}J(O\text{–}H)$ to between -81 and -90 Hz. As the O–H distance increases dramatically in the transition structures, $^{1}J(O\text{–}H)$ decreases significantly in absolute value to -20 Hz in **1** and from -30 to -47 Hz in the complexes with the acids. Once again, the variation in this coupling constant can be readily seen in the plot of $^{1}J(O\text{–}H)$ versus the O–H distance, shown in Figure 8. The correlation coefficient of the second-order trendline is 0.973.

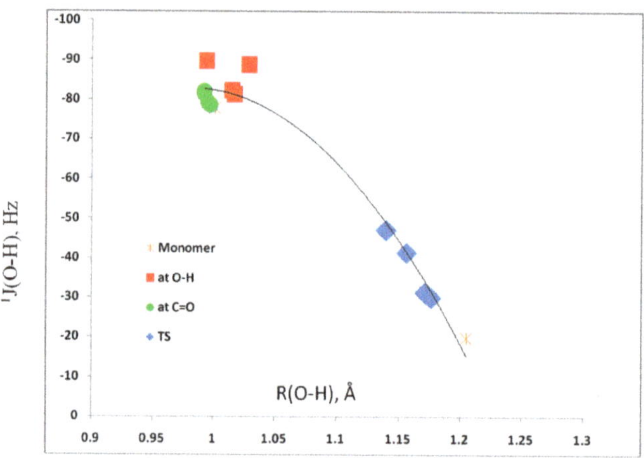

Figure 8. $^{1}J(O\text{–}H)$ versus the O–H distance for complexes **1**:acid.

3. Methods

The structures of the isolated monomer 1-oxo-3-hydroxy-2-propene; the acids LiH, LiF, BeH_2, and BeF_2; and the complexes of 1-oxo-3-hydroxy-2-propene with the acids were optimized at second-order Møller–Plesset perturbation theory (MP2) [52–55] with the aug'-cc-pVTZ basis set [56]. This basis set was derived from the Dunning aug-cc-pVTZ basis set [57,58] by removing diffuse functions from H atoms. Searches were made of the 1-oxo-3-hydroxy-2-propene:acid potential surfaces for equilibrium structures and transition structures. Frequencies were computed to confirm that the optimized structures are indeed equilibrium structures with no imaginary frequencies and that the transition structures have one imaginary frequency along the path that connects two equilibrium structures. Optimization and frequency calculations were performed using the Gaussian 16 program [59]. The binding energies of the equilibrium complexes were computed as $-\Delta E$ for the reaction that forms these complexes from the isolated monomers.

The natural bond orbital (NBO) method [60] was used to obtain the stabilizing charge-transfer interactions using the NBO-6 program [61]. Since MP2 orbitals are nonexistent, the charge-transfer interactions were computed using the B3LYP functional with the aug'-cc-pVTZ basis set at the MP2/aug'-cc-pVTZ geometries so that at least some electron correlation effects could be included. The atoms in molecules (AIM) methodology [62–65] was used to produce the molecular graphs of the complexes, employing the AIMAll program [66]. The molecular graph identifies the location of electron density features of interest, including the electron density (ρ) maxima associated with the various nuclei and saddle points that correspond to bond critical points (BCPs). The zero gradient line that connects a BCP with two nuclei is the bond path.

Spin–spin coupling constants were evaluated using the equation-of-motion coupled cluster singles and doubles (EOM-CCSD) method in the CI (configuration interaction)-like approximation [67,68] with all electrons correlated. For these calculations, the Ahlrichs [69] qzp basis set was placed on ^{13}C, ^{17}O, and ^{19}F atoms, the hybrid basis set developed previously on ^{7}Li and ^{9}Be [70], and the qz2p basis set on the hydrogen-bonded ^{1}H atom. The Dunning cc-pVDZ basis was placed on the remaining ^{1}H atoms. All terms that contribute to the total coupling constant, namely, the paramagnetic spin–orbit (PSO), diamagnetic spin–orbit (DSO), Fermi contact (FC), and spin–dipole (SD) were evaluated. The EOM-CCSD calculations were performed using ACES II [71] on the HPC cluster Owens at the Ohio Supercomputer Center.

4. Conclusions

Ab initio MP2/aug'-cc-pVTZ calculations were carried out to identify and characterize hydrogen-bonded equilibrium structures and transition structures on the 1-oxo-3-hydroxy-2-propene:acid (1:acid) potential energy surfaces, with the acids LiH, LiF, BeH_2, and BeF_2. The results of these calculations support the following statements:

1. Two equilibrium structures, one with the acid interacting with the C=O group and the other with the interaction occurring at the O–H group, exist on all surfaces. These structures are separated by transition structures that present the barriers to the interconversion of the two equilibrium structures.
2. The binding energies of these complexes vary between 65 and 100 kJ·mol^{-1}, with binding at the C=O group preferred by 10 to 20 kJ·mol^{-1}.
3. The barrier to interconverting the equilibrium structures with the acid at the C=O group to the structure with the acid at the O–H group is 12 kJ·mol^{-1} in isolated **1** and increases to between 15 and 21 kJ·mol^{-1} in the complexes. The reverse barriers range from 2 to 6 kJ·mol^{-1}. Thus, only structures with the acid interacting at the C=O group would be experimentally observed, even at low temperatures.
4. Charge-transfer stabilizes the **1**:acid complexes. The greater charge-transfer interactions involve electron donation from an oxygen lone pair orbital on the C=O group to an antibonding pi-type orbital on the O–H group.

5. EOM-CCSD spin–spin coupling constants $^{2h}J(O\text{–}O)$, $^{1h}J(H\text{–}O)$, and $^{1}J(O\text{–}H)$ were computed for all equilibrium and transition structures. Plots of $^{2h}J(O\text{–}O)$ versus the O–O distance, $^{1h}J(H\text{–}O)$ versus the H–O distance across the hydrogen bond, and $^{1}J(O\text{–}H)$ versus the covalent O–H bond distance of **1** exhibit a second-order dependence of the coupling constant on the corresponding distance, with very high correlation coefficients.

Supplementary Materials: The following are available online, Table S1: Structures (Å), total energies (au), and molecular graphs of 1-oxo-3-hydroxy-2-propene:acid complexes; Figure S1: Relationship between electron densities at the O–H hydrogen bonds and interatomic distances; Table S2: Components of spin–spin coupling constants $^{2h}J(O\text{–}O)$, $^{1h}J(H\text{–}O)$, and $^{1}J(O\text{–}H)$ (Hz).

Author Contributions: I.A. and J.E.D.B. carried out the calculations. J.E.D.B., J.E., and I.A. contributed equally to the writing of this paper. All authors have read and agreed to the published version of the manuscript.

Funding: This work was carried out with financial support from the Ministerio de Ciencia, Innovación y Universidades of Spain (Project No. PGC2018–094644-B-C22) and Comunidad Autónoma de Madrid (P2018/EMT–4329 AIRTEC-CM).

Institutional Review Board Statement: Not applicable.

Informed Consent Statement: Not applicable.

Data Availability Statement: The data presented in this study are available in the article and in the Supplementary Material.

Conflicts of Interest: The authors declare no conflict of interest.

Sample Availability: Not applicable.

References

1. Grabowski, S. An estimation of strength of intramolecular hydrogen bonds—Ab initio and AIM studies. *J. Mol. Struct.* **2001**, *562*, 137–143. [CrossRef]
2. Carrington, T.; Miller, W.H. Reaction Surface Hamiltonian for the Dynamics of Reactions in Polyatomic Systems. *J. Chem. Phys.* **1984**, *81*, 3942–3950. [CrossRef]
3. Tew, D.P.; Handy, N.C.; Carter, S. A reaction surface Hamiltonian study of malonaldehyde. *J. Chem. Phys.* **2006**, *125*, 084313. [CrossRef]
4. Fillaux, F.; Nicolaï, B. Proton Transfer in Malonaldehyde: From Reaction Path to Schrödinger's Cat. *Chem. Phys. Lett.* **2005**, *415*, 357–361. [CrossRef]
5. Wong, K.F.; Sonnenberg, J.L.; Paesani, F.; Yamamoto, T.; Vaníček, J.; Zhang, W.; Schlegel, H.B.; Case, D.A.; Cheatham, T.E.; Miller, W.H.; et al. Proton Transfer Studied Using a Combined Ab Initio Reactive Potential Energy Surface with Quantum Path Integral Methodology. *J. Chem. Theory Comput.* **2010**, *6*, 2566–2580. [CrossRef] [PubMed]
6. Karandashev, K.; Xu, Z.-H.; Meuwly, M.; Vaníček, J.; Richardson, J.O. Kinetic isotope effects and how to describe them. *Struct. Dyn.* **2017**, *4*, 061501. [CrossRef] [PubMed]
7. Grosch, A.A.; Van der Lubbe, S.C.C.; Guerra, C.F. Nature of Intramolecular Resonance Assisted Hydrogen Bonding in Malonaldehyde and Its Saturated Analogue. *J. Phys. Chem. A* **2018**, *122*, 1813–1820. [CrossRef]
8. Grabowski, S.J. Intramolecular Hydrogen Bond Energy and Its Decomposition—O–H···O Inyteractions. *Crystals* **2021**, *11*, 5. [CrossRef]
9. Gilli, G.; Bellucci, F.; Ferretti, V.; Bertolasi, V. Evidence for Resonance-Assisted Hydrogen Bonding from Crystal-Structure Correlations on the Enol Form of the β-Diketone Fragment. *J. Am. Chem. Soc.* **1989**, *111*, 1023–1028. [CrossRef]
10. Bertolasi, V.; Nanni, L.; Gilli, P.; Ferretti, V.; Gilli, G.; Issa, Y.M.; Sherif, O.E. Intramolecular N–H···O=C Hydrogen-Bonding Assisted by Resonance-Intercorrelation between Structural and Spectroscopic Data for Six β-Diketo-Arylhydrazones Derived from Benzoylacetone or Acetylacetone. *New J. Chem.* **1994**, *18*, 251–261.
11. Trujillo, C.; Sánchez-Sanz, G.; Alkorta, I.; Elguero, J.; Mó, O.; Yáñez, M. Resonance assisted hydrogen bonds in open-chain and cyclic structures of malonaldehyde enol: A theoretical study. *J. Mol. Struct.* **2013**, *1048*, 138–151. [CrossRef]
12. Sanz, P.; Montero-Campillo, M.M.; Mó, O.; Yáñez, M.; Alkorta, I.; Elguero, J. Intramolecular magnesium bonds in malonaldehyde-like systems: A critical view of the resonance-assisted phenomena. *Theor. Chem. Acc.* **2018**, *137*, 97. [CrossRef]
13. Palusiak, M.; Simon, S.; Solà, M. The proton transfer reaction in malonaldehyde derivatives: Substituent effects and quasi-aromaticity of the proton bridge. *Chem. Phys.* **2007**, *342*, 43–54. [CrossRef]

14. Meyer, K. Über die Tautomerie des Acetessigesters. Über Keto-Enol-Tautomerie III. *Berichte der Deutschen Chemischen Gesellschaft* **1911**, *44*, 2718–2724. [CrossRef]
15. Meyer, K. Über den Zusammenhang zwischen Konstitution und Gleichgewicht bei keto-enol-desmotropen Verbindungen. Über Keto-Enol-Tautomerie VI. *Berichte der Deutschen Chemischen Gesellschaft* **1912**, *45*, 2843–2864. [CrossRef]
16. Bothner-By, A.A.; Harris, R.K. Conformational Preferences in Malonaldehyde and Acetylacetal-dehyde Enols Investigated by Nuclear Magnetic Resonance. *J. Org. Chem.* **1965**, *30*, 254–257. [CrossRef]
17. Emsley, J. The composition, structure and hydrogen bonding of the β-diketones. In *Complex Chemistry. Structure and Bonding*; Springer: Berlin/Heidelberg, Germany, 1984; Volume 57, pp. 147–191.
18. Claramunt, R.M.; López, C.; Lott, S.; María, M.D.S.; Alkorta, I.; Elguero, J. Solid-State NMR Study of the Tautomerism of Acetylacetone Included in a Host Matrix. *Helv. Chim. Acta* **2005**, *88*, 1931–1942. [CrossRef]
19. Jabłoński, M. A Critical Overview of Current Theoretical Methods of Estimating the Energy of Intramolecular Interactions. *Molecules* **2020**, *25*, 5512. [CrossRef] [PubMed]
20. Montero, M.D.A.; Martínez, F.A.; Aucar, G.A. Magnetic descriptors of hydrogen bonds in malonaldehyde and its derivatives. *Phys. Chem. Chem. Phys.* **2019**, *21*, 19742–19754. [CrossRef] [PubMed]
21. Pichierri, F. Effect of fluorine substitution on the proton transfer barrier in malonaldehyde. A density functional theory study. *Chem. Phys. Lett.* **2003**, *376*, 781–787. [CrossRef]
22. Scivetti, I.; Sen, K.; Elena, A.M.; Todorov, I.T. Reactive Molecular Dynamics at Constant Pressure via Nonreactive Force Fields: Extending the Empirical Valence Bond Method to the Isothermal-Isobaric Ensemble. *J. Phys. Chem. A* **2020**, *124*, 7585–7597. [CrossRef] [PubMed]
23. List, N.H.; Dempwolff, A.L.; Dreuw, A.; Norman, P.; Martínez, T.J. Probing competing relaxation pathways in malonaldehyde with transient X-ray absorption spectroscopy. *Chem. Sci.* **2020**, *11*, 4180–4193. [CrossRef]
24. Yagi, K.; Taketsugu, T.; Hirao, K. Generation of full-dimensional potential energy surface of intramolecular hydrogen atom transfer in malonaldehyde and tunneling dynamics. *J. Chem. Phys.* **2001**, *115*, 10647–10655. [CrossRef]
25. Tautermann, C.S.; Voegele, A.F.; Loerting, T.; Liedl, K.R. The Optimum Tunneling Path for the Proton Transfer in Malonaldehyde. *J. Chem. Phys.* **2002**, *117*, 1962–1966. [CrossRef]
26. Mil'Nikov, G.V.; Yagi, K.; Taketsugu, T.; Nakamura, H.; Hirao, K. Simple and accurate method to evaluate tunneling splitting in polyatomic molecules. *J. Chem. Phys.* **2004**, *120*, 5036–5045. [CrossRef] [PubMed]
27. Gulaczyk, I.; Kręglewski, M. Multi-dimensional proton tunneling in 2-methylmalonaldehyde. *J. Mol. Struct.* **2020**, *1220*, 128733. [CrossRef]
28. Alparone, A.; Millefiori, S. Anharmonic vibrational spectroscopic investigation of malonaldehyde. *Chem. Phys.* **2003**, *290*, 15–25. [CrossRef]
29. Baughcum, S.L.; Duerst, R.W.; Rowe, W.F.; Smith, Z.; Wilson, E.B. Microwave spectroscopic study of malonaldehyde (3-hydroxy-2-propenal). 2. Structure, dipole moment, and tunneling. *J. Am. Chem. Soc.* **1981**, *103*, 6296–6303. [CrossRef]
30. Baughcum, S.L.; Smith, Z.; Wilson, E.B.; Duerst, R.W. Microwave spectroscopic study of malonaldehyde. 3. Vibration-rotation interaction and one-dimensional model for proton tunneling. *J. Am. Chem. Soc.* **1984**, *106*, 2260–2265. [CrossRef]
31. Turner, P.; Baughcum, S.L.; Coy, S.L.; Smith, Z. Microwave spectroscopic study of malonaldehyde. 4. Vibration-rotation interaction in parent species. *J. Am. Chem. Soc.* **1984**, *106*, 2265–2267. [CrossRef]
32. Delchev, V.B. Hydrogen Bonded Complexes of Acetylacetone and Methanol: HF and DFT level Study. *Monatsh. Chem.* **2004**, *135*, 249–260. [CrossRef]
33. Grabowski, S.J. Intramolecular O–H O Hydrogen Bonds—The Influence of External Agents on Their Properties. *Pol. J. Chem.* **2007**, *81*, 799–811.
34. Mó, O.; Yáñez, M.; Alkorta, I.; Elguero, J. Modulating the Strength of Hydrogen Bonds through Beryllium Bonds. *J. Chem. Theory Comput.* **2012**, *8*, 2293–2300. [CrossRef] [PubMed]
35. Cremer, D.; Gräfenstein, J. Calculation and analysis of NMR spin–spin coupling constants. *Phys. Chem. Chem. Phys.* **2007**, *9*, 2791–2816. [CrossRef] [PubMed]
36. Pople, J.A.; Schneider, W.G.; Bernstein, H.J. *High Resolution Nuclear Magnetic Resonance*; McGraw-Hill: New York, NY, USA, 1959.
37. Roberts, J.D. *Nuclear Magnetic Resonance*; McGraw-Hill: New York, NY, USA, 1959.
38. Ernst, R.R.; Bodenhausen, G.; Wokaun, A. *Principles of Nuclear Magnetic Resonance in One and Two Dimensions*; Clarendon Press: New York, NY, USA, 1987.
39. Shenderovich, I.G.; Smirnov, S.N.; Denisov, G.S.; Gindin, V.A.; Golubev, N.S.; Dunger, A.; Reibke, R.; Kirpekar, S.; Malkina, O.L.; Limbach, H.-H. Nuclear magnetic resonance of hydrogen bonded clusters between F− and (HF)n: Experiment and theory. *Berichte der Deutschen Chemischen Gesellschaft* **1998**, *102*, 422–428. [CrossRef]
40. Del Amo, J.M.L.; Langer, U.; Torres, V.; Buntkowsky, G.; Vieth, H.-M.; Pérez-Torralba, M.; Sanz, D.; Claramunt, R.M.; Elguero, J.; Limbach, H.-H. NMR Studies of Ultrafast Intramolecular Proton Tautomerism in Crystalline and Amorphous N,N′-Diphenyl-6-aminofulvene-1-aldimine: Solid-State, Kinetic Isotope, and Tunneling Effects. *J. Am. Chem. Soc.* **2008**, *130*, 8620–8632. [CrossRef]
41. Del Bene, J.E.; Perera, S.A.; Bartlett, R.J. Predicted NMR Coupling Constants Across Hydrogen Bonds: A Fingerprint for Specifying Hydrogen Bond Type? *J. Am. Chem. Soc.* **2000**, *122*, 3560–3561. [CrossRef]
42. Del Bene, J.E.; Perera, S.A.; Bartlett, R.J.; Alkorta, I.; Elguero, J. $^{4h}J(^{31}P–^{31}P)$ Coupling Constants through N–H+–N Hydrogen Bonds: A Comparison of Computed ab Initio and Experimental Data. *J. Phys. Chem. A* **2000**, *104*, 7165–7166. [CrossRef]

43. Hoelger, C.-G.; Limbach, H.-H. Localization of Hydrons in Hydrogen Bonds Using Dipolar Solid State NMR Spectroscopy. *J. Phys. Chem.* **1994**, *98*, 11803–11810. [CrossRef]

44. Alkorta, I.; Elguero, J.; Frontera, A. Not Only Hydrogen Bonds: Other Noncovalent Interactions. *Crystals* **2020**, *10*, 180. [CrossRef]

45. Steiner, T. Lengthening of the N–H bond in N–H··· N hydrogen bonds. Preliminary structural data and implications of the bond valence concept. *J. Chem. Soc. Chem. Commun.* **1995**, 1331–1332. [CrossRef]

46. Ramos, M.; Alkorta, I.; Elguero, J.; Golubev, N.S.; Denisov, G.S.; Benedict, H.; Limbach, H.-H. Theoretical Study of the Influence of Electric Fields on Hydrogen-Bonded Acid−Base Complexes. *J. Phys. Chem. A* **1997**, *101*, 9791–9800. [CrossRef]

47. Cremer, D.; Kraka, E. A Description of the Chemical Bond in Terms of Local Properties of Electron Density and Energy, *Croat. Chem. Acta* **1984**, *57*, 1259–1281.

48. Rozas, I.; Alkorta, I.; Elguero, J. Behavior of Ylides Containing N, O, and C Atoms as Hydrogen Bond Acceptors. *J. Am. Chem. Soc.* **2000**, *122*, 11154–11161. [CrossRef]

49. Mata, I.; Alkorta, I.; Molins, E.; Espinosa, E. Universal Features of the Electron Density Distribution in Hydrogen-Bonding Regions: A Comprehensive Study Involving H ··· X (X=H, C, N, O, F, S, Cl, π) Interactions. *Chem. Eur. J.* **2010**, *16*, 2442–2452. [CrossRef]

50. Alkorta, I.; Solimannejad, M.; Provasi, P.F.; Elguero, J. Theoretical Study of Complexes and Fluoride Cation Transfer between N2F+ and Electron Donors. *J. Phys. Chem. A* **2007**, *111*, 7154–7161. [CrossRef]

51. Sánchez-Sanz, G.; Alkorta, I.; Elguero, J. Theoretical study of the HXYH dimers (X, Y = O, S, Se). Hydrogen bonding and chalcogen–chalcogen interactions. *Mol. Phys.* **2011**, *109*, 2453–2552. [CrossRef]

52. Pople, J.A.; Binkley, J.S.; Seeger, R. Theoretical models incorporating electron correlation. *Int. J. Quantum Chem.* **2009**, *10*, 1–19. [CrossRef]

53. Krishnan, R.; Pople, J.A. Approximate fourth-order perturbation theory of the electron correlation energy. *Int. J. Quantum Chem.* **1978**, *14*, 91–100. [CrossRef]

54. Bartlett, R.J.; Silver, D.M. Many-body perturbation theory applied to electron pair correlation energies. I. Closed-shell first-row diatomic hydrides. *J. Chem. Phys.* **1975**, *62*, 3258–3268. [CrossRef]

55. Bartlett, R.J.; Purvis, G.D. Many-body perturbation theory, coupled-pair many-electron theory, and the importance of quadruple excitations for the correlation problem. *Int. J. Quantum Chem.* **1978**, *14*, 561–581. [CrossRef]

56. Del Bene, J.E. Proton affinities of ammonia, water, and hydrogen fluoride and their anions: A quest for the basis-set limit using the Dunning augmented correlation-consistent basis sets. *J. Phys. Chem.* **1993**, *97*, 107–110. [CrossRef]

57. Dunning, T.H. Gaussian Basis Sets for Use in Correlated Molecular Calculations. I. The Atoms Boron through Neon and Hydrogen *J. Chem. Phys.* **1989**, *90*, 1007–1023. [CrossRef]

58. Woon, D.E.; Dunning, T.H. Gaussian basis sets for use in correlated molecular calculations. V. Core-valence basis sets for boron through neon. *J. Chem. Phys.* **1995**, *103*, 4572–4585. [CrossRef]

59. Frisch, M.J.G.W.; Trucks, G.W.; Schlegel, H.B.; Scuseria, G.E.; Robb, M.A.; Cheeseman, J.R.; Scalmani, G.; Barone, V.; Petersson, G.A.; Nakatsuji, H.; et al. *Gaussian 16, Revision A. 03*; Gaussian, Inc.: Wallingford, CT, USA, 2016.

60. Reed, A.E.; Curtiss, L.A.; Weinhold, F. Intermolecular interactions from a natural bond orbital, donor-acceptor viewpoint. *Chem. Rev.* **1988**, *88*, 899–926. [CrossRef]

61. Glendening, E.D.; Badenhoop, J.K.; Reed, A.E.; Carpenter, J.E.; Bohmann, J.A.; Morales, C.M.; Landis, C.R.; Weinhold, F. *NBO 6.0*; University of Wisconsin: Madison, WI, USA, 2013.

62. Bader, R.F.W. A quantum theory of molecular structure and its applications. *Chem. Rev.* **1991**, *91*, 893–928. [CrossRef]

63. Bader, R.F.W. *Atoms in Molecules, a Quantum Theory*; Oxford University Press: Oxford, UK, 1990.

64. Popelier, P.L.A. *Atoms in Molecules. An Introduction*; Prentice Hall: Harlow, UK, 2000.

65. Matta, C.F.; Boyd, R.J. *The Quantum Theory of Atoms in Molecules: From Solid State to DNA and Drug Design*; Wiley-VCH: Weinheim, Germany, 2007.

66. Keith, T.A. *AIMAll (Version 19.10.12)*; TK Gristmill Software: Overland Park, KS, USA, 2019.

67. Perera, S.A.; Nooijen, M.; Bartlett, R.J. Electron correlation effects on the theoretical calculation of nuclear magnetic resonance spin–spin coupling constants. *J. Chem. Phys.* **1996**, *104*, 3290–3305. [CrossRef]

68. Perera, S.A.; Sekino, H.; Bartlett, R.J. Coupled-cluster calculations of indirect nuclear coupling constants: The importance of non-Fermi contact contributions. *J. Chem. Phys.* **1994**, *101*, 2186–2191. [CrossRef]

69. Schäfer, A.; Horn, H.; Ahlrichs, R. Fully optimized contracted Gaussian basis sets for atoms Li to Kr. *J. Chem. Phys.* **1992**, *97*, 2571–2577. [CrossRef]

70. Del Bene, J.E.; Elguero, J.; Alkorta, I.; Yáñez, A.M.; Mó, O. An ab Initio Study of ^{15}N−^{11}B Spin−Spin Coupling Constants for Borazine and Selected Derivatives. *J. Phys. Chem. A* **2006**, *110*, 9959–9966. [CrossRef]

71. Stanton, J.F.; Gauss, J.; Watts, J.D.; Nooijen, J.M.; Oliphant, N.; Perera, S.A.; Szalay, P.S.; Lauderdale, W.J.; Gwaltney, S.R.; Beck, S.; et al. *ACES II*; University of Florida: Gainesville, FL, USA, 1991.

Article

Perturbating Intramolecular Hydrogen Bonds through Substituent Effects or Non-Covalent Interactions

Al Mokhtar Lamsabhi *, Otilia Mó and Manuel Yáñez *

Departamento de Química (Módulo 13, Facultad de Ciencias) and Institute of Advanced Chemical Sciences (IadChem), Universidad Autónoma de Madrid, Campus de Excelencia UAM-CSIC, Cantoblanco, 28049 Madrid, Spain; otilia.mo@uam.es
* Correspondence: mokhtar.lamsabhi@uam.es (A.M.L.); manuel.yanez@uam.es (M.Y.)

Abstract: An analysis of the effects induced by F, Cl, and Br-substituents at the α-position of both, the hydroxyl or the amino group for a series of amino-alcohols, $HOCH_2(CH_2)_nCH_2NH_2$ ($n = 0$–5) on the strength and characteristics of their OH···N or NH···O intramolecular hydrogen bonds (IMHBs) was carried out through the use of high-level G4 ab initio calculations. For the parent unsubstituted amino-alcohols, it is found that the strength of the OH···N IMHB goes through a maximum for $n = 2$, as revealed by the use of appropriate isodesmic reactions, natural bond orbital (NBO) analysis and atoms in molecules (AIM), and non-covalent interaction (NCI) procedures. The corresponding infrared (IR) spectra also reflect the same trends. When the α-position to the hydroxyl group is substituted by halogen atoms, the OH···N IMHB significantly reinforces following the trend H < F < Cl < Br. Conversely, when the substitution takes place at the α-position with respect to the amino group, the result is a weakening of the OH···N IMHB. A totally different scenario is found when the amino-alcohols $HOCH_2(CH_2)_nCH_2NH_2$ ($n = 0$–3) interact with BeF_2. Although the presence of the beryllium derivative dramatically increases the strength of the IMHBs, the possibility for the beryllium atom to interact simultaneously with the O and the N atoms of the amino-alcohol leads to the global minimum of the potential energy surface, with the result that the IMHBs are replaced by two beryllium bonds.

Keywords: intramolecular hydrogen bonds; amino-alcohols; α-substitution; beryllium bonds; calculated infrared spectra

Citation: Lamsabhi, A.M.; Mó, O.; Yáñez, M. Perturbating Intramolecular Hydrogen Bonds through Substituent Effects or Non-Covalent Interactions. *Molecules* **2021**, *26*, 3556. https://doi.org/10.3390/molecules26123556

Academic Editor: Mirosław Jabłoński

Received: 27 May 2021
Accepted: 8 June 2021
Published: 10 June 2021

Publisher's Note: MDPI stays neutral with regard to jurisdictional claims in published maps and institutional affiliations.

1. Introduction

The name "hydrogen bonding" was used for the first time by Linus Pauling in 1929 [1], although the interactions associated with a hydrogen bond had been described for the first time more than a century ago, by W.M. Latimer and W.H. Rodebush, in a paper published in 1920 [2]. Since 1934, the year in which Pauling and Brockway experimentally confirmed that carboxylic acids could indeed form hydrogen bonds [3], hydrogen bonding became one of the most fruitful concepts in chemistry, being behind a huge amount of stabilizing intermolecular interactions [4–15]. In addition, in the 1920s, the existence of intramolecular interactions that, nowadays, we know as intramolecular hydrogen bonds (IHBs), were described in the literature. Indeed, the first paper on the coordination of hydrogen to explain the abnormal solubilities of some benzene derivatives was published in 1924 [16]. The intramolecular hydrogen bond was actually described in a 1926 publication by H.E. Amstrong [17], making reference to a paper of T.M. Lowry on the anomalies of the optical rotatory dispersion of tartaric acid published the same year [18], as a "bigamous hydrogen". Almost a century has elapsed since the publication of these articles, and the presence of IHBs is very often shown to be behind many different phenomena in chemistry, physics, in particular in photophysics, material science, and biochemistry, to the point that IUPAC dedicated a specific paper to the definition of this chemical interaction [19]. Let us mention here a few suitable examples, drawn from a huge number of publications, such as the substituent effects on the IHB of malonaldehyde [20], or the role of IMHBs on: protonation and deprotonation processes [21], NMR chemical shifts [22], bond dissociation enthalpies [23,24], or their influence on the role of aromaticity in chemical reactions [25]. They are also crucial to understand excited-state intramolecular proton transfers [26–28], or the control of photosubstitution phenomena in metal complexes [29], or the control of emissive properties of different organic compounds [30]. IMHBs are also behind the high chemical stability of metal-organic frameworks [31] and the excellent electroluminescence of some organic light-emitting diodes (OLEDs) [32], and related with linear and nonlinear electric properties [33] or with the thermodynamic properties of compounds like pyridinol derivatives [34]. In biochemistry, they are responsible for proton shuttle mechanisms in certain lipase-catalyzed reactions [35]. IMHBs between hydroxyl and amino groups of a serinol function, characteristic of an important lipid, sphingosine, seem to be responsible for its existence, as neutral at the physiological pH [36], IMHBs allow also for designing novel peptide inhibitors [37], and play an important role as far as the molecular conformation of amino acids is concerned [38]. In addition, important properties are associated with the so-called intramolecular charge-inverted hydrogen bonds [39,40].

Since the pioneering studies by microwave spectroscopy on 2-aminoethanol [41] and by infrared spectroscopy on *trans*-8a and -8b-decahydroquinilinol reveal the important role of IMHBs in their stability, the interest in amino-alcohols increased significantly. The first ab initio calculations showed that 2-aminoethanol, 3-aminopropanol, and 4-aminobutanol were indeed stabilized by OH···N IMHBs, this stabilization being larger the longer the chain [42]. Very recent microwave experiments in parallel with MP2 ab initio calculations confirmed that, in both 3-aminopropanol [43] and 4-aminobutanol [44], the ground state is stabilized through the formation of an OH···N IMHB with O-N internuclear distance of 2.856 and 2.954 Å, respectively. In addition, a recent and rather complete study [45] using FTIR measurements, ^{1}H NMR spectroscopy, density functional theory (DFT) calculations, and molecular dynamics (MD) on 3-aminopropan-1-ol, 3-methylaminopropan-1-ol, and 3-dimethylaminopropan-1-ol, unambiguously showed that the methylation at the N atom results in a systematic enhancement of the OH···N IMHB, reflecting a decrease of the s-character of the nitrogen lone pair orbital.

These results clearly indicate that the effects of substituents directly attached to the active sites of the OH···N IMHB have a significant effect in its strength, but, to the best of our knowledge, not much is known on the effects induced by substituents that are at the α-position of both the hydroxyl or the amino group, or what the effects are when any of these two groups are also actively participating in a non-covalent interaction with a second body. Hence, one of the aims of this paper is to investigate, through the use of high-level ab initio calculations, the characteristics of the OH···N IMHB in the series of $HOCHX(CH_2)_nCH_2NH_2$ and $HOCH_2(CH_2)_nCHXNH_2$ ($n = 0$–5) with substituted amino-alcohols being X = H, F, Cl, Br. As mentioned above, there is an alternative way in which the strength of an IMHB can be altered, and this is through non-covalent interactions in which either the proton donor or the proton acceptor is engaged. Among the many possible non-covalent interactions capable of interacting with IMHBs, the so-called beryllium bonds [46] are particularly interesting. A good and rather complete compilation on the coordination chemistry of beryllium was reported recently by Perera et al. [47]. Beryllium bonds have been found to compete with dihydrogen bonds [48], and they are able to modulate the strength of tetrel bonds [49]. They can also be formed when interacting with other electron deficient systems, such as boron derivatives [50] or with noble gases [51]. Although there are previous studies on the mutual interaction of intermolecular hydrogen bonds and beryllium bonds [52,53], very little has been done dealing with IMHBs. In this respect, a recent theoretical study must be mentioned describing how intramolecular hydrogen bonds are able to enhance tetrel bonds [54], or the competition between pnictogen bonds and intramolecular hydrogen bonds in protonated systems [55]. Here, we have decided to investigate this phenomenon when the system interacting with the amino-alcohol is an electron deficient compound. To achieve this goal, we will investigate the characteristics of the complexes between $HOCH_2(CH_2)_nCH_2NH_2$ (n = 0–3) when they interact, either through the amino group or the hydroxyl group, with BeF_2 molecules, yielding the corresponding beryllium bonds [46].

2. Computational Details

In order to ensure the reliability of the theoretical characterization of the IMHB stabilizing the compounds under investigation, we have decided to use the Gaussian-4 (G4) theory [56]. The G4 theory is a high-level ab initio composite method based on DFT optimized geometries and thermochemical corrections obtained at the B3LYP/6-31G(2df,p) level [56]. Correlation effects are accounted for by using the Moller–Plesset perturbation theory up to the fourth-order and CCSD(T) coupled cluster theory. A further correction, to account for the Hartree–Fock basis set limit, is added using a linear two-point extrapolation scheme and quadruple-zeta and quintuple-zeta basis sets.

However, it would be impossible to apply the G4 scheme to characterize all possible conformers associated with these two series of substituted amino-alcohols $HOCHX(CH_2)_n$ CH_2NH_2 and $HOCH_2(CH_2)_nCHXNH_2$ (n = 0–5, X = H, F, Cl, Br) under scrutiny because this number is huge in particular when n $\geq$ 3, and it would be applied only to the most stable ones. In order to make the selection of the most stable conformers, we first located the ensemble of them by using the conformer–rotamer ensemble sampling tool (CREST) recently developed by Grimme [57,58]. This method is based on the semiempirical tight-binding based quantum chemistry method GFN2-xTB [59] in the framework of meta-dynamics (MTD). Once the conformers ensemble is obtained, the script ENSO (Energetic Sorting of Crest ensembles) [60] is applied to classify them in a three-step process. The first one consists of a single point calculation at pbeh-3c/Def2-SVP [61] level of theory of the different structures of the ensemble. In a second step, the conformers in the range of 25 kJ/mol are selected for optimization at the same level of theory. In the last step, the range of energy is reduced to about 8 kJ/mol, the final energy, and the percentage of Boltzmann distribution being obtained from a single point wb97x/Def2-TZVPP calculations [62]. The conformers with a percentage higher than 1% were chosen to carry out the G4 calculations. It should also be mentioned that, for all the complexes including Beryllium bonds, final energies were also calculated at the G4 level, but, in these cases, instead of using the standard geometry optimization procedure within the G4 formalism, the structures were optimized using a larger basis set expansion, aug-cc-pVTZ, that includes diffuse functions, very often critical to correctly describe non-covalent interactions.

The bonding in all the systems investigated was analyzed through the use of four different approaches, namely the quantum theory of atoms in molecules (QTAIM) [63,64], the electron localization function (ELF) [65], the natural bond orbital (NBO) analysis [66], and the non-covalent interaction (NCI) formalism [67]. In the QTAIM approach, a topological analysis of the electron density of the systems permits the location of its critical points and the paths of minimum gradients connecting them, which leads to an unambiguous definition of chemical bonds, and/or ring and cage structures. The ELF procedure allows a partition of the molecular space in monosynaptic and disynaptic (or polysynaptic) basins in which the electrons of the system are distributed. The monosynaptic ones are associated with core electrons and/or lone pairs, whereas the disynaptic (or polysynaptic) correspond to bonding regions. The NBO method is based on the generation of localized hybrid orbitals which would correspond to Lewis-type representations of the molecular structures. The calculation of second-order orbital interaction energies between occupied and empty orbitals permit quantitatively characterizing donations and backdonations among occupied and empty localized molecular orbitals involved in inter and/or intramolecular interactions. The NCI formalism is an alternative analysis of the electron density, based on the fact that regions of small reduced-density gradient at low electronic densities are associated with the presence of non-covalent interactions. This analysis leads to rather visual representations if a colored scale is used to plot the isosurfaces of the reduced density, in both two- and three-dimensional spaces.

3. Results and Discussion

In our theoretical survey, we have investigated, for each one of the species under consideration, the possibility of having O–H$\cdots$N or N–H$\cdots$O IMHBs, though, in almost all cases, the former are stronger than the latter, and, therefore, in most of the cases, the global minima are characterized by the existence of an O–H$\cdots$N IMHB.

The number of stable conformers of HOCHX(CH$_2$)$_n$CH$_2$NH$_2$ and HOCH$_2$(CH$_2$)$_n$CHXNH$_2$ (n = 0–5, X = H, F, Cl, Br) compounds obviously increases as the number of carbon atoms increases along the series, reaching for the larger values of n several hundred. In what follows, we will analyze the six more stable unsubstituted HOCH$_2$(CH$_2$)$_n$CH$_2$NH$_2$ (n = 0–5) amino-alcohols first, and, after that, we will pay attention to the effect of the X substituent at the α-position with respect to the CH$_2$OH group and with respect to the CH$_2$NH$_2$ group.

3.1. HOCH$_2$(CH$_2$)$_n$CH$_2$NH$_2$ (n = 0–5) Compounds

We will start our analysis by looking at the effect that the length of the carbon chain has on the strength of the IMHB between the alcohol and the amino functional groups. As expected, the conformer stabilized by the formation of a O–H$\cdots$N IMHB is more stable (see Figure S1) than that exhibiting a N–H$\cdots$O IMHB because the hydroxy group is a better proton donor but a weaker base than the amino group, so, in what follows, we will discuss only the systems exhibiting a O–H$\cdots$N IMHB.

The structures corresponding to the most stable conformer for each value of n are shown in Figure 1. It can be seen that the longest O–H$\cdots$N IMHB corresponds to the first member of family, 2-amino ethanol (n = 0) which should be the amino-alcohol with the weakest bond along the series. This bond length reaches its minimum for n = 2, which in principle should be the member of the series with the strongest O–H$\cdots$N IMHB, its strength decreasing for larger values of n (n = 0–5).

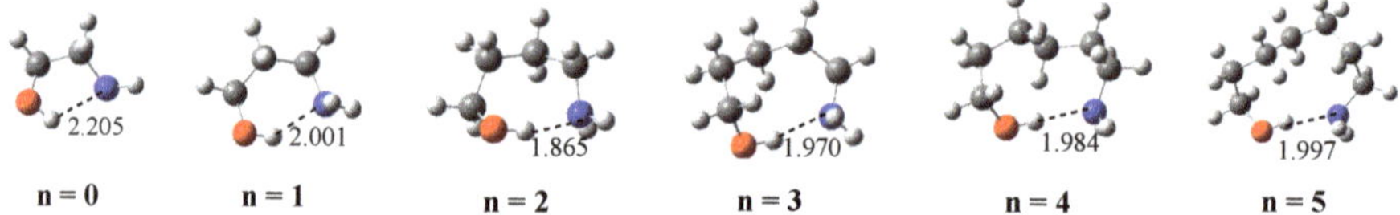

Figure 1. G4 optimized structures for HOCH$_2$(CH$_2$)$_n$CH$_2$NH$_2$ (n = 0–5) amino alcohols, showing the IMHB length in Å.

Unfortunately, although the calculation of the energy of an intermolecular hydrogen bond is straightforward, this is not so when dealing with IMHBs, though some procedures to estimate it have been proposed [68]. We propose, however, the use of the isodesmic reaction [69] (1) as a suitable method to have, at least, a reasonably good estimate of the relative stabilization gained in these compounds when the IMHB is formed. The isodesmic process we have used corresponds to the first reaction shown in Scheme 1.

Scheme 1. Isodesmic reactions to estimate the stabilization produced by the IMHB (reaction (**1**)) and the repulsive interaction between the terminal methyl groups (reaction (**2**)).

Reaction (2) in Scheme 1, in which the carbon chain is fully deployed, was used just to check whether, in the isodesmic reactions proposed, the repulsive interaction between the terminal methyl groups may significantly affect the isodesmic energy obtained with the first reaction. The ideal situation should correspond to that in which the second reaction is thermoneutral. The results obtained for these two reactions at the G4 level of theory are summarized in Table 1. It can be observed that, indeed, the reactions (2) for the set of derivatives investigated are nearly thermoneutral, which is an indication of the reliability of our isodesmic reaction (1) to provide a good estimation of the stabilization energy associated with each IMHB. Not surprisingly, the largest deviation for thermoneutrality of reaction (2) is obtained for the first member of the series, as a consequence of its higher rigidity. Nevertheless, it should be taken into account that, in these isodesmic reactions, and, mainly, when the carbon chain is sufficiently large and very flexible, the number of possible conformers for the three compounds that do not have IMHB is very large, and, in many cases, the energy difference between them is very small, even smaller than 1 kJ·mol^{-1}.

Table 1. G4 calculated enthalpies (kJ·mol^{-1}) for the isodesmic reactions included in Scheme 1.

n	ΔH_{HB}	ΔH_{Int}
0	−8.7	1.8
1	−16.8	−0.6
2	−24.6	−0.5
3	−16.1	0.1
4	−16.6	0.1
5	−17.5	0.3

The values in Table 1 show, in agreement with the IMHB length reported in Figure 1, that the strongest isodesmic stabilization interaction energy corresponds to $n = 2$, being the weakest one corresponding to $n = 0$. These energetic trends are also consistent with the characteristics of the corresponding molecular graphs drawn in Figure 2a, showing that, in all cases, the electron density at the BCP located between the H atom of the hydroxyl group and the N atom of the amino group goes through a maximum for $n = 2$.

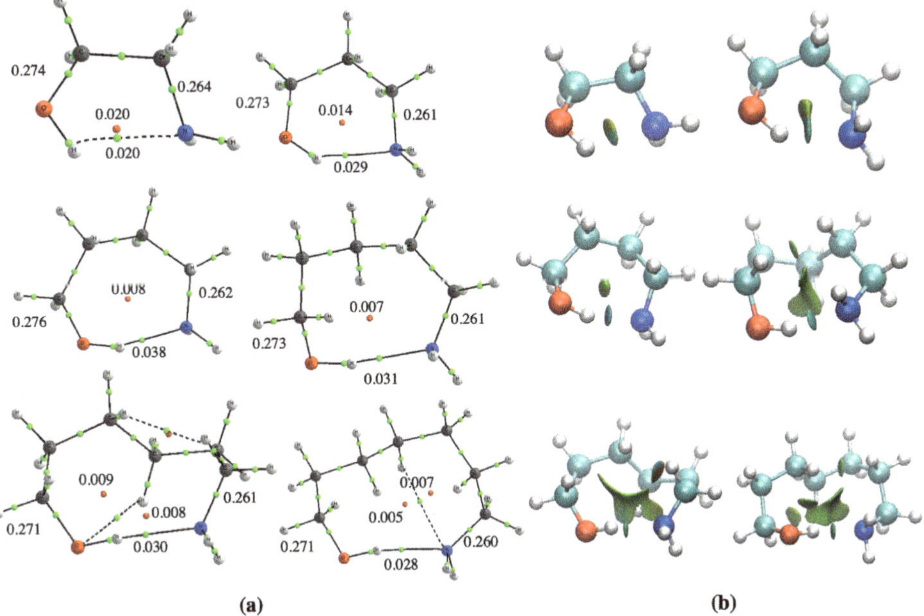

Figure 2. (a) Molecular graphs for HOCH$_2$(CH$_2$)$_n$CH$_2$NH$_2$ ($n = 0$–5) amino alcohols showing the electron density at the BCPs (green dots) and at the RCPs (red dots); (b) 3D-NCI plots for the same compounds showing the isosurfaces associated with the different intramolecular interactions.

The 3D-NCI plots included in Figure 2b also show the existence of an isosurface between the OH and the NH_2 groups, which denotes the existence of a NCI whose strength increases from $n = 0$ (greenish) up to $n = 2$ (blueish). It is also worth noting that, already for $n = 1$, a second greenish lobe appears attached to the blueish one associated to the van der Waals interactions range involving the chain of carbon atoms. Indeed, for $n = 2$, the two isosurfaces are now independent and, for the remaining systems ($n = 3-5$), the extension of these secondary interactions increases with the length of the chain of carbons.

The evolution of the characteristics of the IMHB along the series is nicely reflected in the IR spectra of the different species. As shown in Figure 3, the absorption band associated with the O–H stretching of the alcoholic function, which, for the first compound ($n = 0$), is predicted to appear at 3715 cm^{-1}, is clearly redshifted when moving to larger compounds. This red-shifting is maximum (233 cm^{-1}) for $n = 2$, but even for $n = 5$ the red-shifting with respect to $n = 0$ is significant (141 cm^{-1}), indicating, in agreement with the other indices that the IMHB for $n = 5$, though weaker than that for $n = 2$ is still stronger than for $n = 0$.

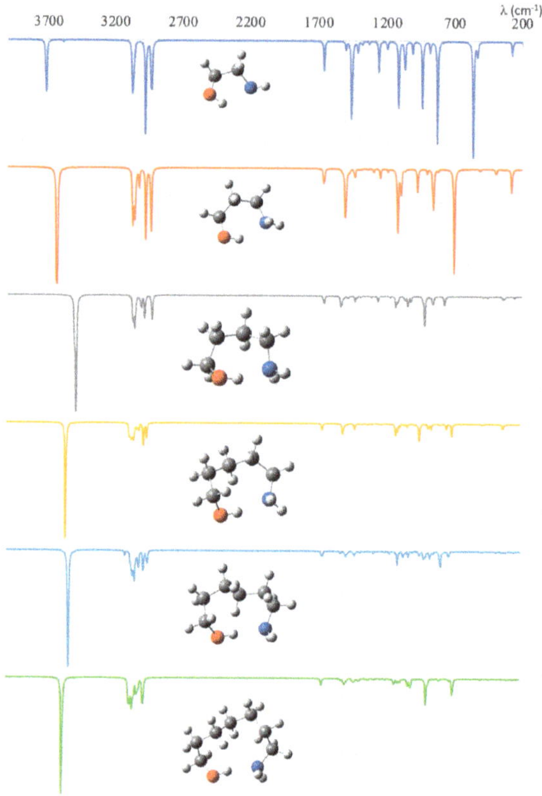

Figure 3. Calculated IR spectra for the $HOCH_2(CH_2)_nCH_2NH_2$ ($n = 0-5$) amino alcohols.

3.2. $HOCHX(CH_2)_nCH_2NH_2$ and $HOCH_2(CH_2)_nCHXNH_2$ ($n = 0-5$, $X = F$, Cl, Br) Derivatives

Let us look now at the effects of halogen substituents at the α-position of both the -CH_2OH and the -CH_2NH_2 functional groups. As indicated above, the number of possible conformers is huge, so, in Figures S2 and S3 of the Supplementary Materials, we present only the optimized geometries of the most stable conformers stabilized by an O–H$\cdots$N or a N–H$\cdots$O IMHB. As it was also the case for the unsubstituted amino alcohols, and regardless of whether the halogen substituent is at the α-position of the CH_2OH group or the CH_2NH_2 group, the conformers with a N–H$\cdots$O IMHB are less stable than those with a O–H$\cdots$N IMHB, with the only exception being the $HOCH_2CHXNH_2$ (X = F, Cl, Br) derivatives (see the first row of Figure S3), where the conformer with a N–H$\cdots$O IMHB is predicted to be, at the G4-level of theory, slightly more stable than the conformer exhibiting a O–H$\cdots$N IMHB regardless of the nature of the substituent X.

Focusing our attention then on the compounds stabilized by O–H$\cdots$N IMHB, it can be observed that, as it was the case for the unsubstituted amino-alcohols, when the halogen substituent (F, Cl or Br) is at the α-position of the CH_2OH group, the IMHB length decreases from $n = 0$ to $n = 2$, where it reaches its minimum value. An NBO analysis of these compounds not only indicates (see Figure 4a) that, as expected, the most significant orbital interactions involve the occupied lone pair of the amino group and the empty antibonding $\sigma^*_{O–H}$ orbital, which necessarily results in a weakening of the O–H bond, but also (see Figure 4b) that this interaction reaches its maximum, regardless of the nature of the substituent (F, Cl or Br) for $n = 2$. It is also very important to note that the same picture shows that this effect increases following the sequence: H < F < Cl < Br, parallel to the energy of the antibonding $\sigma^*_{O–H}$ orbital decreases, favoring the charge donation from the nitrogen lone-pair. Consistently, the electron densities at the O–H BCP decrease, but those at the IMHB BCP (See Figure 4c) increase following the sequence H < F < Cl < Br (see Figure 4c).

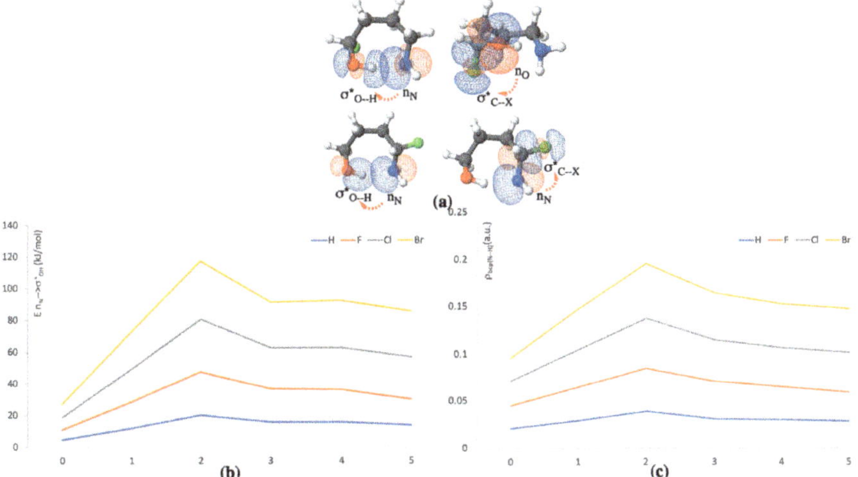

Figure 4. (a) Dominant orbital interactions associated with the formation of O–H$\cdots$N IMHBs when the substituents are at the α-position with respect to the CH_2OH group (first row) and with respect to the CH_2NH_2 group (second row); (b) variation of the energy involved in these orbital interactions as a function of n for the different substituents considered; (c) variation of the electron density at the IMHB BCP as a function of n for the different substituents considered.

The effects on the IMHB characteristics by introducing the halogen substituent (F, Cl or Br) at the α-position of the CH_2NH_2 group are just the opposite as those just discussed for α-substitution with respect to the CH_2OH group. Indeed, the presence of an electronegative atom at α-position of the CH_2NH_2 group results in a reduction of the intrinsic basicity of the amino group, which accordingly becomes a poorer donor towards the $\sigma^*_{O–H}$ orbital, leading to O–H$\cdots$N IMHB weaker than in the unsubstituted compounds. Accordingly, this IMHB becomes about 10% longer. Again, the substituent effect increases as H < F < Cl < Br, and, therefore, whereas for substituents α to the CH_2OH group the strongest O–H$\cdots$N IMHB is observed for the Br derivative, when the substituent is α to the CH_2NH_2 group, the Br derivative exhibits the weaker IMHB. This behavior is in perfect agreement with the values of the electron densities at the BCP associated with the IMHB, as shown in Table 2 for the particular case $n = 2$ being taken as a suitable example.

Table 2. Electron densities (a.u.) at the O–H···N IMHB BCP in $HOCHX(CH_2)_2CH_2NH_2$ and $HOCH_2(CH_2)_2CHXNH_2$ (X = H, F, Cl, Br).

X	$HOCHX(CH_2)_2CH_2NH_2$	$HOCH_2(CH_2)_2CHXNH_2$
H	0.038	0.038
F	0.046	0.032
Cl	0.053	0.031
Br	0.057	0.030

These different trends are also nicely reflected in the characteristics of the corresponding IR spectra. We are going to illustrate this point again using the case $n = 2$ as a suitable example. As shown in Figure 5a, for the case in which the substitution takes place at the α-position with respect to the CH_2OH group, the absorption band associated with the O–H stretching frequency is clearly red shifted upon F, Cl, and Br substitution. If the substitution takes place at the α-position with respect to the CH_2NH_2 group (see Figure 5b), the O–H stretching band is now blue shifted when H is replaced by F, Cl, Br.

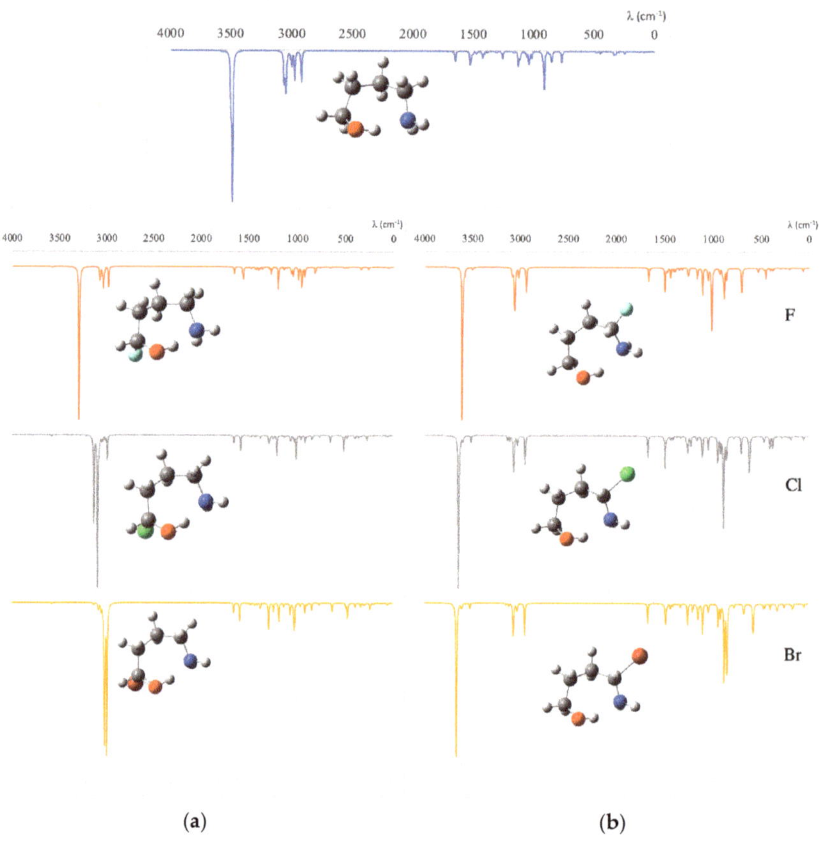

(a)   (b)

Figure 5. Calculated IR spectra for $HOCH_2(CH_2)_2CH_2NH_2$ and for compounds: (**a**) $HOCHX(CH_2)_2CH_2NH_2$ (X = H, F, Cl, Br) and (**b**) $HOCH_2(CH_2)_2CHXNH_2$ (X = F, Cl, Br).

3.3. Complexes between HOCH$_2$(CH$_2$)$_n$CH$_2$NH$_2$ (n = 0–3) and BeF$_2$

In this section, we will examine the effect that the interaction of the amino-alcohols HOCH$_2$(CH$_2$)$_n$CH$_2$NH$_2$ (n = 0–3) with a strong Lewis acid as BeF$_2$ will have on the IMHBs stabilizing these compounds and characterized in Section 3.1. However, the presence of the beryllium derivative opens up new scenarios in which new IMHBs enter into play as well as the possibility of having beryllium bonds replacing the IMHBs which characterize the isolated amino-alcohols.

Indeed, the association of BeF$_2$ to the alcohol functional group (see the first column of Figure 6) leads to the amino-alcohols moiety structures being very similar to those exhibited by the isolated amino-alcohol, though the bond length of the OH···N IMHB is much shorter than the one found in Section 3.1 for the isolated system.

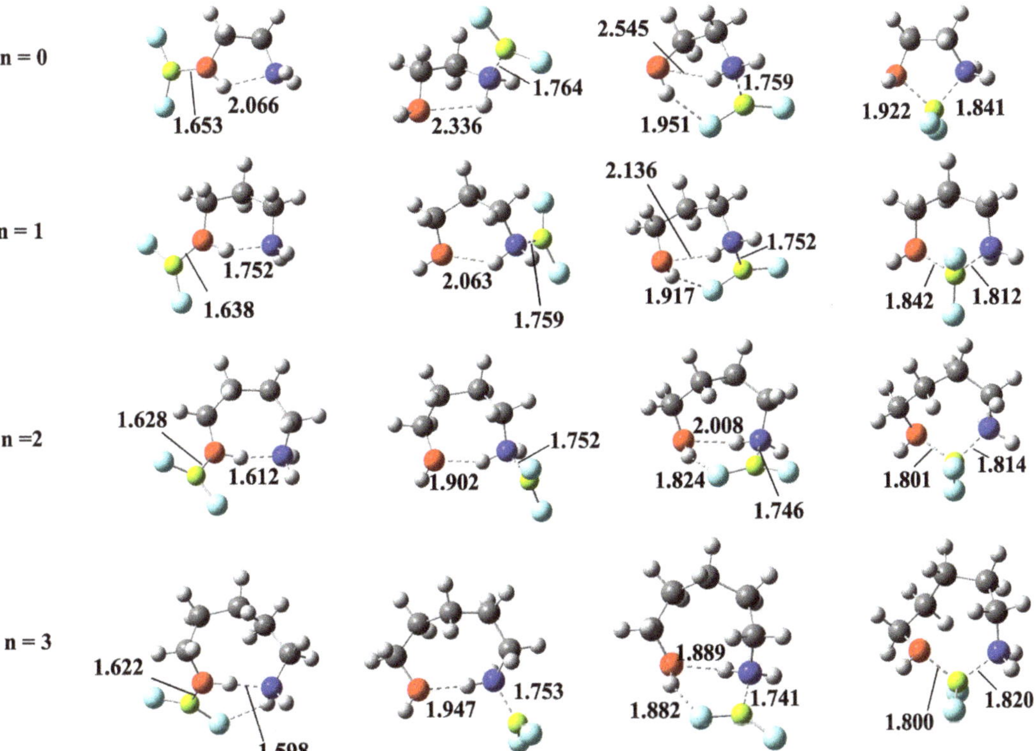

Figure 6. B3LYP/aug-cc-pVTZ optimized structures for the complexes between HOCH$_2$(CH$_2$)$_n$CH$_2$NH$_2$ (n = 0–3) amino-alcohols with BeF$_2$ stabilized through OH···N IMHBs (first column), NH···O IMHBs (second column), NH···O and OH···F IMHBs, simultaneously (third column) and through the beryllium bonds involving both the OH and the NH$_2$ groups (fourth column). The length of all these non-covalent interactions is in Å.

This reinforcement of the OH···N IMHB is the obvious consequence of the significant electron density transfer from the oxygen atom of the hydroxy group to the empty orbitals of the Be atom to form the corresponding beryllium bond, which necessarily increases the proton donor character of the OH group. Indeed, the second-order orbital interaction energies between the nitrogen lone-pair and the σ*$_{O-H}$ antibonding orbital, responsible for the formation of the O–H···N IMHB, are about four times larger for the BeF$_2$ complexes than for the isolated amino-alcohol as shown in Table S1 of the Supplementary Materials. The effect is qualitatively similar when BeF$_2$ interacts with the amino group (second column of Figure 6), which, upon beryllium association, also becomes a stronger proton donor, reinforcing the N–H⋯O IMHB, but necessarily to a lesser extent than when the group involved in the alcohol function, with the only exception of the amino-ethanol, in which case the conformer with a N–H⋯O IMHB is predicted to be 9 kJ·mol^{-1} more stable than the one stabilized by the formation of a O–H⋯N IMHB (see Table 3).

Table 3. Relative stability (kJ·mol^{-1}) of the complexes between $HOCH_2(CH_2)_nCH_2NH_2$ (n = 0–3) amino-alcohols and BeF_2, stabilized by the different IMHBs indicated in the first row, or through the bridge resulting from the simultaneous interaction of the Be atom with the O and N atoms of the amino-alcohol.

n	O–H⋯N	N–H⋯O	N–H⋯O, O–H⋯F	Bridged
0	22.2	13.2	8.9	0.0
1	23.4	30.8	19.9	0.0
2	9.7	32.0	24.7	0.0
3	8.7	33.4	27.2	0.0

It is also worth noting that, when the complex is stabilized through the formation of a N–H⋯O IMHB, the possible formation of a second IMHB by the interaction of the O–H group with one of the fluorine atoms attached to beryllium is open, and, indeed, as shown in Table 3, this new conformer is systematically (from 6 to 11 kJ·mol^{-1}) more stable (compare the third and the fourth column of Table 3). Nevertheless, the most important finding is that, in all cases, the global minimum (fifth column of Table 3) corresponds to a complex in which no IMHBs are formed because the interaction of the O and N atoms of the amino-alcohol with beryllium atom of BeF_2 molecule forming the corresponding beryllium bonds is energetically more favorable. The propensity of Be to exhibit a tetrahedral coordination has been previously reported in the literature when competing with intermolecular hydrogen bonds [47,70]. Here, we find that this tendency of Be to behave like a "tetrahedral proton" [47,70] is also observed when competing with IMHBs. In addition, importantly, the enthalpy gap between these bridged structures and the most stable conformers exhibiting a O–H⋯N IMHB is big enough (8.7 to 22.2 kJ·mol^{-1}) as to conclude, using a Boltzmann distribution function, that, at room temperature, practically 100% of the complexes are those stabilized through the formation of beryllium bonds.

An inspection of the molecular graphs of these complexes (see Figure 7) confirms the tendency of Be to be tetracoordinated, the electron density at the N–Be beryllium bond being systematically larger than that at the O–Be beryllium bond, since the amino group is a better electron donor than the hydroxyl one. On the other hand, the intrinsic stability of the complex, measured by the corresponding binding energy, increases from the amino-ethanol to the amino-pentanol, but, for the amino-butanol, goes through a little sinkhole that is consistent with the fact that, as we have discussed before, the stabilization produced in the isolated compound by the O–H⋯N IMHB is maximum for amino-butanol.

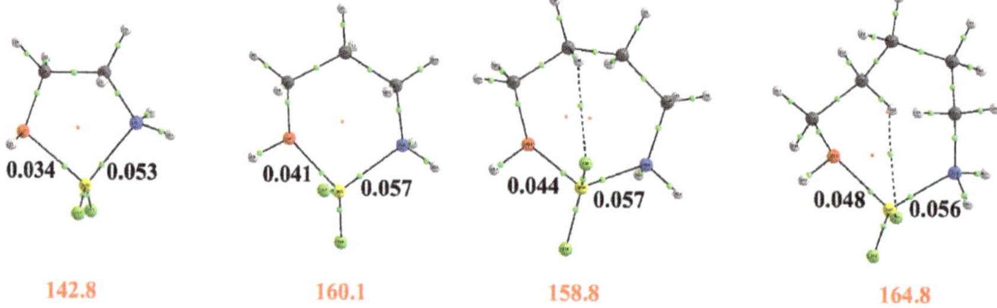

Figure 7. Molecular graphs for the HOCH2(CH2)nCH2NH2–BeF2 (n = 0–3) bridged conformers showing the electron densities (a.u.) at the BCPs between Be and the two basic sites of the amino-alcohol. The values in red correspond to the binding energies (kJ·mol^{-1}) defined as the energy for the reaction: $HOCH_2(CH_2)_nCH_2NH_2–BeF_2 \rightarrow HOCH_2(CH_2)_nCH_2NH_2 + BeF_2$.

Finally, it is interesting to highlight some of the peculiarities of the IR spectra of these complexes, showed in Figure 8.

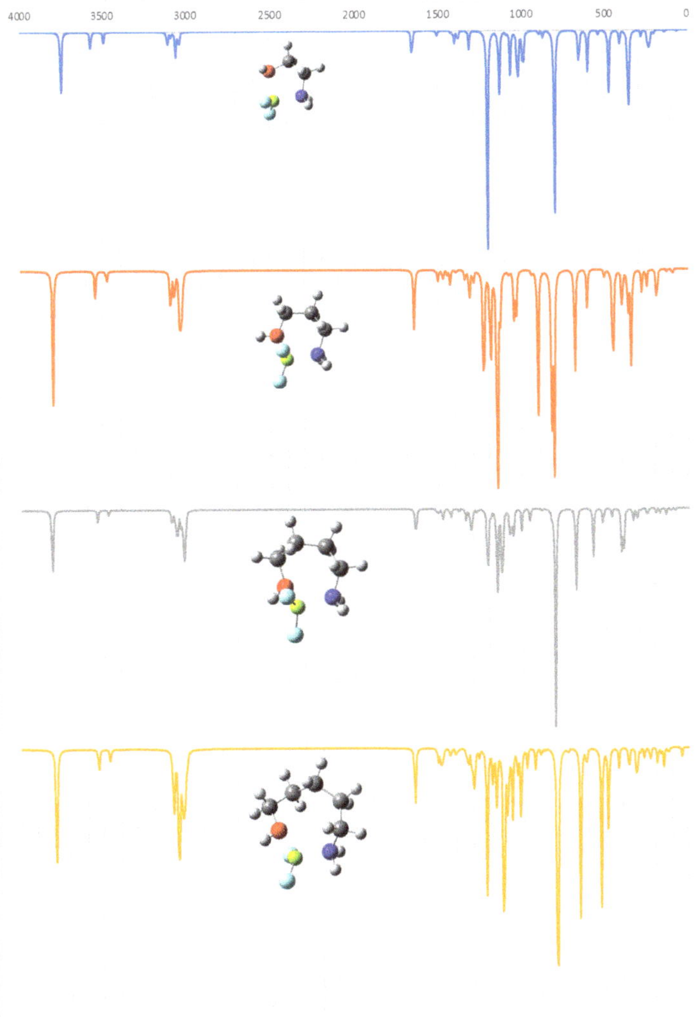

Figure 8. Calculated IR spectra for the bridged $HOCH_2(CH_2)_nCH_2NH_2$-BeF_2 (n = 0–3).

The first conspicuous fact is that the stretching O–H absorption band appears significantly blue-shifted with respect to those in Figure 5a, as it corresponds to a free OH group that, in these systems, does not participate in hydrogen bonding. Indeed, the OH stretching frequency calculated for the bridge complexes in Figure 8 go from 3767 to 3832 cm^{-1}, values rather similar to that obtained for methanol (3831 cm^{-1}) at the same level of theory. However, for the amino-ethanol, the blue-shifting, with respect to the isomer in Figure 5a, is only 200 cm^{-1} because of the rather weak O–H$\cdots$N IMHB in the isomer of Figure 5a. For amino-butanol and amino-pentanol, the shifting is greater than 1100 cm^{-1}. Another common finger-print of the IR spectra of these complexes is the presence of a rather intense band always around 800 cm^{-1} corresponding to the symmetric stretching of the BeF_2 moiety.

4. Conclusions

From our G4 calculations, we can conclude that, when F, Cl, or Br atoms are introduced as substituents at the α-position of both the hydroxyl or the amino group of amino-alcohols, there is a significant change in the strength of both the O–H$\cdots$N or the N–H$\cdots$O IMHBs that can be formed in these compounds. Substitution at the α-position of the hydroxyl group results in a significant reinforcement of the O–H$\cdots$N, which follows the trend H < F < Cl < Br, the effects on the N–H$\cdots$O IMHB being the opposite. Consistently, substitution at the α-position of the amino group also results in a reinforcement of the N–H$\cdots$O IMHB, but the effect is weaker than the one affecting the O–H$\cdots$N IMHB. Accordingly, the global minima for the unsubstituted and the substituted amino-alcohols are stabilized by the formation of O–H$\cdots$N IMHBs. In addition, for both sets, the strength of the O–H$\cdots$N IMHB, as shown by the characteristics of the corresponding molecular graphs, the NBO and NCI analyses, and the stabilization energy estimated through the use of isodesmic reactions, goes through a maximum for n = 2 (amino-butanol).

The scenario changes completely when the $HOCH_2(CH_2)_nCH_2NH_2$ (n = 0–3) amino-alcohols interact with a typical electro-deficient compound such as BeF_2. The interaction of BeF_2 with either the hydroxyl or the amino group of the amino-alcohol changes the strength of the IMHBs dramatically. In the first case, the O–H$\cdots$N IMHB becomes much stronger, whereas, in the second case, it is the N–H$\cdots$O IMHB that becomes very much reinforced, though the effect is still weaker than the one observed for the O–H$\cdots$N IMHB. In this second case, interaction of BeF_2 with the amino group, a new possibility is open through the formation of an additional IMHB between the O–H group of the alcoholic function and one of the F atoms of the BeF_2 molecule. However, it is the possibility for the beryllium atom to interact simultaneously with the O and the N atoms of the amino-alcohol that leads to the global minimum of the potential energy surface, with the result that the IMHBs are replaced by two beryllium bonds. These complexes stabilized by the formation of two beryllium bonds, O$\cdots$Be and N$\cdots$Be, present very peculiar IR spectra, in which the OH stretching band appears in some cases more than 1100 cm^{-1} blue-shifted with respect to the α-substituted amino-alcohols, whereas a very intense band, around 800 cm^{-1}, corresponding to the symmetric stretching of the BeF_2 moiety, is another finger-print of these spectra.

Supplementary Materials: The following are available: Figure S1: Relative stability of the $HOCH_2(CH_2)_nCH_2NH_2$ (n = 0–2) aminoalcohols stabilized by the formation of an O–H$\cdots$N IMHB or a N-H$\cdots$O IMHB, Figure S2: Molecular structures of the substituted HOCHX(CH2)nCH2NH2 (n = 0–5, X = H, F, Cl, Br) aminoalcohols, Figure S3: Molecular structures of the substituted HOCH2(CH2)nCHXNH2 (n = 0–5, X = F, Cl, Br) aminoalcohols, Table S1: Second-order orbital interactions energies between the nitrogen lone-pair and the s*OH antibonding orbital for the $HOCH_2(CH_2)_nCH_2NH_2$ (n = 0–3) amino-alcohols and their O$\cdots$BeF$_2$ complexes.

Author Contributions: Conceptualization, M.Y. and A.M.L.; methodology, O.M. and A.M.L.; software, A.M.L.; validation, A.M.L., O.M. and M.Y.; formal analysis, A.M.L., O.M. and M.Y.; investigation, A.M.L. and M.Y.; resources, O.M. and M.Y.; data curation, A.M.L.; writing—original draft preparation, M.Y.; writing—review and editing, A.M.L., O.M. and M.Y.; visualization, O.M.; supervision, M.Y. and A.M.L.; project administration, O.M. and M.Y.; funding acquisition, O.M., A.M.L. and M.Y. All authors have read and agreed to the published version of the manuscript.

Funding: This work was carried out with financial support from the Ministerio de Ciencia, Innovación y Universidades of Spain (Project No. PGC2018-094644-B-C21 and PID2019-110091GB-I00 (MICINN), and the PRIES-CM project Ref: Y2020/EMT-6290 from the Comunidad Autónoma de Madrid.

Institutional Review Board Statement: Not applicable.

Informed Consent Statement: Not applicable.

Data Availability Statement: The data presented in this study are available in the article and in the Supplementary Material.

Acknowledgments: This work was carried out with the financial support from the project PGC2018-094644-B-C21 and PID2019-110091GB-I00 (MICINN) of the Ministerio de Ciencia, Innovación y Universidades of Spain and the PRIES-CM project Ref: Y2020/EMT-6290 from the Comunidad Autónoma de Madrid. The authors would also like to thank the Centro de Computación Científica of the UAM (CCC-UAM) for the generous allocation of computer time and for their continued technical support.

Conflicts of Interest: The authors declare no conflict of interest.

Sample Availability: Samples of the compounds are not available from the authors.

References

1. Pauling, L. The Shared-Electron Chemical Bond. *Proc. Natl. Acad. Sci. USA* **1928**, *14*, 359–362. [CrossRef] [PubMed]
2. Latimer, W.M.; Rodebush, W.H. Polarity and Ionization from the Standpoint of the Lewis Theory of Valence. *J. Am. Chem. Soc.* **1920**, *42*, 1419–1433. [CrossRef]
3. Pauling, L.; Brockway, L.O. The Structure of the Carboxyl Group: I. The Investigation of Formic Acid by the Diffraction of Electrons. *Proc. Natl. Acad. Sci. USA* **1934**, *20*, 336–340. [CrossRef] [PubMed]
4. Pimentel, G.C.; McClelland, A.L. *The Hydrogen Bond*; W.H. Freeman: San Francisco, CA, USA, 1960.
5. Kollman, P.A.; Allen, L.C. Theory of hydrogen bond.Electronic structure and properties of water dimer. *J. Chem. Phys.* **1969**, *51*, 3286. [CrossRef]
6. Witkowski, A. Far-Infrared Spectrum of the Hydrogen Bond in Acetic Acid Dimers. *J. Chem. Phys.* **1970**, *52*, 4403. [CrossRef]
7. Allen, L.C. Simple model of hydrogen bonding. *J. Am. Chem. Soc.* **1975**, *97*, 6921–6940. [CrossRef]
8. Jeffrey, G.A.; Saenger, W. *Hydrogen Bonding in Biological Structures*; Springer Science and Business Media LLC: Berlin, Germany, 1991.
9. Jeffrey, G.A. *An Introduction to Hydrogen Bonding*; Oxford University Press: New York, NY, USA, 1997.
10. Desiraju, G.R.; Steiner, T. *The Weak Hydrogen Bond in Structural Chemistry and Biology*; Oxford University Press: Oxford, UK, 1999.
11. Grabowski, S.J. *Hydrogen Bonding—New Insights*; Springer: Dordrecht, The Netherlands, 2006; Volume 3.
12. Han, K.-L.; Zhao, G.-J. *Hydrogen Bonding and Transfer in the Excited State*; John Wiley& Sons: Chichester, UK, 2011.
13. Kuo, S.-W. *Hydrogen Bonding in Polymeric Materials*; Wiley-VCH: Weinheim, Germany, 2018.
14. Han, K.; Zhao, G. *Hydrogen-Bonding Research in Photochemistry, Photobiology, and Optoelectronic Materials*; World Scientific Pub. Co. Pte. Lt.: Singapore, 2018.
15. Grabowski, S.J. *Understanding Hydrogen Bonds: Theoretical and Experimental Views*; CPI Group, UK Ltd.: Croydon, UK, 2020.
16. Sidwick, N.V.; Callow, R.K. Abnormal benzene derivatives. *J. Chem. Soc.* **1924**, *125*, 527–538. [CrossRef]
17. Amstrong, H.E. Bigamous Hydrogen—A Protest. *Nature* **1926**, *117*, 553–554. [CrossRef]
18. Lowry, T.M. Optical Rotatory Dispersion. *Nature* **1926**, *117*, 271–275. [CrossRef]
19. Arunan, E.; Desiraju, G.R.; Klein, R.A.; Sadlej, J.; Scheiner, S.; Alkorta, I.; Clary, D.; Crabtree, R.H.; Dannenberg, J.J.; Hobza, P.; et al. Definition of the hydrogen bond (IUPAC Recommendations 2011). *Pure Appl. Chem.* **2011**, *83*, 1637–1641. [CrossRef]
20. Buemi, G.; Zuccarello, F. DFT study of the intramolecular hydrogen bonds in the amino and nitro-derivatives of malonaldehyde. *Chem. Phys.* **2004**, *306*, 115–129. [CrossRef]
21. Gonzalez, L.; Mó, O.; Yáñez, M. Protonation and Deprotonation of Thiomalonaldehyde. The Role of the Intramolecular Hydrogen Bond. In *Recent Theoretical and Experimental Advances in Hydrogen Bonded Clusters*; Springer Science and Business Media LLC: Berlin, Germany, 2000; pp. 393–402.
22. West-Nielsen, M.; Dominiak, P.; Wozniak, K.; Hansen, P.E. Strong intramolecular hydrogen bonding involving nitro- and acetyl groups. Deuterium isotope effects on chemical shifts. *J. Mol. Struct.* **2006**, *789*, 81–91. [CrossRef]
23. Nazarparvar, E.; Zahedi, M.; Klein, E. Density Functional Theory (B3LYP) Study of Substituent Effects on O–H Bond Dissociation Enthalpies of trans-Resveratrol Derivatives and the Role of Intramolecular Hydrogen Bonds. *J. Org. Chem.* **2012**, *77*, 10093–10104. [CrossRef]
24. Nazarparvar, E.; Zahedi, M.; Klein, E. Theoretical study of the substituent effects on O–H BDE of trans-resveratrol derivatives in water and benzene: NBO analysis of intramolecular hydrogen bonds. *Struct. Chem.* **2014**, *26*, 47–59. [CrossRef]
25. Galvao, T.L.P.; Rocha, I.M.; da Silva, M.D.M.C.R.; da Silva, M.A.V.R. Energetic Study of 4(3H)-Pyrimidinone: Aromaticity of Reactions, Hydrogen Bond Rules, and Support for an Anomeric Effect. *J. Phys. Chem. A* **2014**, *118*, 3360–3366. [CrossRef] [PubMed]
26. Liu, Z.-Y.; Hu, J.-W.; Huang, T.-H.; Chen, K.-Y.; Chou, P.-T. Excited-state intramolecular proton transfer in the kinetic-control regime. *Phys. Chem. Chem. Phys.* **2020**, *22*, 22271–22278. [CrossRef]
27. Laner, J.N.; Junior, H.d.C.S.; Rodembusch, F.S.; Moreira, E.C. New insights on the ESIPT process based on solid-state data and state-of-the-art computational methods. *Phys. Chem. Chem. Phys.* **2021**, *23*, 1146–1155. [CrossRef] [PubMed]
28. Guo, Q.; Ji, D.; Zhao, J. Theoretical insights into photochemical behavior and ESIPT mechanism for 2,6-dimethyl phenyl derivatives. *Chem. Phys. Lett.* **2021**, *767*, 138377. [CrossRef]
29. Hirahara, M.; Nakano, H.; Uchida, K.; Yamamoto, R.; Umemura, Y. Intramolecular Hydrogen Bonding: A Key Factor Controlling the Photosubstitution of Ruthenium Complexes. *Inorg. Chem.* **2020**, *59*, 11273–11286. [CrossRef] [PubMed]

30. Listkowski, A.; Masiera, N.; Kijak, M.; Luboradzki, R.; Leśniewska, B.; Waluk, J. Controlling Emissive Properties by Intramolecular Hydrogen Bonds: Alkyl and Aryl meso-Substituted Porphycenes. *Chem. A Eur. J.* **2021**, *27*, 6324–6333. [CrossRef] [PubMed]

31. Wang, K.; Wang, Q.; Wang, X.; Wang, M.; Wang, Q.; Shen, H.-M.; Yang, Y.-F.; She, Y.-B. Intramolecular hydrogen bond-induced high chemical stability of metal–organic frameworks. *Inorg. Chem. Front.* **2020**, *7*, 3548–3554. [CrossRef]

32. Li, W.; Chasing, P.; Benchaphanthawee, W.; Nalaoh, P.; Chawanpunyawat, T.; Kaiyasuan, C.; Kungwan, N.; Namuangruk, S.; Sudyoadsuk, T.; Promarak, V. Intramolecular hydrogen bond—Enhanced electroluminescence performance of hybridized local and charge transfer (HLCT) excited-state blue-emissive materials. *J. Mater. Chem. C* **2021**, *9*, 497–507. [CrossRef]

33. Avramopoulos, A.; Jabłoński, M.; Papadopoulos, M.G.; Sadlej, A.J. Linear and nonlinear electric properties and their dependence on the conformation and intramolecular H-bonding: A model study. *Chem. Phys.* **2006**, *328*, 33–44. [CrossRef]

34. Fallas, J.A.; Gonzalez, L.; Corral, I. Density functional theory rationalization of the substituent effects in trifluoromethyl-pyridinol derivatives. *Tetrahedron* **2009**, *65*, 232–239. [CrossRef]

35. Syrén, P.-O.; Le Joubioux, F.; Henda, Y.B.; Maugard, T.; Hult, K.; Graber, M. Proton Shuttle Mechanism in the Transition State of Lipase-Catalyzed N-Acylation of Amino Alcohols. *ChemCatChem* **2013**, *5*, 1842–1853. [CrossRef]

36. Loru, D.; Pena, I.; Alonso, J.L.; Sanz, M.E. Intramolecular interactions in the polarheadgroup of sphingosine: Serinol. *Chem. Commun.* **2016**, *52*, 3615–3618. [CrossRef]

37. Yang, X.; Zhong, J.; Zhang, Q.; Qian, J.; Song, K.; Ruan, C.; Xu, J.; Ding, K.; Zhang, J. Rational Design and Structure Validation of a Novel Peptide Inhibitor of the Adenomatous-Polyposis-Coli (APC)–Rho-Guanine-Nucleotide-Exchange-Factor-4 (Asef) Interaction. *J. Med. Chem.* **2018**, *61*, 8017–8028. [CrossRef]

38. Giubertoni, G.; Sofronov, O.O.; Bakker, H.J. Effect of intramolecular hydrogen-bond formation on the molecular conformation of amino acids. *Commun. Chem.* **2020**, *3*, 1–6. [CrossRef]

39. Jabłoński, M. Binding of X–H to the lone-pair vacancy: Charge-inverted hydrogen bond. *Chem. Phys. Lett.* **2009**, *477*, 374–376. [CrossRef]

40. Jabłoński, M. QTAIM-Based Comparison of Agostic Bonds and Intramolecular Charge-Inverted Hydrogen Bonds. *J. Phys. Chem. A* **2015**, *119*, 4993–5008. [CrossRef]

41. Penn, R.E.; Curl, R.F. Microwave Spectrum of 2-Aminoethanol: Structural Effects of the Hydrogen Bond. *J. Chem. Phys.* **1971**, *55*, 651–658. [CrossRef]

42. Kelterer, A.-M.; Ramek, M. Intramolecular hydrogen bonding in 2-aminoethanol, 3-aminopropanol and 4-aminobutanol. *J. Mol. Struct. Theochem* **1991**, *232*, 189–201. [CrossRef]

43. Khalil, A.S.; Lavrich, R.J. Intramolecular hydrogen bond stabilized conformation of 3-aminopropanol. *J. Mol. Spectrosc.* **2020**, *370*, 111279. [CrossRef]

44. Khalil, A.S.; Duguay, T.M.; Lavrich, R.J. Conformation and hydrogen bonding in 4-Aminobutanol. *J. Mol. Struct.* **2017**, *1138*, 12–16. [CrossRef]

45. Batista, P.R.; Karas, L.J.; Viesser, R.V.; De Oliveira, C.C.; Gonçalves, M.B.; Tormena, C.F.; Rittner, R.; Ducati, L.C.; De Oliveira, P.R. Dealing with Hydrogen Bonding on the Conformational Preference of 1,3-Aminopropanols: Experimental and Molecular Dynamics Approaches. *J. Phys. Chem. A* **2019**, *123*, 8583–8594. [CrossRef]

46. Yáñez, M.; Sanz, P.; Mó, O.; Alkorta, I.; Elguero, J. Beryllium Bonds, do they exist? *J. Chem Theor. Comput.* **2009**, *5*, 2763–2771. [CrossRef]

47. Perera, L.C.; Raymond, O.; Henderson, W.; Brothers, P.J.; Plieger, P.G. Advances in beryllium coordination chemistry. *Coord. Chem. Rev.* **2017**, *352*, 264–290. [CrossRef]

48. Li, Q.Z.; Liu, X.F.; Li, R.; Cheng, J.B.; Li, W.Z. Competition between dihydrogen bond and beryllium bond in complexes between HBeH and HArF: A huge blue shift of distant H-Ar stretch. *Spectrochim. Acta Part A Mol. Biomol. Spectrosc.* **2012**, *90*, 135–140. [CrossRef]

49. Liu, M.; Yang, L.; Li, Q.; Cheng, J.; Xiao, B.; Yu, X. Modulating the strength of tetrel bonding through beryllium bonding. *J. Mol. Model.* **2016**, *22*, 10. [CrossRef] [PubMed]

50. Arnold, T.; Braunschweig, H.; Ewing, W.; Kramer, T.; Mies, J.; Schuster, J.K. Beryllium bis(diazaborolyl): Old neighbors finally shake hands. *Chem. Commun.* **2014**, *51*, 737–740. [CrossRef]

51. Zhang, Q.; Chen, M.; Zhou, M.; Andrada, D.M.; Frenking, G. Experimental and Theoretical Studies of the Infrared Spectra and Bonding Properties of NgBeCO3 and a Comparison with NgBeO (Ng = He, Ne, Ar, Kr, Xe). *J. Phys. Chem. A* **2014**, *119*, 2543–2552. [CrossRef]

52. Albrecht, L.; Boyd, R.J.; Mó, O.; Yáñez, M. Cooperativity between hydrogen bonds and beryllium bonds in $(H_2O)_n BeX_2$ (n = 1–3, X = H, F) complexes. A new perspective. *Phys. Chem. Chem. Phys.* **2012**, *14*, 14540–14547. [CrossRef]

53. Mó, O.; Yáñez, M.; Alkorta, I.; Elguero, J. Modulating the Strength of Hydrogen Bonds through Beryllium Bonds. *J. Chem. Theory Comput.* **2012**, *8*, 2293–2300. [CrossRef] [PubMed]

54. Trujillo, C.; Alkorta, I.; Elguero, J.; Sánchez-Sanz, G. Cooperative Effects in Weak Interactions: Enhancement of Tetrel Bonds by Intramolecular Hydrogen Bonds. *Molecules* **2019**, *24*, 308. [CrossRef]

55. Sánchez-Sanz, G.; Trujillo, C.; Alkorta, I.; Elguero, J. Competition between intramolecular hydrogen and pnictogen bonds in protonated systems. *Theor. Chem. Accounts* **2016**, *135*, 140. [CrossRef]

56. Curtiss, L.A.; Redfern, P.C.; Raghavachari, K. Gaussian-4 theory. *J. Chem. Phys.* **2007**, *126*, 084108. [CrossRef]

57. Pracht, P.; Bohle, F.; Grimme, S. Automated exploration of the low-energy chemical space with fast quantum chemical methods. *Phys. Chem. Chem. Phys.* **2020**, *22*, 7169–7192. [CrossRef] [PubMed]
58. Grimme, S. Exploration of Chemical Compound, Conformer, and Reaction Space with Meta-Dynamics Simulations Based on Tight-Binding Quantum Chemical Calculations. *J. Chem. Theory Comput.* **2019**, *15*, 2847–2862. [CrossRef]
59. Bannwarth, C.; Ehlert, S.; Grimme, S. GFN2-xTB-An Accurate and Broadly Parametrized Self-Consistent Tight-Binding Quantum Chemical Method with Multipole Electrostatics and Density-Dependent Dispersion Contributions. *J. Chem. Theory Comput.* **2019**, *15*, 1652–1671. [CrossRef] [PubMed]
60. Grimme, S.; Bannwarth, C.; Dohm, S.; Hansen, A.; Pisarek, J.; Pracht, P.; Seibert, J.; Neese, F. Fully automated quantum-chemistry-based computation of spin–spin-coupled nuclear magnetic resonance spectra. *Angew. Chem. Int. Ed.* **2017**, *56*, 14763–14769. [CrossRef]
61. Grimme, S.; Brandenburg, J.G.; Bannwarth, C.; Hansen, A. Consistent structures and interactions by density functional theory with small atomic orbital basis sets. *J. Chem. Phys.* **2015**, *143*, 054107. [CrossRef]
62. Chai, J.-D.; Head-Gordon, M. Systematic optimization of long-range corrected hybrid density functionals. *J. Chem. Phys.* **2008**, *128*, 084106. [CrossRef] [PubMed]
63. Bader, R.F.W. *Atoms in Molecules. A Quantum Theory*; Clarendon Press: Oxford, UK, 1990.
64. Popelier, P.L.A. Quantum Chemical Topology: On Bonds and Potentials. In *Intermolecular Forces and Clusters I. Structure and Bonding*; Wales, D.J., Ed.; Springer: Berlin/Heidelberg, Germany, 2005; Volume 115.
65. Savin, A.; Nesper, R.; Wengert, S.; Fässler, T.F. ELF: The Electron Localization Function. *Angew. Chem. Int. Ed.* **1997**, *36*, 1808–1832. [CrossRef]
66. Reed, A.E.; Curtiss, L.A.; Weinhold, F. Intermolecular interactions from a natural bond orbital, donor-acceptor viewpoint. *Chem. Rev.* **1988**, *88*, 899–926. [CrossRef]
67. Contreras-García, J.; Johnson, E.R.; Keinan, S.; Chaudret, R.; Piquemal, J.-P.; Beratan, D.N.; Yang, W. NCIPLOT: A Program for Plotting Noncovalent Interaction Regions. *J. Chem. Theory Comput.* **2011**, *7*, 625–632. [CrossRef] [PubMed]
68. Jabłoński, M.; Kaczmarek, A.; Sadlej, A.J. Estimates of the Energy of Intramolecular Hydrogen Bonds. *J. Phys. Chem. A* **2006**, *110*, 10890–10898. [CrossRef]
69. Muller, P. Glossary of Terms Used in Physical Organic Chemistry. *Pure Appl. Chem.* **1994**, *66*, 1077–1184. [CrossRef]
70. McCleskey, T.M.; Ehler, D.S.; Keizer, T.S.; Asthagiri, D.N.; Pratt, L.R.; Michalczyk, R.; Scott, B.L. Beryllium displacement of H+ from strong hydrogen bonds. *Angew. Chem. Int. Edit.* **2007**, *46*, 2669–2671. [CrossRef] [PubMed]

molecules

Article

On the Relationship between Hydrogen Bond Strength and the Formation Energy in Resonance-Assisted Hydrogen Bonds

José Manuel Guevara-Vela [1], Miguel Gallegos [2], Mónica A. Valentín-Rodríguez [3], Aurora Costales [2], Tomás Rocha-Rinza [1] and Ángel Martín Pendás [2,*]

[1] Institute of Chemistry, National Autonomous University of Mexico, Circuito Exterior, Ciudad Universitaria, Delegación Coyoacán, Mexico City C.P. 04510, Mexico; jmguevarav@gmail.com (J.M.G.-V.); tomasrocharinza@gmail.com (T.R.-R.)
[2] Department of Analytical and Physical Chemistry, University of Oviedo, 33006 Oviedo, Spain; gallegosmiguel@uniovi.es (M.G.); costalesmaria@uniovi.es (A.C.)
[3] Instituto de Física Fundamental, Consejo Superior de Investigaciones Científicas (IFF-CSIC), Serrano 123, 28006 Madrid, Spain; monicavr@iff.csic.es
* Correspondence: ampendas@uniovi.es

Abstract: Resonance-assisted hydrogen bonds (RAHB) are intramolecular contacts that are characterised by being particularly energetic. This fact is often attributed to the delocalisation of π electrons in the system. In the present article, we assess this thesis via the examination of the effect of electron-withdrawing and electron-donating groups, namely $-F$, $-Cl$, $-Br$, $-CF_3$, $-N(CH_3)_2$, $-OCH_3$, $-NHCOCH_3$ on the strength of the RAHB in malondialdehyde by using the Quantum Theory of Atoms in Molecules (QTAIM) and the Interacting Quantum Atoms (IQA) analyses. We show that the influence of the investigated substituents on the strength of the investigated RAHBs depends largely on its position within the π skeleton. We also examine the relationship between the formation energy of the RAHB and the hydrogen bond interaction energy as defined by the IQA method of wave function analysis. We demonstrate that these substituents can have different effects on the formation and interaction energies, casting doubts regarding the use of different parameters as indicators of the RAHB formation energies. Finally, we also demonstrate how the energy density can offer an estimation of the IQA interaction energy, and therefore of the HB strength, at a reduced computational cost for these important interactions. We expected that the results reported herein will provide a valuable understanding in the assessment of the energetics of RAHB and other intramolecular interactions.

Keywords: hydrogen bond; interacting quantum atoms; resonance-assisted hydrogen bond

Citation: Guevara-Vela, J.M.; Gallegos, M.; Valentín-Rodríguez, M.A.; Costales, A.; Rocha-Rinza, T.; Martín Pendás, Á. On the Relationship between Hydrogen Bond Strength and the Formation Energy in Resonance-Assisted Hydrogen Bonds. *Molecules* **2021**, *26*, 4196. https://doi.org/10.3390/molecules26144196

Academic Editor: Mirosław Jabłoński

Received: 4 June 2021
Accepted: 7 July 2021
Published: 10 July 2021

Publisher's Note: MDPI stays neutral with regard to jurisdictional claims in published maps and institutional affiliations.

1. Introduction

The structure and observable properties of condensed phases depend greatly on noncovalent interactions (NCI). The hydrogen bond (HB) is arguably the most important of these contacts, as it involved in many crucial phenomena in chemistry and biology, e.g., the association between DNA strands [1], enzymatic catalysis [2,3], molecular recognition [4,5], and protein folding [6]. HBs can result from the interaction of moieties in either different (intermolecular) or the same (intramolecular) molecules. Regarding the latter case, the formation of intramolecular HBs might produce profound changes in molecular properties and structure with far-reaching consequences. Indeed, intramolecular HBs can, for example, (i) alter membrane permeability, water solubility, and lipophilicity of molecules relevant in medicinal chemistry [7] or (ii) produce noticeable variations in the photochemical properties of molecules, such as shifts in the photoabsorption energy [8] or substantial variations in their photoisomerisation processes [9,10].

Resonance-assisted hydrogen bonds (RAHB) are very energetic HBs that are characterised by the connection between the proton donor and acceptor groups throughout conjugated double bonds. From their inception in the crystallographic work of Gill and coworkers [11–15], the concept of RAHB has been successfully adopted by the chemical community to explain phenomena in diverse fields such as physical [16–18] and organic chemistry [19–21] and in nuclear magnetic resonance [22,23].

Concerning the energetics of RAHBs, some authors consider the difference in energy between the associated closed and open conformers, also known as formation energy, as a measure of the strength of RAHBs [24]. Nevertheless, a problem that arises with this approach to study these interactions, and, as a matter of fact, with any other intramolecular contact, is that the rupture of the YH···X bond cannot occur without changing the structure of the molecule. Indeed, the above-mentioned method has the drawback that it involves the energy of (i) the HB itself and (ii) those corresponding to the changes taking place elsewhere in the molecule. In this context, different procedures to compute the interaction energy of RAHBs and other intramolecular non-covalent interactions have arisen [25,26]. Some of the most used methods are those that rely on the theoretical framework of the Quantum Theory of Atoms in Molecules (QTAIM) [27,28], such as the interacting quantum atoms (IQA) method of wave function analysis. The IQA approach involves an energy partition scheme that separates the total energy of an electronic system in intra- and interatomic terms [29,30]. Importantly, IQA allows for the univocal calculation of intramolecular interaction energies without requiring the definition of non-interacting fragments as opposed to traditional Energy Decomposition Analysis (EDA) methods [31]. IQA has been employed in the study of different intramolecular interactions in general [32–36] including, as a case of particular interest to this investigation, RAHBs [37–40].

In this article, we carried out electronic structure calculations as well as QTAIM and IQA analyses for a series of malondialdehydes substituted at different positions of the conjugated π system with an electron-withdrawing (EWG: $-F$, $-Cl$, $-Br$ and $-CF_3$) or an electron-donating (EDG: $-N(CH_3)_2$, $-OCH_3$ and $-NCOCH_3$) group [41]. These calculations enabled us (i) to study the effect of substituents in the strength of the RAHB depending on its relative position and (ii) to compare the values of the IQA interaction energies and other parameters from different EDA analyses on one hand, and the corresponding RAHB formation energies on the other. Our results show that in a considerable fraction of the examined cases, the formation energy deviates markedly from the IQA results. Therefore, such approaches to assess the strength of RAHBs are unable to properly differentiate contacts of this nature with very different energetic features. In contrast, the empirical formula proposed by Espinosa and coworkers [42,43] seems to adequately discern between RAHBs of different strengths. Overall, we expect the present investigation to yield novel insights about the different methods to compute the strength of RAHB and other relevant non-covalent intramolecular interactions.

2. Theoretical Framework

The QTAIM provides a division of space based on the topology of the electronic density. This method of wave function analysis enables the recovery of important chemical concepts, such as atoms, functional groups, atomic charges, and bond orders from either electronic structure calculations or X-ray experiments [44]. Consequently, QTAIM has been applied in the study of a wide variety of chemical and physical problems, such as the examination of different bonds [45,46], adsorption [47–49], electrical conductivity [50–52], and catalysis [53–55].

The traditional implementation of the IQA energy partition uses the atoms of QTAIM as a starting point to divide the total energy of an electronic system into the sum of self energies for each atom and interaction energies between the atoms in the system [29,30],

$$E = \sum_A E_{\text{self}}^A + \sum_{A>B} E_{\text{int}}^{AB}. \tag{1}$$

E_{self}^A in Equation (1) is the energy corresponding to atom A, which includes its kinetic energy, the electron–nucleus attraction and the interelectronic repulsion within atom A.

E_{int}^{AB} is the total interaction energy between atoms A and B and comprises all the possible combinations of the interaction terms between the nucleus and electrons of A on one hand, with the nucleus and electrons of B on the other.

We can also reorganise the terms included in E_{int}^{AB} in order to obtain an expression that gives us additional information about the nature of the interaction between A and B,

$$E_{\text{int}}^{AB} = V_{\text{cl}}^{AB} + V_{\text{xc}}^{AB}. \tag{2}$$

where V_{cl}^{AB} corresponds to the ionic part of the interaction energy while V_{xc}^{AB} is a term related with the covalency of the bond [56].

2.1. Models to Estimate the Energies of Intramolecular Hydrogen Bonds

The work dedicated to the estimation of the strength of intramolecular HBs has been very extensive, as reflected in the excellent review on the subject by Jabłoński [24]. Specifically, we will mainly focus on two indirect measurements. The first approach is based on the differences between the open and the closed conformations, referred to hereafter as the Open-Closed Method (OCM):

$$E_{\text{HB}}^{\text{intra}} \approx E_{\text{form}} = E_{\text{closed}} - E_{\text{open}}. \tag{3}$$

This methodology, albeit popular, presents two important drawbacks. First, it is not clear what geometry should be used as "open" [24]. For instance, Schuster has argued that the optimal open conformation for comparison purposes would be the one wherein minimal changes occur with respect to the closed conformation, even if its geometry is not a local minimum of the potential energy hypersurface [57]. We chose to use a different approach from that put forward by Schuster, and we considered optimised structures for both closed and open conformations of the systems under study. The other important drawback of the OCM method is that it combines changes taking place in other parts of the molecule with the energy corresponding to the HB itself [38]. Thus, stabilising and destabilising contributions, which result from other effects apart from the HB can be misattributed to this interaction. For example, steric destabilisation elsewhere in the molecule could be discounted from an examined intramolecular HB energy because it might occur that the HB is strong enough to compensate such unfavourable steric effects.

The second approximation, proposed by Espinosa et al. and denoted hereafter as Espinosa's Method (EM), is based on the topology of the electronic density, specifically, on the correlation of the potential electron energy density at the bond critical point, $V(\mathbf{r}_{\text{bcp}})$, associated with a given HB and its corresponding energy according to the empirical expression [42,43],

$$E_{\text{HB}}^{\text{intra}} \approx E_{\text{HB}} = \frac{1}{2} V(\mathbf{r}_{\text{bcp}}). \tag{4}$$

This equation was put forward for the study of intermolecular HBs and has proven to be a suitable estimator for the formation energies as computed by the IQA approach in small and medium-sized water clusters, accounting for the relative order for the different types of HB contacts in these systems [58,59]. Nevertheless, some authors have questioned the uncritical use of EM for intramolecular HBs [60–62].

2.2. Computational Details

We carried out electronic structure calculations for a series of derivatives of malondialdehyde in their open and closed configurations (Figure 1), where one of the hydrogens in the three carbon atoms of the conjugated skeleton is replaced by an EWG or an EDG, namely $-F$, $-Cl$, $-Br$, $-CF_3$, $-OCH_3$, $-N(CH_3)_2$, $-NHCOCH_3$ or $-NO_2$ (Figure 1). Thus, we computed the open and closed conformations for eight substitutions in three different positions resulting in 48 different structures wherein the RAHB is present in

24 of them. The conformers were chosen to minimise the differences between open and closed configurations and also to avoid secondary interactions, e.g., contacts between carbonyl and C-H groups. All the geometries were optimised with the aid of the B3LYP functional [63,64], along with the aug-cc-pVTZ basis set [65–68], as implemented in the GAUSSIAN09 package [69]. This combination of exchange-correlation functional and basis set has yielded good results concerning the study of intramolecular hydrogen-bonded systems [40]. Harmonic frequency calculations were done in order to confirm that the optimised structures are indeed local minima. The QTAIM analyses were carried out with the help of the AIMALL program [70]. The IQA energy partitions were carried out with our in-house PROMOLDEN code [71] using β-spheres with radii between 0.1 and 0.3 Bohr along with restricted angular Lebedev quadratures. We partitioned the exchange-correlation energy in accordance with Equation (1) via scaling techniques [72] previously used in conjunction with QTAIM.

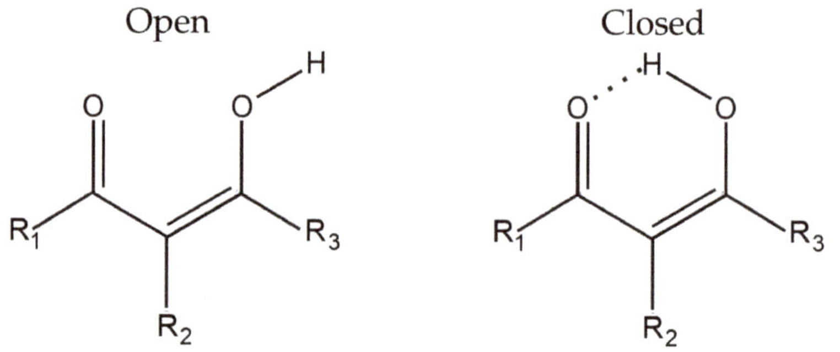

Figure 1. Malondialdehyde structure in its open (**left**) and closed (**right**) conformations. The R_n (n = 1, 2, 3) symbols indicate the different positions available for substitution by the $-F$, $-Cl$, $-Br$, $-CF_3$, $-OCH_3$, $-N(CH_3)_2$, $-NHCOCH_3$, and $-NO_2$ groups.

3. Results

We present the main results of this investigation in three parts. First, the effect of the monosubstitution in malondialdehyde by the EWG and EDG considered in this work. Second, we will shed some light on the origin of the strong positional dependency of the substitution. Furthermore, third, we will compare the IQA results with the estimations put forward in Equations (3) and (4).

3.1. Influence of Substitution on RAHB Energetics

The IQA methodology partitions the energy of an electronic system into intra- and interatomic terms, a condition that allows the study of individual interactions within a molecule. An important characteristic of IQA is that this partition is carried out without using any reference system or empirical data. These features make IQA arguably the gold standard among the different methodologies, geometric or energetic alike, to study intramolecular interactions, including RAHBs. Figure 2 shows the excellent correlation between IQA interaction energies and the intramolecular hydrogen bond distance for the examined RAHBs.

The values for the interaction energy corresponding to the O···H contact with respect to those of malondialdehyde are reported in Table 1. The same chart reports the dissection of the IQA interaction energies into classical and exchange-correlation components. The relevance of the former over the latter contributions is conspicuous for malondialdehyde and the investigated EWGs and EDGs.

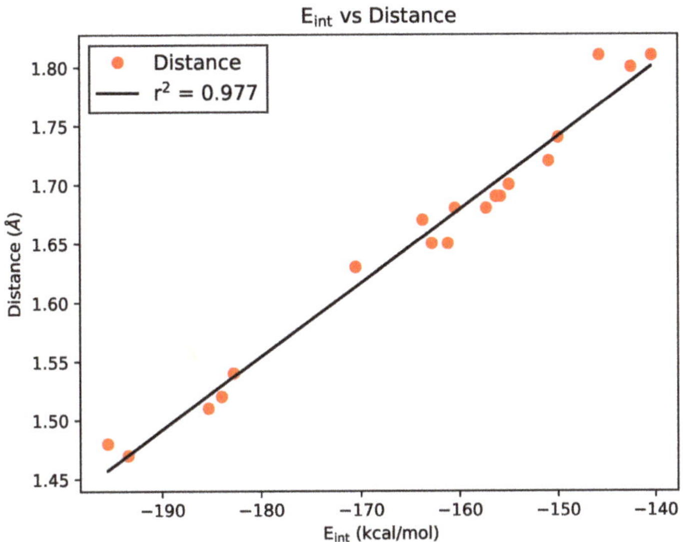

Figure 2. Correlation between IQA interaction energies in kcal/mol and OH···H distances in angstroms.

Table 1. IQA interaction energies (E_{int}), as well as its classical (V_{cl}) and exchange-correlation (V_{xc}) parts, for the investigated O···H RAHB contacts with respect to those in malondialdehyde (Table S1). The values are reported in kcal·mol^{-1}.

	R_1			R_2			R_3		
–R	E_{int}	V_{cl}	V_{xc}	E_{int}	V_{cl}	V_{xc}	E_{int}	V_{cl}	V_{xc}
–CF$_3$	8.78	6.22	2.57	−1.43	−0.90	−0.53	−3.06	−2.25	−0.80
–F	13.99	8.11	5.88	9.77	6.97	2.80	−33.71	−21.09	−12.62
–Cl	17.24	11.36	5.88	3.91	2.68	1.24	−24.20	−15.19	−9.02
–Br	19.35	12.93	6.41	3.53	2.45	1.08	−25.55	−15.70	−9.85
–N(CH$_3$)$_2$	−10.76	−9.02	−1.74	3.45	2.35	1.10	−35.77	−23.84	−11.92
–OCH$_3$	−3.99	−4.22	0.23	4.77	3.33	1.44	−22.97	−15.63	−7.34
–NCOCH$_3$	−0.71	−1.73	1.03	2.47	1.81	0.67	−23.01	−15.57	−7.43
–NO$_2$	18.20	12.11	6.09	−4.50	−2.84	−1.66	−7.75	−5.09	−2.67

We note that the monosubstitution in different positions can either weaken or strengthen the associated RAHB. For example, the CF$_3$ group weakens the RAHB when it is bonded directly to the carbonyl group, but it has the opposite effect in positions 2 and 3. Concerning the halogens, they decrease the intensity of the O···H interaction when they are located in positions 1 and 2. On the other hand, they decrease the magnitude of the interaction energy when they are bonded to the enolic carbon. Mesomeric structures suggest that the influence of EWGs via resonance would be more noticeable on the RAHB strength when the EWG is bonded to the α carbon (Figure 3). The effect of the EWG are therefore more likely interpreted to occur via inductive effects. Furthermore, the influence of these groups is most obvious when they are close to the HB donor. Nevertheless, Table 1 shows that the exchange-correlation contribution to bonding also increases when EWG is bonded to the β carbon, and thus resonance effects cannot be completely neglected.

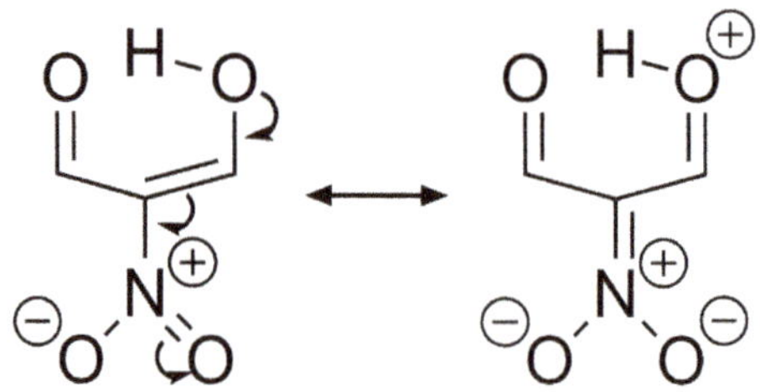

Figure 3. Resonance effect of an electron withdrawing substituent at position 2 in the examined RAHBs.

With respect to the examined EDG, we note that these groups have a minimal effect (a slight reduction) when they are located at position 2. Notwithstanding when they are at position 1 and especially at position 3, they notably increase the RAHB interaction energy. This effect can be understood in terms of the mesomeric structures shown in Figure 4. Interestingly, the substitution in position 3 has the most conspicuous influence effect, leading in all cases to a strengthening of the interaction.

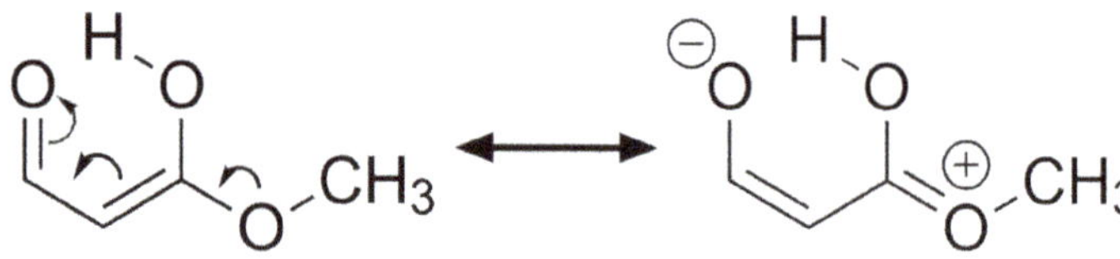

Figure 4. Resonance effect of an electron donating substituent at position 3 in the examined RAHBs.

The above-mentioned effects can be used as guidelines by synthetic chemists to modulate the strength of RAHBs, via the electron withdrawing or donating features of a given substituent, together with its position in the conjugated system.

3.2. Comparison between IQA, OCM, and EM Methods

Table 2 reports the different assessments of the RAHB energy considered in this paper, namely, IQA, OCM, and EM. Figure 5 shows the relationship between (i) the IQA interaction energy and (ii) the formation energy computed with the OCM method together with the HB energy calculated using EM. As we can appreciate from the left side of Figure 5, the IQA interaction energy is not correlated with the E_{form} results yielded by the OCM approach. The fact that these results are unconnected can be associated with the main thesis of this work: the breaking of an RAHB can trigger a rearrangement in the electronic density, which is unrelated to the energetic features of the investigated RAHB [38]. Indeed, these unavoidable modifications take place in molecular regions unrelated to the RAHB. This fact makes the OCM (E_{form}) unsuitable as a parameter for the assessment of the formation energies of the RAHBs under consideration.

Table 2. IQA interaction energies (E_{int}), formation energies (E_{form}) computed by means of the OCM (expression (3)), and H-bond interaction energies (E_{HB}) estimated via Equation (4) for the investigated O···H RAHB contacts. The values are reported in kcal·mol^{-1}.

	R_1			R_2			R_3		
–R	E_{int}	E_{form}	E_{HB}	E_{int}	E_{form}	E_{HB}	E_{int}	E_{form}	E_{HB}
–CF$_3$	8.78	1.08	2.51	−1.43	−1.09	−1.06	−3.06	1.35	−1.34
–F	13.99	3.07	6.34	9.77	2.50	3.27	−33.71	−0.36	−15.04
–Cl	17.24	3.61	6.16	3.91	1.55	1.09	−24.20	0.30	−10.61
–Br	19.35	3.44	6.65	3.53	0.95	0.84	−25.55	1.33	−11.57
–N(CH$_3$)$_2$	−10.76	1.52	−2.47	3.45	1.11	0.86	−35.77	−0.37	−14.64
–OCH$_3$	−3.99	−0.50	−0.15	4.77	1.55	1.64	−22.97	−0.27	−8.71
–NCOCH$_3$	−0.71	0.67	0.73	2.47	2.46	0.07	−23.01	5.24	−8.87
–NO$_2$	18.20	5.12	6.38	−4.50	0.53	−2.28	−7.75	7.01	−3.35

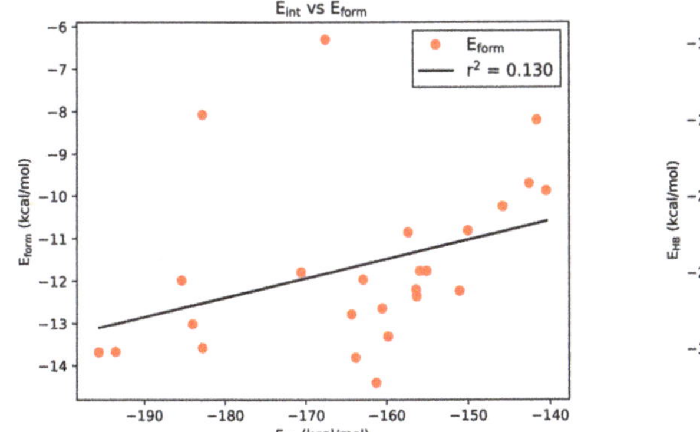
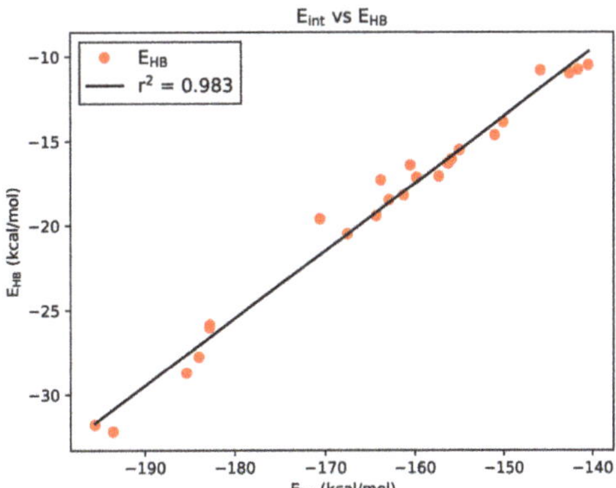

Figure 5. Correlation of IQA interaction energies with (**left**) OCM values of E_{form} and (**right**) EM results of E_{HB}.

Contrary to this fact, the correlation between the values of E_{int} and those corresponding to Equation (4) are excellent. In all cases, IQA interaction energies and Espinosa's empirical formula produce indeed the same relative strengths for the studied systems. This observation indicates that for typical RAHBs, Equation (4) is able to qualitatively recover the interplay between the π-skeleton and the O···H−O moiety.

The good agreement between E_{int} and E_{HB} should not be interpreted as an uncritical approval to the use of Equation (4) for the estimation of relative RAHB strengths. Certainly, different authors have pointed out a series of deficiencies for this empirical formulation. Here we mention two of these potential problems. First, the values resulting from Equation (4) are always negative, a circumstance that always points to an attractive interaction. Nevertheless, certain C−H···O contacts are repulsive in nature [62]. The use of Formula (3) to describe these contacts would produce a qualitatively incorrect result. Second, Equation (4) is not transferable to other contacts, such as H···F, where a different scaling factor from that used in Equation (4) needs to be used [73].

We finally state what we consider the limits for the reasonable application of EM on the study of intramolecular HBs in general and RAHB in particular. Given that Equation (4) is unable to distinguish between attractive and repulsive interactions, it should be only applied when no doubt can arise regarding the attractive nature of the contact. Additionally, although E_{int} and E_{HB} follow the same strength order, their magnitudes are not comparable. Therefore, the EM should primarily be used to study intramolecular contacts where the interaction mainly involve the same atomic species. This situation might be the case, for instance, in different substitutions in an aromatic ring adjacent to the O−H···O group [74] or changes in the protonation degree in an intramolecular HB [75].

4. Conclusions

We investigate the effect of different electron-withdrawing and -donor groups on the energetics of the resonance assisted hydrogen bond in malondialdehyde. Our data indicate that classical contributions are far more important than exchange-correlation components as opposed to the notion that the stability of RAHBs occur mainly due to the delocalisation of π electrons. These groups exert a marked influence on the RAHB interaction energy, which in turn depends notably on the position of the EWG and EDG. Notably, both types of groups considerably strengthen the RAHB when they are bonded to the β carbon atom of malondialdehyde. We also addressed different methodologies to assess the interaction energy of RAHBs. In this regard, we showed how the examination of the energy density offers a good estimation of the IQA interaction energy and therefore of the RAHB energetics at a reduced computational cost.

Supplementary Materials: The following are available online, Table S1: IQA interaction energies for the investigated systems; Table S2: IQA interaction, formation, and HB interaction energies. Figures S1–S8: Structures of all compounds.

Author Contributions: All authors contributed equally to this work. All authors have read and agreed to the published version of the manuscript.

Funding: We gratefully acknowledge financial support from CONACyT/Mexico (grant 253776), PAPIIT/UNAM (project IN205118), and the Spanish MICINN (grant PGC2018-095953-B-I00). We are also grateful to DGTIC/UNAM (project LANCAD-UNAM-DGTIC-250) for computer time. Miguel Gallegos specifically acknowledges the Spanish MICIU for the predoctoral grant FPU19/02903.

Institutional Review Board Statement: Not applicable.

Informed Consent Statement: Not applicable.

Data Availability Statement: Not applicable.

Conflicts of Interest: The authors declare no conflict of interest.

References

1. Guerra, C.F.; Bickelhaupt, F.M.; Snijders, J.G.; Baerends, E.J. Hydrogen bonding in DNA base pairs: Reconciliation of theory and experiment. *J. Am. Chem. Soc.* **2000**, *122*, 4117–4128. [CrossRef]
2. Cleland, W.; Kreevoy, M. Low-barrier hydrogen bonds and enzymic catalysis. *Science* **1994**, *264*, 1887–1890. [CrossRef] [PubMed]
3. Guo, H.; Salahub, D.R. Cooperative hydrogen bonding and enzyme catalysis. *Angew. Chem. Int. Ed.* **1998**, *37*, 2985–2990. [CrossRef]
4. Fersht, A.R. The hydrogen bond in molecular recognition. *Trends Biochem. Sci.* **1987**, *12*, 301–304. [CrossRef]
5. Etter, M.C.; Reutzel, S.M. Hydrogen bond directed cocrystallization and molecular recognition properties of acyclic imides. *J. Am. Chem. Soc.* **1991**, *113*, 2586–2598. [CrossRef]
6. Ben-Naim, A. The role of hydrogen bonds in protein folding and protein association. *J. Phys. Chem.* **1991**, *95*, 1437–1444. [CrossRef]
7. Kuhn, B.; Mohr, P.; Stahl, M. Intramolecular hydrogen bonding in medicinal chemistry. *J. Med. Chem.* **2010**, *53*, 2601–2611. [CrossRef]
8. González, D.; Neilands, O.; Rezende, M.C. The solvatochromic behaviour of 2-and 4-pyridiniophenoxides. *J. Chem. Soc. Per. Trans.* **1999**, *2*, 713–718. [CrossRef]
9. Lewis, F.D.; Stern, C.L.; Yoon, B.A. Effects of inter-and intramolecular hydrogen bonding upon the structure and photoisomerization of 3-(2-pyridyl) propenamides. *J. Am. Chem. Soc.* **1992**, *114*, 3131–3133. [CrossRef]
10. Cui, G.; Lan, Z.; Thiel, W. Intramolecular hydrogen bonding plays a crucial role in the photophysics and photochemistry of the GFP chromophore. *J. Am. Chem. Soc.* **2012**, *134*, 1662–1672. [CrossRef]
11. Gilli, G.; Bellucci, F.; Ferretti, V.; Bertolasi, V. Evidence for resonance-assisted hydrogen bonding from crystal-structure correlations on the enol form of the. beta.-diketone fragment. *J. Am. Chem. Soc.* **1989**, *111*, 1023–1028. [CrossRef]
12. Bertolasi, V.; Gilli, P.; Ferretti, V.; Gilli, G. Evidence for resonance-assisted hydrogen bonding. 2. Intercorrelation between crystal structure and spectroscopic parameters in eight intramolecularly hydrogen bonded 1,3-diaryl-1,3-propanedione enols. *J. Am. Chem. Soc.* **1991**, *113*, 4917–4925. [CrossRef]
13. Gilli, G.; Bertolasi, V.; Ferretti, V.; Gilli, P. Resonance-assisted hydrogen bonding. III. Formation of intermolecular hydrogen-bonded chains in crystals of β-diketone enols and its relevance to molecular association. *Acta Crystallogr. B Struct. Sci.* **1993**, *49*, 564–576.

14. Gilli, P.; Bertolasi, V.; Ferretti, V.; Gilli, G. Evidence for resonance-assisted hydrogen bonding. 4. Covalent nature of the strong homonuclear hydrogen bond. Study of the O–H–O system by crystal structure correlation methods. *J. Am. Chem. Soc.* **1994**, *116*, 909–915. [CrossRef]
15. Gilli, P.; Bertolasi, V.; Pretto, L.; Ferretti, V.; Gilli, G. Covalent versus Electrostatic Nature of the Strong Hydrogen Bond: Discrimination among Single, Double, and Asymmetric Single-Well Hydrogen Bonds by Variable-Temperature X-ray Crystallographic Methods in β-Diketone Enol RAHB Systems. *J. Am. Chem. Soc.* **2004**, *126*, 3845–3855. [CrossRef] [PubMed]
16. Grabowski, S.J. π-Electron delocalisation for intramolecular resonance assisted hydrogen bonds. *J. Phys. Org. Chem.* **2003**, *16*, 797–802. [CrossRef]
17. Sanz, P.; Mó, O.; Yáñez, M.; Elguero, J. Resonance-Assisted Hydrogen Bonds: A Critical Examination. Structure and Stability of the Enols of β-Diketones and β-Enaminones. *J. Phys. Chem. A* **2007**, *111*, 3585–3591. [CrossRef]
18. Lin, X.; Zhang, H.; Jiang, X.; Wu, W.; Mo, Y. The Origin of the Non-Additivity in Resonance-Assisted Hydrogen Bond Systems. *J. Phys. Chem. A* **2017**, *121*, 8535–8541. [CrossRef]
19. Chin, J.; Kim, D.C.; Kim, H.-J.; Panosyan, F.B.; Kim, K.M. Chiral Shift Reagent for Amino Acids Based on Resonance-Assisted Hydrogen Bonding. *Org. Lett.* **2004**, *6*, 2591–2593. [CrossRef]
20. Zubatyuk, R.I.; Volovenko, Y.M.; Shishkin, O.V.; Gorb, L.; Leszczynski, J. Aromaticity-Controlled Tautomerism and Resonance-Assisted Hydrogen Bonding in Heterocyclic Enaminone-Iminoenol Systems. *J. Org. Chem.* **2007**, *72*, 725–735. [CrossRef]
21. Kim, H.; Nguyen, Y.; Yen, C.P.-H.; Chagal, L.; Lough, A.J.; Kim, B.M.; Chin, J. Stereospecific Synthesis of C2 Symmetric Diamines from the Mother Diamine by Resonance-Assisted Hydrogen-Bond Directed Diaza-Cope Rearrangement. *J. Am. Chem. Soc.* **2008**, *130*, 12184–12191. [CrossRef] [PubMed]
22. Marković, R.; Shirazi, A.; Džambaski, Z.; Baranac, M.; Minić, D. Configurational isomerization of push-pull thiazolidinone derivatives controlled by intermolecular and intramolecular RAHB: 1 H NMR dynamic investigation of concentration and temperature effects. *J. Phys. Org. Chem.* **2004**, *17*, 118–123. [CrossRef]
23. Dračínský, M.; Čechová, L.; Hodgkinson, P.; Procházková, E.; Janeba, Z. Resonance-assisted stabilisation of hydrogen bonds probed by NMR spectroscopy and path integral molecular dynamics. *Chem. Commun.* **2015**, *51*, 13986–13989. [CrossRef]
24. Jabłoński, M. A Critical Overview of Current Theoretical Methods of Estimating the Energy of Intramolecular Interactions. *Molecules* **2020**, *25*, 5512. [CrossRef] [PubMed]
25. Jabłoński, M.; Kaczmarek, A.; Sadlej, A.J. Estimates of the Energy of Intramolecular Hydrogen Bonds. *J. Phys. Chem. A* **2006**, *110*, 10890–10898. [CrossRef] [PubMed]
26. Rusinska-Roszak, D. Intramolecular O–H⋯O=C Hydrogen Bond Energy via the Molecular Tailoring Approach to RAHB Structures. *J. Phys. Chem. A* **2015**, *119*, 3674–3687. [CrossRef]
27. Grabowski, S. An estimation of strength of intramolecular hydrogen bonds—Ab initio and AIM studies. *J. Mol. Struct.* **2001**, *562*, 137–143. [CrossRef]
28. Fuster, F.; Grabowski, S.J. Intramolecular Hydrogen Bonds: The QTAIM and ELF Characteristics. *J. Phys. Chem. A* **2011**, *115*, 10078–10086. [CrossRef]
29. Blanco, M.A.; Martín Pendás, A.; Francisco, E. Interacting Quantum Atoms: A Correlated Energy Decomposition Scheme Based on the Quantum Theory of Atoms in Molecules. *J. Chem. Theory Comput.* **2005**, *1*, 1096–1109. [CrossRef]
30. Francisco, E.; Martín Pendás, A.; Blanco, M.A. A Molecular Energy Decomposition Scheme for Atoms in Molecules. *J. Chem. Theory Comput.* **2005**, *2*, 90–102. [CrossRef]
31. von Hopffgarten, M.; Frenking, G. Energy decomposition analysis. *Wiley Interdiscip. Rev. Comput. Mol. Sci.* **2011**, *2*, 43–62. [CrossRef]
32. Tognetti, V.; Joubert, L. On the physical role of exchange in the formation of an intramolecular bond path between two electronegative atoms. *J. Chem. Phys.* **2013**, *138*, 024102. [CrossRef] [PubMed]
33. Bartashevich, E.V.; Martín Pendás, Á.; Tsirelson, V.G. An anatomy of intramolecular atomic interactions in halogen-substituted trinitromethanes. *Phys. Chem. Chem. Phys.* **2014**, *16*, 16780–16789. [CrossRef] [PubMed]
34. Yahia-Ouahmed, M.; Tognetti, V.; Joubert, L. Halogen–halogen interactions in perhalogenated ethanes: An interacting quantum atoms study. *Comput. Theor. Chem.* **2015**, *1053*, 254–262. [CrossRef]
35. Yahia-Ouahmed, M.; Tognetti, V.; Joubert, L. Intramolecular halogen bonding: An interacting quantum atoms study. *Theor. Chem. Acc.* **2016**, *135*, 45. [CrossRef]
36. Thacker, J.C.R.; Popelier, P.L.A. Fluorine Gauche Effect Explained by Electrostatic Polarization Instead of Hyperconjugation: An Interacting Quantum Atoms (IQA) and Relative Energy Gradient (REG) Study. *J. Phys. Chem. A* **2018**, *122*, 1439–1450. [CrossRef] [PubMed]
37. Ebrahimi, S.; Dabbagh, H.A.; Eskandari, K. Nature of intramolecular interactions of vitamin C in view of interacting quantum atoms: The role of hydrogen bond cooperativity on geometry. *Phys. Chem. Chem. Phys.* **2016**, *18*, 18278–18288. [CrossRef]
38. Guevara-Vela, J.M.; Romero-Montalvo, E.; Costales, A.; Martín Pendás, Á.; Rocha-Rinza, T. The nature of resonance-assisted hydrogen bonds: A quantum chemical topology perspective. *Phys. Chem. Chem. Phys.* **2016**, *18*, 26383–26390. [CrossRef]
39. Romero-Montalvo, E.; Guevara-Vela, J.M.; Costales, A.; Martín Pendás, Á.; Rocha-Rinza, T. Cooperative and anticooperative effects in resonance assisted hydrogen bonds in merged structures of malondialdehyde. *Phys. Chem. Chem. Phys.* **2017**, *19*, 97–107. [CrossRef]

40. Guevara-Vela, J.M.; Romero-Montalvo, E.; del Río Lima, A.; Martín Pendás, Á.; Hernández-Rodríguez, M.; Rocha Rinza, T. Hydrogen-Bond Weakening through π Systems: Resonance-Impaired Hydrogen Bonds (RIHB). *Chem. Eur. J.* **2017**, *23*, 16605–16611. [CrossRef]

41. Domingo, L.R.; Pérez, P.; Contreras, R. Electronic Contributions to the σpParameter of the Hammett Equation. *J. Org. Chem.* **2003**, *68*, 6060–6062. [CrossRef] [PubMed]

42. Espinosa, E.; Molins, E.; Lecomte, C. Hydrogen bond strengths revealed by topological analyses of experimentally observed electron densities. *Chem. Phys. Lett.* **1998**, *285*, 170–173. [CrossRef]

43. Espinosa, E.; Molins, E. Retrieving interaction potentials from the topology of the electron density distribution: The case of hydrogen bonds. *J. Chem. Phys.* **2000**, *113*, 5686–5694. [CrossRef]

44. Bader, R.F.W. *Atoms in Molecules. A Quantum Theory*; Oxford University Press: Oxford, UK, 1995.

45. Grabowski, S.J.; Casanova, D.; Formoso, E.; Ugalde, J.M. Tetravalent Oxygen and Sulphur Centres Mediated by Carborane Superacid: Theoretical Analysis. *ChemPhysChem* **2019**, *20*, 2443–2450. [CrossRef]

46. Radoske, T.; Kloditz, R.; Fichter, S.; März, J.; Kaden, P.; Patzschke, M.; Schmidt, M.; Stumpf, T.; Walter, O.; Ikeda-Ohno, A. Systematic comparison of the structure of homoleptic tetradentate N2O2-type Schiff base complexes of tetravalent f-elements (M(iv) = Ce, Th, U, Np, and Pu) in solid state and in solution. *Dalton Trans.* **2020**, *49*, 17559–17570. [CrossRef]

47. Marana, N.L.; Casassa, S.M.; Sambrano, J.R. Adsorption of NH3 with Different Coverages on Single-Walled ZnO Nanotube: DFT and QTAIM Study. *J. Phys. Chem. C* **2017**, *121*, 8109–8119. [CrossRef]

48. Malček, M.; Bučinský, L.; Teixeira, F.; Cordeiro, M.N.D.S. Detection of simple inorganic and organic molecules over Cu-decorated circumcoronene: A combined DFT and QTAIM study. *Phys. Chem. Chem. Phys.* **2018**, *20*, 16021–16032. [CrossRef]

49. Ohno, T.; Kubicki, J.D. Adsorption of Organic Acids and Phosphate to an Iron (Oxyhydr)oxide Mineral: A Combined Experimental and Density Functional Theory Study. *J. Phys. Chem. A* **2020**, *124*, 3249–3260. [CrossRef]

50. Martín Pendás, Á.; Guevara-Vela, J.M.; Crespo, D.M.; Costales, A.; Francisco, E. An unexpected bridge between chemical bonding indicators and electrical conductivity through the localization tensor. *Phys. Chem. Chem. Phys.* **2017**, *19*, 1790–1797. [CrossRef]

51. Astakhov, A.A.; Tsirelson, V.G. Spatially resolved characterization of electron localization and delocalization in molecules: Extending the Kohn-Resta approach. *Int. J. Quantum Chem.* **2018**, *118*, e25600. [CrossRef]

52. Gil-Guerrero, S.; Ramos-Berdullas, N.; Martín Pendás, Á.; Francisco, E.; Mandado, M. Anti-ohmic single molecule electron transport: Is it feasible? *Nanoscale Adv.* **2019**, *1*, 1901–1913. [CrossRef]

53. Teixeira, F.; Mosquera, R.; Melo, A.; Freire, C.; Cordeiro, M.N.D.S. Driving Forces in the Sharpless Epoxidation Reaction: A Coupled AIMD/QTAIM Study. *Inorg. Chem.* **2017**, *56*, 2124–2134. [CrossRef]

54. Hooper, T.N.; Garçon, M.; White, A.J.P.; Crimmin, M.R. Room temperature catalytic carbon–hydrogen bond alumination of unactivated arenes: Mechanism and selectivity. *Chem. Sci.* **2018**, *9*, 5435–5440. [CrossRef]

55. Escofet, I.; Armengol-Relats, H.; Bruss, H.; Besora, M.; Echavarren, A.M. On the Structure of Intermediates in Enyne Gold(I)-Catalyzed Cyclizations: Formation of trans-Fused Bicyclo[5.1.0]octanes as a Case Study. *Chem. Eur. J.* **2020**, *26*, 15738–15745. [CrossRef]

56. Martín Pendás, A.; Casals-Sainz, J.L.; Francisco, E. On Electrostatics, Covalency, and Chemical Dashes: Physical Interactions versus Chemical Bonds. *Chem. Eur. J.* **2018**, *25*, 309–314. [CrossRef] [PubMed]

57. Schuster, P. *The Hydrogen Bond: Recent Developments in Theory and Experiments*; North-Holland Pub. Co. Distributor: Amsterdam, The Netherlands; American Elsevier Pub. Co.: New York, NY, USA, 1976.

58. Guevara-Vela, J.M.; Romero-Montalvo, E.; Gómez, V.A.M.; Chávez-Calvillo, R.; García-Revilla, M.; Francisco, E.; Martín Pendás, Á.; Rocha-Rinza, T. Hydrogen bond cooperativity and anticooperativity within the water hexamer. *Phys. Chem. Chem. Phys.* **2016**, *18*, 19557–19566. [CrossRef] [PubMed]

59. Castor-Villegas, V.M.; Guevara-Vela, J.M.; Narváez, W.E.V.; Martín Pendás, Á.; Rocha-Rinza, T.; Fernández-Alarcón, A. On the strength of hydrogen bonding within water clusters on the coordination limit. *J. Comput. Chem.* **2020**, *41*, 2266–2277. [CrossRef]

60. Gatti, C.; May, E.; Destro, R.; Cargnoni, F. Fundamental Properties and Nature of CH··O Interactions in Crystals on the Basis of Experimental and Theoretical Charge Densities. The Case of 3, 4-Bis(dimethylamino)-3-cyclobutene-1, 2-dione (DMACB) Crystal. *J. Phys. Chem. A* **2002**, *106*, 2707–2720. [CrossRef]

61. Nikolaienko, T.Y.; Bulavin, L.A.; Hovorun, D.M. Bridging QTAIM with vibrational spectroscopy: The energy of intramolecular hydrogen bonds in DNA-related biomolecules. *Phys. Chem. Chem. Phys.* **2012**, *14*, 7441. [CrossRef] [PubMed]

62. Jabłoński, M.; Monaco, G. Different Zeroes of Interaction Energies As the Cause of Opposite Results on the Stabilizing Nature of C–H··O Intramolecular Interactions. *J. Chem. Inf. Model.* **2013**, *53*, 1661–1675. [CrossRef]

63. Lee, C.; Yang, W.; Parr, R.G. Development of the Colle-Salvetti correlation-energy formula into a functional of the electron density. *Phys. Rev. B* **1988**, *37*, 785–789. [CrossRef]

64. Becke, A.D. Density-functional thermochemistry. III. The role of exact exchange. *J. Chem. Phys.* **1993**, *98*, 5648–5652. [CrossRef]

65. Dunning, T.H. Gaussian basis sets for use in correlated molecular calculations. I. The atoms boron through neon and hydrogen. *J. Chem. Phys.* **1989**, *90*, 1007–1023. [CrossRef]

66. Kendall, R.A.; Dunning, T.H.; Harrison, R.J. Electron affinities of the first-row atoms revisited. Systematic basis sets and wave functions. *J. Chem. Phys.* **1992**, *96*, 6796–6806. [CrossRef]

67. Woon, D.E.; Dunning, T.H. Gaussian basis sets for use in correlated molecular calculations. III. The atoms aluminum through argon. *J. Chem. Phys.* **1993**, *98*, 1358–1371. [CrossRef]

68. Wilson, A.K.; Woon, D.E.; Peterson, K.A.; Dunning, T.H. Gaussian basis sets for use in correlated molecular calculations. IX. The atoms gallium through krypton. *J. Chem. Phys.* **1999**, *110*, 7667–7676. [CrossRef]
69. Frisch, M.J.; Trucks, G.W.; Schlegel, H.B.; Scuseria, G.E.; Robb, M.A.; Cheeseman, J.R.; Scalmani, G.; Barone, V.; Mennucci, B.; Petersson, G.A.; et al. *Gaussian 09 Revision E.01*; Gaussian Inc.: Wallingford, CT, USA, 2009.
70. Keith, A.T. *AIMALL (Version 19.02.13)*; TK Gristmill Software: Overland Park, KS, USA, 2019.
71. Martín Pendás, Á.; Francisco, E. PROMOLDEN. A QTAIM/IQA code. Unpublished work.
72. Francisco, E.; Casals-Sainz, J.L.; Rocha-Rinza, T.; Martín Pendás, Á. Partitioning the DFT exchange-correlation energy in line with the interacting quantum atoms approach. *Theor. Chem. Acc.* **2016**, *135*, 170. [CrossRef]
73. Mata, I.; Alkorta, I.; Espinosa, E.; Molins, E. Relationships between interaction energy, intermolecular distance and electron density properties in hydrogen bonded complexes under external electric fields. *Chem. Phys. Lett.* **2011**, *507*, 185–189. [CrossRef]
74. Pareras, G.; Palusiak, M.; Duran, M.; Solà, M.; Simon, S. Tuning the Strength of the Resonance-Assisted Hydrogen Bond in o-Hydroxybenzaldehyde by Substitution in the Aromatic Ring. *J. Phys. Chem. A* **2018**, *122*, 2279–2287. [CrossRef]
75. Alkorta, I.; Elguero, J.; Mó, O.; Yáñez, M.; Bene, J.E.D. Are resonance-assisted hydrogen bonds 'resonance assisted'? A theoretical NMR study. *Chem. Phys. Lett.* **2005**, *411*, 411–415. [CrossRef]

Article

Focal Point Evaluation of Energies for Tautomers and Isomers for 3-hydroxy-2-butenamide: Evaluation of Competing Internal Hydrogen Bonds of Types -OH … O=, -OH … N, -NH … O=, and CH … X (X=O and N)

Zikri Altun [1], **Erdi Ata Bleda** [1] **and Carl Trindle** [2,*]

1   Physics Department, Marmara University, Göztepe Kampus, Istanbul 34772, Turkey; zikalt@marmara.edu.tr (Z.A.); ata.bleda@marmara.edu.tr (E.A.B.)
2   Chemistry Department, University of Virginia, Charlottesville, VA 22902, USA
*   Correspondence: cot@virginia.edu; Tel.: +1-4347709197

Abstract: The title compound is a small molecule with many structural variations; it can illustrate a variety of internal hydrogen bonds, among other noncovalent interactions. Here we examine structures displaying hydrogen bonding between carbonyl oxygen and hydroxyl H; between carbonyl oxygen and amino H; hydroxyl H and amino N; hydroxyl O and amino H. We also consider H-bonding in its tautomer 2-oxopropanamide. By extrapolation algorithms applied to Hartree-Fock and correlation energies as estimated in HF, MP2, and CCSD calculations using the cc-pVNZ correlation-consistent basis sets (N = 2, 3, and 4) we obtain reliable relative energies of the isomeric forms. Assuming that such energy differences may be attributed to the presence of the various types of hydrogen bonding, we attempt to infer relative strengths of types of H-bonding. The Atoms in Molecules theory of Bader and the Local Vibrational Modes analysis of Cremer and Kraka are applied to this task. Hydrogen bonds are ranked by relative strength as measured by local stretching force constants, with the stronger =O … HO- > NH … O= > -OH … N well separated from a cluster > NH … O= ≈ >NH … OH ≈ CH … O= of comparable and intermediate strength. Weaker but still significant interactions are of type CH … N which is stronger than CH … OH.

Keywords: intramolecular hydrogen bonding; high-accuracy extrapolation methods; QTAIM; non-covalent interactions; local vibrational modes

Citation: Altun, Z.; Bleda, E.A.; Trindle, C. Focal Point Evaluation of Energies for Tautomers and Isomers for 3-hydroxy-2-butenamide: Evaluation of Competing Internal Hydrogen Bonds of Types -OH … O=, -OH … N, -NH … O=, and CH … X (X=O and N). *Molecules* **2021**, *26*, 2623. https://doi.org/10.3390/molecules26092623

Academic Editors: Mirosław Jabłoński and Esmail Alikhani

Received: 26 March 2021
Accepted: 27 April 2021
Published: 30 April 2021

Publisher's Note: MDPI stays neutral with regard to jurisdictional claims in published maps and institutional affiliations.

1. Introduction

The concept of hydrogen bonding has evolved considerably over its century of history [1,2]. The anomalous properties of water prompted the first suspicion of an attractive interaction between a hydrogen atom in one water molecule with the oxygen of another (HOH … OH_2). Similar phenomena involving N and F were soon recognized as well. H bonding has been recognized as a primarily electrostatic phenomenon expressed by its effect on vibrational frequencies (most often red-shifting the OH stretch), molecular structure (short X … HY distances, where X and Y are electronegative atoms F, O, N etc.), and characteristic NMR parameters (owing to charge shifts) to mention the most prominent. The concept of H bonding has been expanded from its original context so to include a number of surprising interactions, involving atoms other than oxygen, fluorine, and nitrogen [3,4]; "strong" H-bonding [5]; and "anomalous" (blue-shifting) H bonding [6]. Further expansion of the concept also has been recognized [7].

The computational modeling of hydrogen bonding was reviewed in 1997 [8] and 1999 [9], in 2006 [10], and in 2009 [11]. The definition of hydrogen bonding has been formalized by IUPAC [12]. IUPAC's criteria for recognition of H bonding in a structure X-H … YZ include: origins of bonding (largely electrostatic, but also including contributions from charge transfer and dispersion); polarization of the XH bond; geometry of the structure

(XH ... Y near linearity); distortions from reference structures (extension of the XH bond, with impact on vibrational frequencies); effects on NMR spectra including deshielding of H and coupling of X to Y. In the language of the Atoms in Molecules theory [13], a hydrogen bond is associated with presence of a bond path H ... Y including a bond critical point (BCP) between H and Y. We discuss this criterion below.

Recent collections of studies of intramolecular H bonding include the Molecules special issue from 2017 edited by S. Scheiner [14] and a follow-up 2019 special issue of Molecules edited by G. Sanchez [15]. The current special issue gives primary attention to computational studies. Interest in evaluating the strength of H bonds continues [16,17] and is an important part of this special issue of Molecules.

Intramolecular hydrogen bonding has a significant effect specifically on conformational preference in systems of importance in biochemistry, as various types of H bonding can occur in various conformations and isomers. The intramolecular H bonds in histidine have been described by Yannacome, Sethlo, and Kraka (YSK) [18]. To enhance insight into the relative strength of H bonds in context these authors brought together descriptors from Bader's theory of atoms in molecules (AIM) and the reduced density gradient from analysis of noncovalent interactions (NCI), both of which are described below. According to YSK two local measures of bond strength, the density at a BCP and the local stretching force constant are related by a power law. An analogous l link between a local stretching force constant (and hence that density) and bond energy seems eminently reasonable [19,20]. One particularly simple connection, between bond strength and the potential energy density at a Bond Critical Point, has been proposed [21]. However there is a conceptual issue in relating a strictly local property (such as the density at a specific point) to what is not a strictly local property, the energy associated with molecular rearrangement, especially dissociation.

The IUPAC AIM bond path and critical point criteria [12] for the existence of a hydrogen bond is subject to interpretation. Sometimes a path and its critical point seem much at variance with intuitive notions of bonding [22–26]. Bader recognized this awkward conflict, [27] and a proposal that these topological entities be called "line" CPs has been offered. [28] The issue is particularly troubled for weak interactions, which vary in directionality depending on the extent of electrostatic character [29].

In this work we study acetoacetamide and variants of its tautomer 1-amino-1- hydroxyacetone. Bauer and Wilcox addressed the issue of relative stability of the similar but simpler systems malonaldehyde and acetyl acetone [30], concluding that the enol form was the more stable. Intramolecular hydrogen bonding in many hydroxycarbonyl systems has been studied by Afonin and Vashchenko [31]. All the systems they reported have =O ... HO- interactions. The keto-enol isomerization of many substituted diketones has been studied by Belova et al. [32] In the unsubstituted system propane-1,3-dione which we can call (H, H), the enol form was favored in internal energy by 14.6 kJ/mol; ΔG = 6.2 kJ/mol according to the CBS-4 thermochemical scheme [33]. That scheme favors the enol of the (CH_3, CH_3) species acetylacetone by a Gibbs energy of 11.0 kJ/mol [26]. MP2 and B3LYP calculations by Belova et al. [32] favor the enol form by ca. 10 kJ/mol. Their systems include (R1, R2) = (NH_2, CH_3) of central interest here. The DFT model B3LYP/aug-cc-pVTZ placed a keto form above an enol form by 13 kJ/mol; the O ... O distance in the enol was found to be 2.55 Å and the O ... H distance was 1.638 Å in the enol. The enol form was evidently the structure we label TWO; see below. The authors addressed the issue of H-bond strength in their systems, recognizing the long-established link between H bonding and the OH stretching frequency shift to the red relative to the value for a OH group not involved in H bonding. Strong correlations between O ... O shortening, O ... H shortening and OH lengthening are noted.

2. Results

In the following sections we define the systems under study, describe the extrapolation techniques by which we obtain accurate relative energies, and lay out results of AIM analysis of their charge distributions. Complementary information on non-covalent interactions is provided by graphical representation of the reduced density gradient. We interpret the vibrational spectra for all species, and recover the force constants for significant local modes by the LMODEA algorithms. Those force constants are found to be useful indices of relative intramolecular bond strength. All these analyses are described in detail in the section on methods and software.

2.1. Systems under Study

In the following enumeration of structures under study, actual reaction paths are not described; there is instead a series of logical connections between related species.

We begin with the tautomer ONE (acetoacetamide) which has a saturated link and two carbonyl groups. Species ONE-A and ONE-B obtained by rotation of the amido and acyl groups revert spontaneously to species ONE as shown in Figure 1, and are not assigned a relative energy.

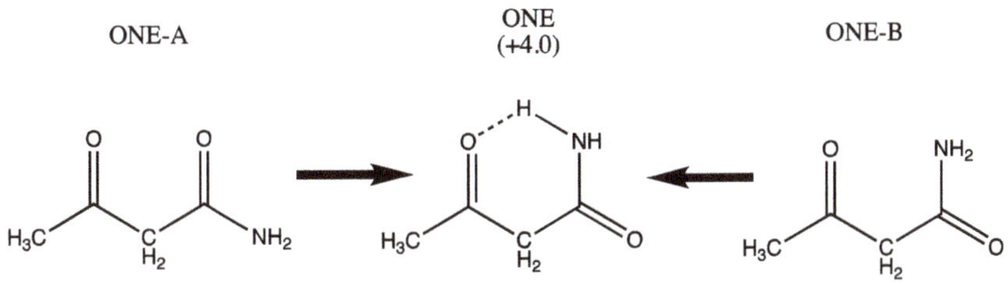

Figure 1. Diketone variant ONE with NH … O hydrogen bonding. Species ONE-A and ONE-B (which lacks the NH … O = H bond) proceed spontaneously upon optimization to ONE, which lies about 4.0 kJ/mol above the most stable isomer TWO.

All other systems TWO to ELEVEN (shown in Figures 2–5) have an unsaturated link and a hydroxyl group at either the methyl-substituted C (CH_3-C-OH) or at the amino-substituted C group (NH_2-C-OH). The most stable of all species, TWO, has an evident OH … O = hydrogen bond. Rotating the CH_3-C-O-H dihedral forms THREE, which lacks that hydrogen bond (Figure 2). Simple torsion of THREE around the CC-amide bond produces an unstable form (unlabeled) which upon optimization can establish either a NH donor -OH acceptor H bond (FOUR) or an N acceptor OH-donor H bond (FIVE).

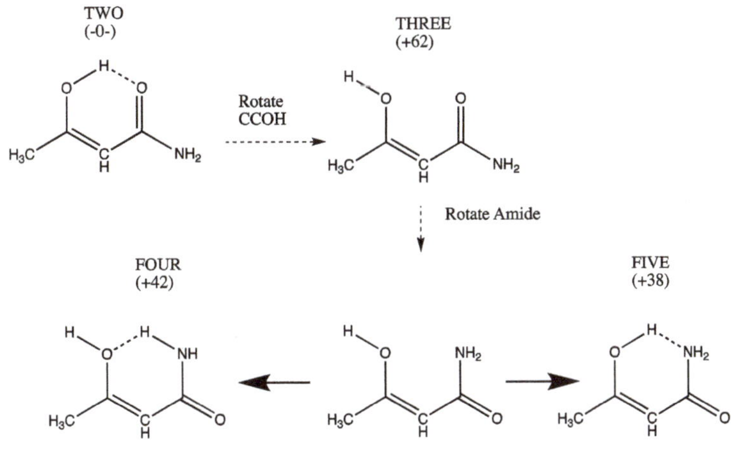

Figure 2. Enolic form THREE can assume forms FOUR with NH...O and FIVE with OH...N H-bonding.

SEVEN
(+38)

SIX
(+43)

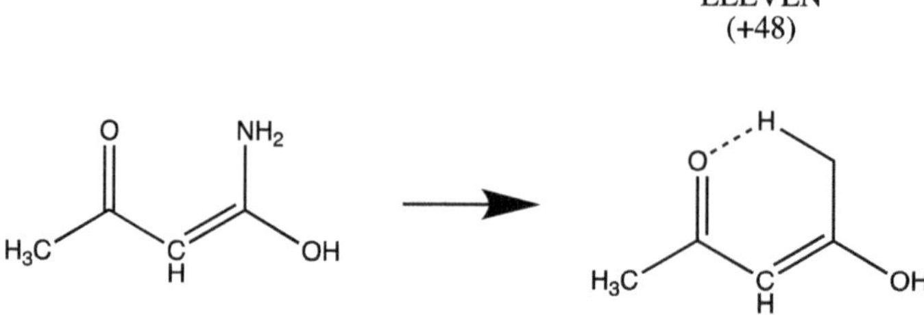

Rotate
Amide

Figure 3. SIX, which may exhibit CH...O= hydrogen bonding, can form SEVEN by a rotation of the amide group. SEVEN may contain CH...N hydrogen bonding.

EIGHT
(+101)

Rotate NCOH   TWO

Rotate Acyl

Exchange
OH and NH₂

NINE
(+91)

TEN
(+61)

Figure 4. A high energy isomer EIGHT with no hydrogen bonding can achieve OH . . . O = hydrogen bonding; the resultant species rearranges spontaneously to stable TWO. By rotation of the acyl fragment EIGHT can produce NINE which may have CH . . . OH hydrogen bonding. Swapping OH and NH_2 produces TEN which may contain a CH . . . NH_2 interaction.

ELEVEN
(+48)

Figure 5. Swapping amino and hydroxyl groups in EIGHT produces a form (unlabeled) which spontaneously forms ELEVEN which has NH . . . O= hydrogen bonding.

Beginning with THREE, exchanging methyl and hydroxyl produces SIX. A subsequent rotation of the amide produces SEVEN (Figure 3).

We turn to structures with an unsaturated link and one carbonyl group at the methyl-substituted C. The system EIGHT which lacks H bonding occupies a high-energy relative minimum (Figure 4). If the OH group is positioned to donate a H bond to the carbonyl oxygen, the system spontaneously rearranged to species TWO. Rotating the acyl group produces species NINE. Exchange of amino and hydroxyl groups in NINE produces TEN. NINE and TEN may be stabilized by CH . . . OH and CH . . . N interactions.

On the other hand, exchange of amino and hydroxyl groups in EIGHT produces a species lacking H bonding (unlabeled). Upon optimization it rearranges spontaneously to ELEVEN which is stabilized by > NH . . . O= H bonding (Figure 5).

2.2. Extrapolation of Accurate Relative Energies

The numerical values (kJ/mol) in Figures 1–5 are the result of the extrapolations described in the methods section. Estimates of electronic energies from the G4 thermochemical scheme agree within 1–2 kJ/mol with our extrapolated values. Detailed tables of energies obtained with RHF, MP2, and CCSD Hamiltonians in Dunning basis sets cc-pVNZ with N = 2, 3, and 4 are provided in Supplemental Information. A broad overview is set forth in Table 1.

Table 1. Species exhibiting categories of interaction and their relative energies (kJ/mol.).

No H Bonding	CH . . . O or CH . . . N	OH . . . O, OH . . . N, NH . . . O
THREE (+62) EIGHT (+101)	SIX (+43) SEVEN (+38) NINE (+91) TEN (61)	ONE (4.0) TWO (0.0) FOUR (+42) FIVE (+38) ELEVEN (+48)

There is a rough clustering of H bonding types. Species ONE and TWO with NH . . . O= and OH . . . O= hydrogen bonding are most stable, while systems THREE and EIGHT with little or no H bonding are relatively unstable. However systems with presumably much weaker CH . . . O or CH . . . N interactions (SIX, SEVEN, and TEN) are not entirely separated from systems incorporating XH . . . Y (X, Y = O and N) which are generally thought to be stronger.

It appears that the relative stability, a global property for each of the molecules in question, is not simply explained as a consequence only of differences in local H bonding. Our task, to describe the hydrogen bonds, needs more local analysis.

2.3. Atoms in Molecules (AIM) Characterization of Interactions

The Quantum Theory of Atoms-in-Molecules, which defines local properties of the charge distribution at significant points, has been widely employed to characterize bonding of many kinds. (See further discussion in the Methods and Software section below.) Table 2 collects the electron density and its Laplacian at bond critical points, with the associated kinetic energy density G and the potential energy V. The total charge $Q(H)$ in the basin containing H atom and the delocalization index $\delta(H, B)$ are shown as well. The delocalization index is a measure of the number of electrons shared between two basins, and is related to the extent of covalent bonding and, indirectly, to bond strength. Kraka and co-workers [34] define an empirical bond order n derived from the density at a BCP.

$$n = 0.54 \, \rho^{0.32} \tag{1}$$

Table 2. Topological data for critical points associated with non-covalent interaction in species ONE–ELEVEN.

Species	Interaction	ρ	$\nabla^2\rho$	H	V	Q(H)	Order	DI
ONE	C=O … HN	0.0241	0.0931	0.00640	−0.0177	0.4636	0.164	0.0642
TWO	-OH … O=	0.0615	0.1158	−0.01860	−0.0661	0.3273	0.221	0.1254
THREE	NCI	0.0110	0.0564		−0.0080	0.5876		
FOUR	HO … HN	0.0218	0.0974	0.00390	−0.0167	0.4485	0.158	0.0541
FIVE	−OH … NH$_2$	0.0324	0.0870	−0.00220	−0.0262	0.3661	0.180	0.0831
SIX	CH … O=	0.0202	0.0782	0.00300	−0.0135	0.1164	0.155	0.0715
SEVEN	CH … NH2	0.0140	0.0517	0.00230	−0.0084	0.0604	0.138	0.0531
EIGHT	NCI	0.0133	0.0560		−0.0081	0.5978		
NINE	CH … OH	0.0114	0.0532	0.00260	−0.0081	0.0140	0.129	0.0384
TEN	CH … N	0.0137	0.0511	0.00220	−0.0083	0.0434	0.142	0.0276
ELEVEN	=O … HN	0.0391	0.1192	−0.00260	−0.0350	0.5022	0.191	0.1003

ρ = density; $\nabla^2\rho$ = Laplacian of density; **G** = Kinetic energy density; **V** = potential energy density; **Q(H)** = integral of charge in H atom basin. Order = Kraka definition of bond order and DI = delocalization index for H atom and its partner in the species' putative hydrogen bond. Analysis by AIMALL of density computed with ωB97XD/cc-pVTZ.

The positive values of the Laplacian (Table 2) indicate that the interaction is between closed shells. The values of bond order indicate that the strongest H bond is the link OH … O of TWO at 0.221, while the bond orders descend from 0.191 to 0.180 to 0.164 to 0.158 for ELEVEN, FIVE, ONE, and FOUR. These all have O to N hydrogen bonds. CH … X interactions are found in SIX (0.155), TEN (0.142), SEVEN (0.138), and NINE (0.129). OH seems to be a weaker H-bond acceptor than carbonyl =O. We do not assign a bond order to THREE and EIGHT, despite the presence of appreciable density at BCPs between O atoms. See further discussion on THREE and EIGHT below and in the section on non-covalent interactions.

An energy density diagnostic adopted by Cremer and Kraka [34] identifies interactions as mainly electrostatic (if H = G + V > 0) or mainly covalent (if H < 0). [35–40] By this criterion the only covalent interactions are for TWO (-OH … O=), FIVE (-OH … NH) and ELEVEN (=O … HN). These also have the shortest X … Y distances (2.538, 2.631, and 2.617 Å). ONE, which has an =O … HN- interaction has crossed over to be predominantly electrostatic. This may be attributed to its greater O … N distance, 2.789 Å, since the electrostatic interaction is of longer range than the covalent interaction which depends on orbital overlap.

H > O for the systems with no plausible H bonding (THREE and EIGHT) and those for which CH … X hydrogen bonding is conceivable (SIX, SEVEN, NINE, and TEN). It is notable that ONE and FOUR with =O … HN and HO … HN interactions are to be considered electrostatic according to the H diagnostic.

Several images representative of interactions as characterized by AIM appear in Figure 6 The complete set is to be found in Supplemental Information.

The diagrams display the "bond paths" (solid and dashed lines) and the bond (or line) critical points (green) for each species, and the ring critical points as well (red). The paths and BCPs close a ring; the location of the ring critical point is related to the strength of the ring closure. In the weaker ring closing interactions, the RCP approaches the BCP, as in THREE and EIGHT. For CH … X interactions, the RCP is further removed from the BCP, and for stronger hydrogen bonds the separation is even greater.

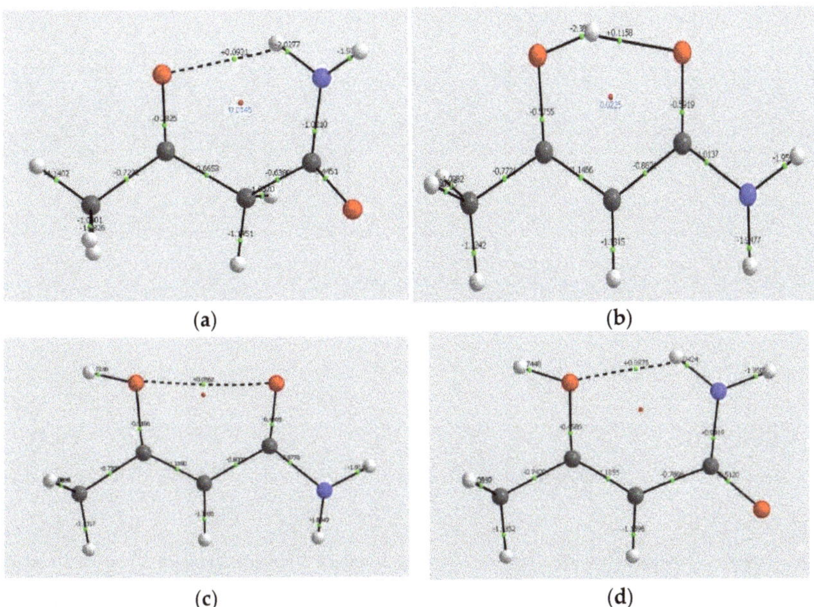

Figure 6. Atoms in Molecules analysis of ωB97XD/cc-pVTZ density. Ring CPs in red, BCPs in green. Laplacian values at BCPs are shown. (**a**) Species ONE, with NH . . . O= hydrogen bonding; (**b**) Species TWO, with OH . . . O= hydrogen bonding; (**c**) Species THREE with repulsive noncovalent interaction; (**d**) Species FOUR, with NH . . . OH.

Our emphasis has been on the BCPs associated with hydrogen bonding and the properties of density at those points. What are we to make of THREE and EIGHT, with O . . . O paths and a substantial positive Laplacian considerable density at the (path) critical point? In common with the other systems, the Laplacians show that the interaction depletes density at the CP, as is characteristic of interactions between closed shells. To deal with such cases we turn to the reduced density gradient, which diagnoses non-covalent interactions.

2.4. Non-Covalent Interaction (NCI) Characterization of Interactions

Isosurfaces for the reduced density gradient characterizing the noncovalent interactions in all species are shown in Figure 7. The color coding identifies the O . . . O interaction in THREE and EIGHT (entirely green) as repulsive, despite the substantial density at the BCPs. In cases TWO and ELEVEN the portion of the NCI isosurface enclosing the BCP is blue (attractive), and the portion enclosing the RCP is green. The NCI enclosing surfaces for weakly interacting CH . . . O and CH . . . N systems are of more subtly varying hue. All these qualitative observations comport with our understanding of the relative strength of the hydrogen bonding.

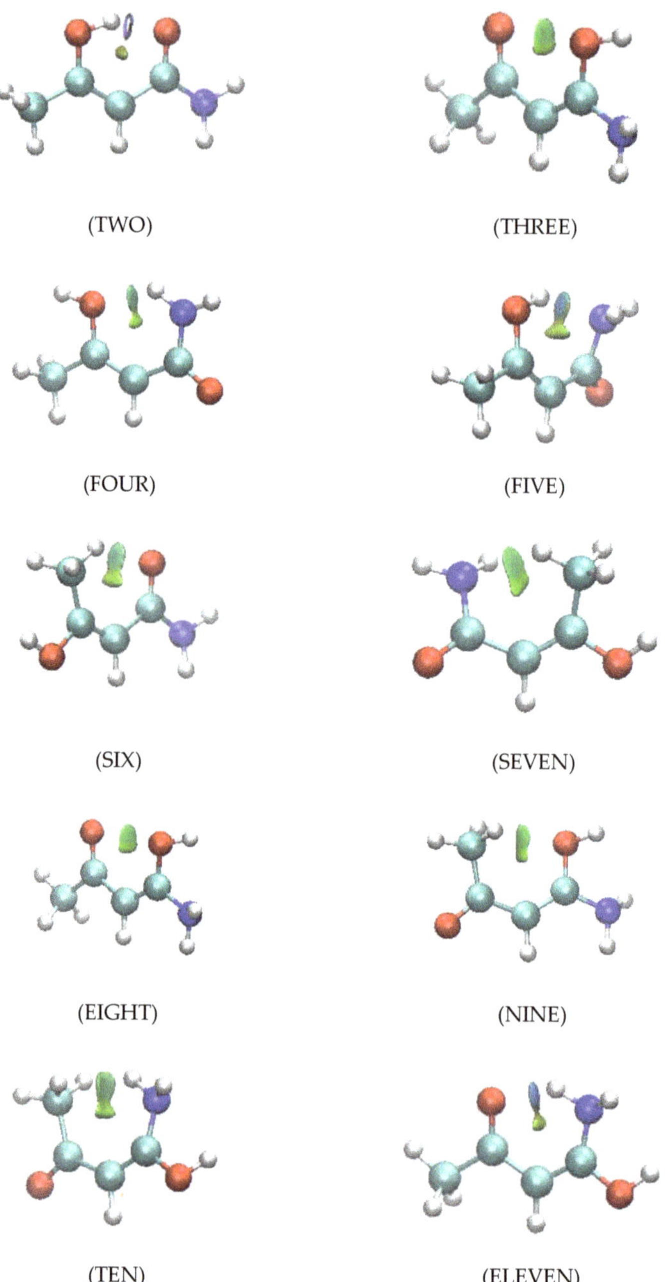

Figure 7. Panels display the isosurfaces for the noncovalent interactions in species TWO through ELEVEN. Species ONE is discussed in detail in the Methods and Software section below. Color coding identify regions of repulsive interaction as green and the attractive regions as blue.

2.5. Expression of Hydrogen Bonding in MP2/cc-pVTZ Computed Harmonic Vibrations

Hydrogen bonding is often expressed in the vibrational spectrum. Here we discuss the canonical frequencies which correspond to normal modes. These are in principle delocalized combinations of local modes, but in some cases the local modes are well isolated. These include OH, NH_2, and CH_3 group modes. See Table 3.

Table 3. anonical harmonic frequencies (in cm^{-1}) for normal modes of all systems, computed by MP2/cc-pVTZ.

Species	NH$_2$ Stretches		Double Bond Stretches		OH Stretches
ONE	3725a	3566s	1790 [1]	1757 [2]	None
TWO	3777a	3630s	1725	1710	3087
THREE	3747a	3609s	1782	1748	3836
FOUR	3777a	3626s	1773	1736	3853
FIVE	3659s	3537s	1780	1704	3571
SIX	3766a	3622s	1771	1742	3842
SEVEN	3726a	3591s	1731	1772	3850
EIGHT	3679a	3559s	1763	1714	3807
NINE	3673a	3554s	1712	1750	3803
TEN	3726a	3591s	1772	1731	3850
ELEVEN	3655a	3347s	1685	1742	3818

Species	CH$_3$ Stretches [4]			CCC Bend [5]	CO Stretch
ONE	3208s	3157a	3076b	236	None
TWO	3198s	3164a	3084b	246	1356
THREE	3202s	3155a	3074b	178	1385
FOUR	3198a	3143s	3063b	226	1347
FIVE	3203s	3162a	3084b	226	1317
SIX	3213s	3135a	3062b	226	1348, 1267
SEVEN	3213s	3132a	3063b	193	1351, 1273
EIGHT	3202s	3155a	3074b	178	1467
NINE	3205s	3175a	3089b	207	1435
TEN	3214a	3132s	3063b	194	1273
ELEVEN	3204s	3161a	3079b	242	1367

[1] Predominantly C=O. [2] Predominantly C=C. NH$_2$ stretches are either antisymmetric (a) or symmetric (s) combinations of local NH motions [4] CH stretches descend in symmetry from the C_{3v} limit; the A_1 mode can be recognized as "breathing" breathing (b) and components of the E pair are the symmetric (s) and antisymmetric (a) modes. [5] The C-C-C bend serves as a surrogate for the hydrogen bond stretch.

OH Mode: The reference OH stretches–uninfluenced by H bonding–fall in the set (THREE, SEVEN, and ELEVEN) and have values 3836, 3850, and 3842 respectively. EIGHT and TEN have OH modes at 3807, which may be associated with coupling of OH at a carbon also bearing NH_2, both being uninvolved in intramolecular hydrogen bonding. The most drastically shifted OH stretches are for TWO to 3087 and FIVE to 3571, suggesting that the OH … O= interaction in TWO is stronger than the OH … NH_2 interaction in FIVE, and that OH is not strongly engaged in hydrogen bonding in any other system.

NH$_2$ Modes: NH stretching modes for most systems cluster in the range 3720–3780 for the asymmetric combination and 3600–3630 for the symmetric combination. High values for the differences Δv between asymmetric and symmetric stretching frequencies correspond to NH participation in H bonding for (especially) system ELEVEN ($\Delta v = 308$) and to a lesser extent for system ONE ($\Delta v = 159$). ONE and ELEVEN have the H-bonding structure (NH … O=) in common, but the larger shift in ELEVEN is easily attributed to the shorter N … O distance in ELEVEN (2.617 Å) than is found for ONE (2.789 Å). The next largest difference is for FOUR (Δv=151), which has an NH … OH interaction.

The NH stretching modes in EIGHT ($\Delta v = 120$) and NINE ($\Delta v = 119$) have minimal differences. In both cases the NH_2 group is isolated from H bonding. The remaining systems have splitting ranging from 132 to 151. TWO ($\Delta v = 147$) and SIX ($\Delta v = 144$) have comparable splitting; for each, NH_2 is attached to a carbonyl carbon. FIVE (132), SEVEN (135), and TEN (136) fall in a narrow range. In each of these, N is a hydrogen bond acceptor. THREE (with $\Delta v = 138$) is unique.

CH stretches in the methyl group: Methyl CH stretches in a C_{3v} environment include the A_1 all-in-phase "breathing" mode and an E set of out-of-phase motions. In this low symmetry setting we can identify motions corresponding to those in high symmetry. The mode analogous to the A_1 breathing is lowest in frequency, ranging from 3060 to 3090 for species ONE through ELEVEN. The former E stretching combinations, which split into in-phase and out-of-phase modes in the low symmetry environment, appear in ranges 3130–3180 and 3060–3090 respectively. There seems to be no pattern in these values indicating whether the methyl group can participate in CH . . . X interaction.

CCC bend: For THREE and EIGHT, for which no H bonding is recognizable and a repulsive NCI zone lies between the C=O and OH oxygens, the C-C=C bending modes have the lowest a frequency, 178. For other systems the C-C=C bend is our rough surrogate for hydrogen bond stretching. The highest bending mode frequencies are found for ELEVEN (242), SEVEN (247), TWO (246) and ONE (242). These have NH . . . O=, CH . . . NH_2, OH . . . O=, and weakened NH . . . O= interactions respectively. Mixing with the methyl internal rotation sometimes makes the isolation of the CCC bend difficult; strong coupling is evident for SEVEN and other species with CH . . . X interaction.

Coupling of local modes within canonical normal modes complicates the interpretation of hydrogen bonding by inspection of vibrations. For example, the C=O and C=C stretching modes are strongly coupled in FOUR, but are weakly coupled in NINE and TEN. Furthermore, the COH bend is often strongly coupled to the CO stretch. as well. We expect that extraction of local modes from the canonical normal modes will simplify the discussion of H bonding and vibrations. This is accomplished by the Local Mode Analysis [41,42].

2.6. Local Mode Analysis

The Local Mode Analysis allowed identification of force constants for several diatomic stretching motions for all species ONE through ELEVEN. These included the two C=O stretches for ONE (Table 4) and the single C=O stretches for TWO through ELEVEN (Table 5). CO and OH stretches were also defined for TWO through ELEVEN. The three methyl CH stretches and the two NH stretches were chosen for all species. For systems which can plausibly be assigned hydrogen bonding structures, the X . . . HY stretches were included in the analysis.

Table 4. For species ONE, local force constants (millidynes/Å) and associated frequencies (cm^{-1}).

Species	C=O	C=O′	NH	NH′	C-H	C-H′	C-H″	H . . . O
ONE	11.7	11.7	7.4	7.0	5.5	5.3	5.3	0.161
	1700	1699	3654	3564	3181	3116	3112	537

Table 5. For species TWO–ELEVEN, local force constants (millidynes/Å) and associated frequencies (cm^{-1}).

Species	C=O	C-O	NH	NH$'$	C-H	C-H$'$	C-H$''$	O-H	X ... Y[1]
TWO	9.7	6.5	7.4	7.0	5.5	5.4	5.4	4.3	0.323
	1552	1273	3651	3670	3169	3130	3130	2801	760
THREE	11.7	6.2	7.3	7.3	5.5	5.3	5.3	8.2	NCI
	1701	1237	3641	3630	3174	3103	3103	3829	
FOUR	11.4	5.6	7.6	7.6	5.5	5.3	5.3	8.3	0.150
	1678	1177	3666	3661	3178	3099	3099	3848	764
FIVE	11.8	6.0	7.1	7.0	5.5	5.4	5.4	6.7	0.201
	1708	1214	3578	3560	3176	3129	3127	3475	712
SIX	11.3	5.8	7.4	7.4	5.6	5.3	5.3	8.2	0.142
	1671	1202	3657	3644	3186	3100	3100	3836	732
SEVEN	11.6	5.8	7.2	7.3	5.6	5.3	5.3	8.2	0.109
	1694	1195	3616	3631	3191	3098	3098	3844	612
EIGHT	11.6	5.8	7.0	7.2	5.5	5.3	5.3	8.0	NCI
	1694	1253	3557	3610	3176	3113	3113	3757	
NINE	11.2	6.0	7.0	7.2	5.5	5.4	5.4	8.0	0.093
	1666	1219	3553	3606	3172	3127	3127	3789	711
TEN	11.2	5.8	6.8	6.8	5.2	5.0	5.2	7.8	0.115
	1664	1197	3511	3598	3176	3022	2995	3728	688
ELEVEN	9.9	6.0	7.2	5.7	5.2	5.2	5.2	8.1	0.231
	1566	1219	3594	3212	3176	3076	3076	3806	804

[1] Atoms linked by the hydrogen bond. Local force constants are based on computed values for TWO and ELEVEN, which provided parameters for a power law fit to BCP densities. Local frequencies are inferred from scaling to the square root of local force constants.

Local CH stretches span a narrow range of frequencies, ca. 3100 to 3200 cm^{-1}. There seems to be no great impact on these values even in systems that may have CH ... O or CH ... N interactions.

C=O stretches extend from 1650 to near 1700. The exception is species TWO at 1550, which has strong =O ... HO hydrogen bonding. OH stretches which do not participate on hydrogen bonding have frequencies near 3800 to 3850 (THREE, SEVEN, EIGHT and NINE) while TWO and FIVE have seriously reduced OH stretching frequencies, in keeping with their participation in hydrogen bonding.

Species TWO, THREE, and SIX have both local NH force constants above 7.2 and both local frequencies above 3600. These systems do not involve the NH$_2$ group in H bonding. ONE and ELEVEN engage one NH bond in hydrogen bonding which is reflected in one low NH stretching frequency (3564 and 3212). FOUR is a puzzling exception, with near-identical NH force contants. TEN shows both NH force constants of 6.8 and frequencies near 3500, and FIVE has its two modes with force constants near 7.1 and frequencies near 3579. EIGHT and NINE have one local mode with low frequency (below 3600) with the other mode higher than 3600. These values are consistent with an interaction between substituents OH and NH$_2$ on an unsaturated carbon.

Yannacone et al. [18] established an empirical power law relation between density at a bond critical point and local force constants, of the form

$$ln\,(k_{HBOND}) = A\,ln(\rho_{BCP}) + B, \qquad (2)$$

We fit the computed local force constants for TWO and ELEVEN to this form, finding A = 0.740 and B = 0.934, and inferred the remaining values reported in Table 4. The sequence of H bond strength begins with the strongest interactions, found in TWO > ELEVEN > FIVE which have k_{LOCAL} above 0.200. The central cluster includes ONE, FOUR, and SIX with NH ... O=, >NH ... OH, and CH ... O= interactions. The weakest interactions (in descending order) are CH ... N, CH ... N, and CH ... OH. It is interesting to see that the strongest CH ... X interaction (with X a carbonyl oxygen) is comparable to NH ... O interactions.

3. Discussion

We have described the intramolecular interactions of conformations of acetoacetamide and isomers of its tautomer 2-oxopropanolamine by accurate *ab intio* calculations of relative energies and analysis of their density by AIM and NCI theories. Recasting the vibrational modes by the Cremer-Kraka local modes analysis allow reliable assignments of relative strengths of hydrogen bonds XH . . . Y with (X, Y) including (O, O), (O, N), (N, C), and (O, C) (Figure 8). Rank ordering of bond strength by measures bond order n, electron density ρ and potential energy density V, and also local force constants k produce an agreed sequence of bond strengths. In specific, ordering according to a bond energy BE estimate based on the potential energy density at the bond critical point agrees with ranking based on local force constants.

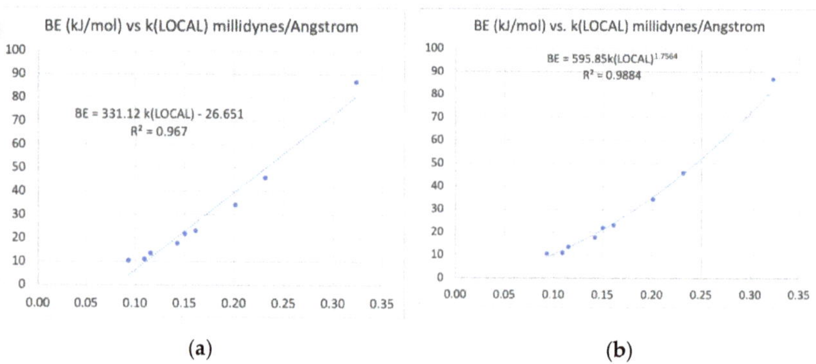

Figure 8. (a) Linear correlation between the local force constants k_{LOCAL} and the Espinosa binding energy (BE) defined as BE = $-$V/2 where V is the potential energy density evaluated by QTAIM at the H bond's Critical Point. (b) Correlation diagram with a power law fit. The rank order of H bond strengths agrees with the ordering predicted by k_{LOCAL} and other measures, including the Kraka bond order *n* and the density itself at the BCPs. A linear fit is imperfect, but illustrates a strong relation between variables.

All measures agree that the H bond strength decreases in magnitude in the order shown in Figure 9, which shows the values of the local force constants for H bonds.

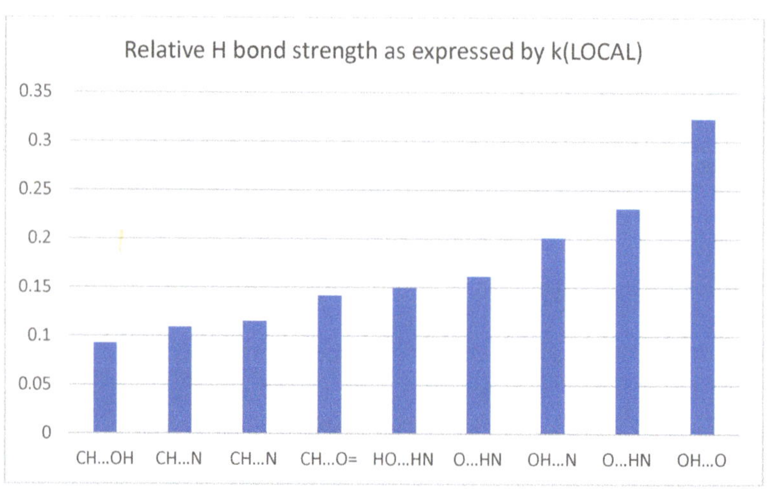

Figure 9. Local force constants (millidynes/Angstrom) for hydrogen bonds.

We expect that the analysis can be extended to many molecules of biological importance and to the description of other kinds of noncovalent interactions.

4. Materials and Methods

We employed extrapolation techniques to arrive at accurate relative energies. The Atoms in Molecules analysis of the computed density, and its extension to the reduced gradient description of non-covalent interactions allowed one perspective on the relative strength of our system's hydrogen bonds. An alternative view is provided by analysis of the vibrational spectrum and its canonical (normal) modes. Recovery of local vibrational modes and their associated frequencies allows more direct discussion properties of the hydrogen bonds.

4.1. Extrapolation Techniques

A well-established method for obtaining highly accurate energies (including energy differences among conformational isomers) is the extrapolation of energies obtained in a sequence of more and more flexible basis sets [35]. Such techniques depend on the behavior of energy values obtained in a sequence of basis sets embracing a series of values of the angular momentum. The Dunning basis sets of general form cc-pVLZ, with maximum angular momentum $L = 2, 3, 4, \ldots$ are most often used [36,37]. The Hartree-Fock energy is known to scale exponentially, while the correlation energy follows a L^{-3} inverse-cubic scaling for sufficiently large L [38]. Simple schemes based on this behavior are often usefully accurate. We employ the formulas shown below.

$$E_{HF}(\infty) = \frac{(E_{HF}(L-2) - E_{HF}(L-1))^2}{(E_{HF}(L-2) - 2E_{HF}(L-1) - E_{HF}(L))} \tag{3}$$

$$E_{CORR}(\infty) = \frac{\left(L^3 E_{CORR}(L) - (L-1)^3 E_{CORR}(L-1)\right)}{\left(4L^3 - (L-1)^3\right)} \tag{4}$$

Here L denotes the maximum angular momentum in the extrapolation, so if we use cc-pVQZ as the largest basis, $E_{HF}(L)$ and $E_{CORR}(L)$ refer to the energies obtained with that basis, $L - 1$ to cc-pVTZ, and $L - 2$ to cc-pVDZ.

A very recent refinement in the extrapolation of the correlation energy has been reported by Lesiuk and Jeziorski [39] which employs E_{CORR} for L and $L - 1$. Defining the constant a by

$$a = L^4(E_{CORR}(L) - E_{CORR}(L-1)) \tag{5}$$

The limit of the correlation energy is

$$E(\infty) = E_{CORR}(L) + a\left[\frac{\pi^4}{90} - \sum_{l=1}^{L} l^{-4}\right] \tag{6}$$

We found that the two-point Zeta-extrapolation lowered the final estimate of total electronic energy by 15 millihartrees. However, the Zeta-extrapolation had only a very small impact (<0.2 kJ/mol.) on relative energies obtained with the simpler scheme.

We do not employ any correction intended to overcome basis set superposition error, since it appears that the complete basis set limit of energies is only very slightly altered by its inclusion [35]. The fact that we are studying intramolecular effects may further discount its significance.

Extrapolation is also an important part of well-established thermochemical schemes, which include the CBS series [33,40], the Gn series [41], and more demanding schemes such as Wn [42,43] and HEAT [44]. These methods include empirical corrections lacking in our calculations. To complement our extrapolations we performed G4 calculations, finding that the G4 estimates of relative electronic energies of species in question were in disagreement with our values by no more than 1–2 kJ/mol.

4.2. Atoms in Molecules

The atoms in Molecules (AIM) theory of Bader [13,27,45] is constituted of a description of the molecular charge density ρ, its gradient and Hessian, and related quantities including kinetic and potential energy densities. Extreme values of the density demark significant regions in the molecular charge distribution. Points with $\nabla\rho = 0$ may be described as quantum atoms, bond critical points (BCPs), ring critical points (RCPs) and cage critical points (CCPs) depending on the sign structure of the set of eigenvalues of the Hessian matrix of the density at those points. A locus of points with zero density gradient connecting atom centers through a BCP is termed a bond path. These paths very often correspond to intuitive ideas of chemical bonds. However serious complications are often encountered in the chemical interpretation of bond paths and critical points for weak interactions, as mentioned already [22–28].

The AIM analysis was applied to H bonding from the beginning. Koch and Popelier [46] developed criteria for judging H-bonding in AIM context. They include:

1. Topology: There should appear a bond path between atoms considered to be linked by a H bond. This path should contain a BCP.
2. Density: At this point the density should have a "reasonable" value, about an order of magnitude smaller than the values for covalent bonds or ca 0.01 atomic units, with a range up to 0.035 atomic units. There is a correlation between the BCP density and *inter*molecular attraction.
3. Laplacian: $\nabla^2\rho$ must be positive, with values in an approximate range 0.02 to 0.10 atomic units. This sign suggests a depletion of charge at the BCP, and is characteristic of an interaction between closed shells [13,21].
4. H atom charge: q(H) involved in a hydrogen bond XH . . . Y should be substantially smaller than q(H) for H atoms not participating in a H bond but with the same X atom in XH. Here q(H) is the charge in the basin containing the nuclear attractor for H atom.

Further criteria include Atomic interpenetration; Destabilization of the H atom; Diminished atomic polarizability for the H atom; and Decreased volume of the H atom. In this work we confine our attention to the first four criteria, as has been the practice in most studies of AIM-based description of intramolecular H bonding in conformational isomers.

The delocalization index δ(A, B), in contrast to many of the parameters of AIM theory, refers to the two-electron density matrix, and is a measure of electrons shared between two basins A and B. A perfectly covalent bond would have a value for δ(A, B) near the integer associated with a Lewis structure, but an ionic component to bonding reduces the value of δ(A, B) [47,48]. The parameter has been used to discuss the strengths of intermolecular hydrogen bonds [48].

4.3. Non-Covalent Interactions and the Reduced Density Gradient

The measure $s(r)$ for non-covalent interaction (NCI) was introduced by Johnson et al. [49,50] The reduced (dimensionless) density gradient s (also called RDG) is defined as

$$s(r) = \frac{|\nabla\rho(r)|}{2(3\pi^2)^{1/3}\rho(r)r^{4/3}} \tag{7}$$

The RDG has been applied to the description of H-bond strength [51]. A useful description of the RDG, its graphical characterization, and its interpretation is given by Contreras-Garcia, et al. [52].

Two NCI graphic realizations are useful. A two-dimensional plot of s vs. the density (given the sign of the second eigenvalue of the Hessian **H**) will produce a broad sweep from one extreme at small density but large s through an extreme and large s and small density and then to a second extreme of small density and small s. (Figure 6) This is characteristic of an exponentially-decaying density. Superimposed on this sweep may be spikes in s extending to values near zero at modest values of density. The high-density (but low s)

regions near nuclei are far to the right or left of the range of densities shown. The low-density region is located primarily in Cartesian space close to the density tails for which s is large. However, s can be small precisely where the density is disturbed by non-covalent interactions, such as van der Waals/dispersion, electrostatic forces, and hydrogen bonds. Figure 10 is the plot for TWO, the most stable species, which contains a strong -OH ... O= hydrogen bond with its corresponding critical point, and a closed ring marked by a ring critical point. Two such spikes are shown, at about +0.015 and −0.025 signed density units. Generally spikes fall into three types: (i) negative values of the signed density indicative of attractive interactions, such as dipole-dipole or H-bonding, (ii) positive signed density indicating non-bonding interactions, such as electrostatic or steric repulsion in the ring/cage, and (iii) values near zero indicating very weak interactions, such as van der Waals interaction.

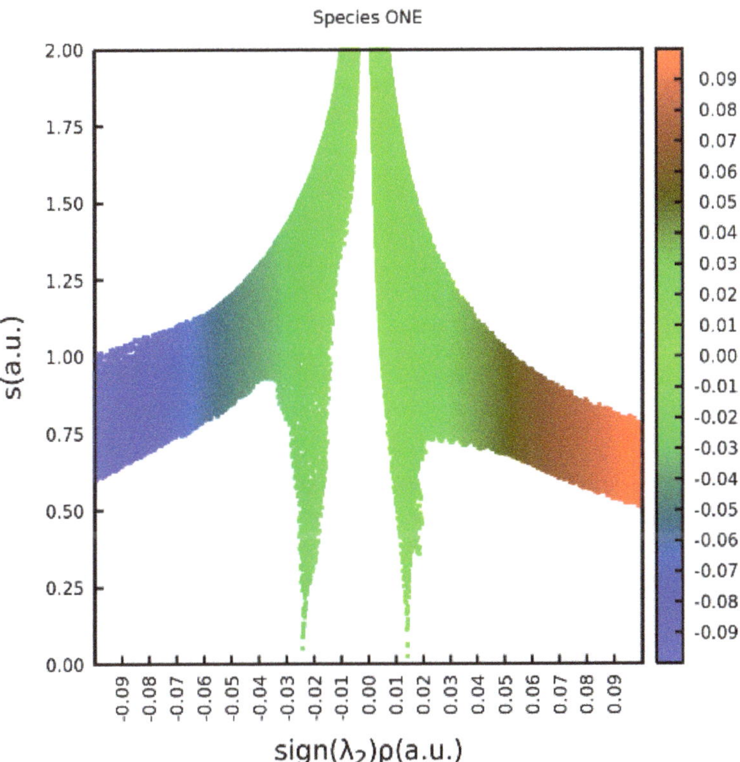

Figure 10. Reduced density gradient s vs. the density signed by the second eigenvalue of the Hessian for species ONE. A positive sign indicates a repulsive non-covalent interaction (often found in the center of a closed ring), and a negative value denotes attraction (H bonding in this case).

Figure 11 shows the isosurface obtained by mapping data from Figure 10 into three-dimensional space. The blue region shows an attractive interaction (according to the reduced density gradient criterion) at the Bond Critical Point along the Path O ... H. At a Ring Critical Point a green isosurface encloses a region of repulsive non-covalent interaction.

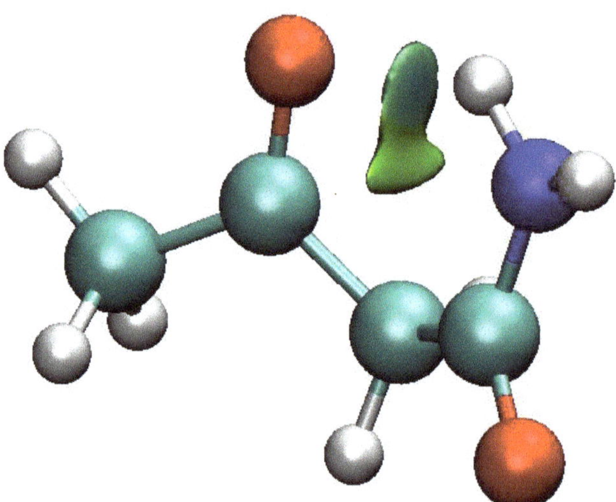

Figure 11. The NCI isosurfaces for ONE make evident the repulsive noncovalent interaction at the RCP interior to the ring (green coded) and the attractive noncovalent interaction at the BCP for the hydrogen bond (blue).

4.4. Canonical Vibrational Spectra and Local Mode Analysis

Computed vibrational frequencies are generally recovered as eigenvalues of the matrix of second derivatives of the energy with respect to mass-weighted displacement coordinates, in the formulation developed by Wilson and coworkers [53]. The associated eigenvectors, normal modes, are generally delocalized. In the study of the strength of hydrogen bonding discussion is eased by the introduction of local force constants without reference to atomic masses. A means of recovery of local force constants from vibrational data was developed by Cremer and coworkers [54]. An overall review is provided by Kraka, Zou, and Tao [55]. The hydrogen bond, among many other kinds of noncovalent interaction has been analyzed [56–58].

4.5. Software

Gaussian 09 [59] and 16 [60] produced optimized structures and the extrapolation sequence of RHF, MP2, and coupled-cluster energies. Atoms-in-Molecules results were obtained with AIMALL [61]. NCIPLOT was obtained by download from the Contreras-Garcia group [62] and LMODEA software for the local modes analysis [63] was provided by the Kraka group.

Supplementary Materials: The following are available online, which includes: Electronic energies used in extrapolation; Cartesian coordinates of all species; AIM figures with bond paths, line critical points and associated values of the Laplacian of the electron density.

Author Contributions: Conceptualization, C.T. and Z.A.; methodology, C.T.; software, Z.A. and E.A.B.; validation, C.T., Z.A. and E.A.B.; formal analysis, C.T.; investigation, C.T., Z.A. and E.A.B.; resources, C.T., Z.A. and E.A.B.; data curation, C.T.; writing—original draft preparation, C.T.; writing—review and editing, C.T. and Z.A.; visualization, C.T. and Z.A.; supervision and project administration, C.T. All authors have read and agreed to the published version of the manuscript.

Funding: Our thanks to Elfi Kraka and Marek Freindorf of Texas Christian University for supplying the LMODEA software. Thanks also to the Body foundation which made possible the acquisition of software and computational resources used in the project. E.A.B. is particularly grateful to BAPKO of Marmara University providing computational resources through Grant No. FEN-A-071015-0477.

Institutional Review Board Statement: Not applicable.

Informed Consent Statement: Not applicable.

Data Availability Statement: Not applicable.

Conflicts of Interest: The authors declare no conflict of interest.

Sample Availability: Samples of the compounds Not applicable; no chemical material is used in this research.

References

1. Needham, P. Hydrogen bonding: Homing in on a tricky chemical concept. *Stud. Hist. Philos. Sci.* **2013**, *44*, 51–65. [CrossRef]
2. Jeffrey, G.A. *An Introduction to Hydrogen Bonding*; Oxford U Press: New York, NY, USA; Oxford, UK, 1997.
3. Taylor, R.; Kennard, O. Crystallographic Evidence for the Existence of CH . . . O, CH . . . N, and CH . . . Cl Hydrogen Bonds. *J. Am. Chem. Soc.* **1982**, *104*, 5063–5070. [CrossRef]
4. Desiraju, G.R. The CH . . . O Hydrogen Bond: Structural Implications and Supramolecular Design. *Acc. Chem. Res.* **1996**, *29*, 441–449. [CrossRef] [PubMed]
5. Perrin, C. Strong hydrogen bonds in chemistry and biology. *Annu. Rev. Phys. Chem.* **1997**, *48*, 511–544. [CrossRef] [PubMed]
6. Hobza, P.; Havlas, Z. Blue-Shifting Hydrogen. *Bonds Chem. Rev.* **2000**, *100*, 4253–4264. [CrossRef] [PubMed]
7. Alkorta, I.; Rozas, I.; José Elguera, J. Non-conventional hydrogen bonds. *Chem. Soc. Rev.* **1998**, *27*, 163–170. [CrossRef]
8. Hadži, D. (Ed.) *Theoretical Treatments of Hydrogen Bonding*; Wiley Research Series in Theoretical Chemistry; John Wiley and Sons, Ltd.: Chichester, UK, 1997.
9. Schuster, P.; Wolschann, P. Hydrogen Bonding: From Small Clusters to Biopolymers. *Mon. Chem.* **1999**, *130*, 947–960.
10. Grabowski, S.J. (Ed.) *Hydrogen Bonding—New Insights*; Springer: Dordrecht, The Netherlands, 2006.
11. Gilli, G.; Gilli, P. *The Nature of the Hydrogen Bond. Outline of a Comprehensive Hydrogen Bond Theory*; Oxford U Press: Oxford, UK, 2009.
12. Arunan, E.; Desiraju, G.R.; Klein, R.A.; Sadlej, J.; Scheiner, S.; Alkorta, I.; Clary, D.C.; Crabtree, R.H.; Dannenberg, J.J.; Hobza, P.; et al. Definition of the Hydrogen Bond. *J. Pure Appl. Chem.* **2011**, *83*, 1637–1641. [CrossRef]
13. Bader, R.F.W. *Atoms in Molecules. A Quantum Theory*; Clarendon: Oxford, UK, 1990.
14. Scheiner, S. Special Issue: Intramolecular Hydrogen Bonding 2017. *Molecules* **2017**, *22*, 1521. [CrossRef] [PubMed]
15. Sanchez, G. Introduction to "Intramolecular Hydrogen Bonding 2018". *Molecules* **2019**, *24*, 2858. [CrossRef] [PubMed]
16. Afonin, A.V.; Vashchenko, A.V. Benchmark calculations of intramolecular hydrogen bond energy based on molecular tailoring and function-based approaches: Developing hybrid approach. *Int. J. Quantum. Chem.* **2019**, *119*, 26001. [CrossRef]
17. Emamian, S.; Lu, T.; Kruse, H.; Emamian, H. Exploring Nature and Predicting Strength of Hydrogen Bonds: A Correlation Analysis between Atoms-in-Molecules Descriptors, Binding Energies, and Energy Components of Symmetry-Adapted Perturbation Theory. *J. Comp. Chem.* **2019**, *40*, 2868–2881. [CrossRef] [PubMed]
18. Yannacone, S.F.; Sethio, D.; Kraka, E. Quantitative assessment of intramolecular hydrogen bonds in neutral histidine. *Theor. Chem. Acc.* **2020**, *139*, 125. [CrossRef]
19. Zavitsas, A.A. Quantitative relationship between bond dissociation energies, infrared stretching frequencies, and force constants in polyatomic Molecules. *J. Phys. Chem.* **1987**, *91*, 5573–5577. [CrossRef]
20. Cremer, D.; Kraka, E. Characterization of CF Bonds with Multiple-Bond Character: Bond Lengths, Stretching Force Constants, and Bond Dissociation Energies. *Chem. Phys. Chem.* **2009**, *10*, 686–698.
21. Espinosa, J.E.; Molins, E.; LeComte, C. Hydrogen bond strengths revealed by topological analyses of experimentally observed electron densities. *Chem. Phys. Lett.* **1998**, *285*, 170–173. [CrossRef]
22. Jablónski, M.A. Critical Overview of Current Theoretical Methods of Estimating the Energy of Intramolecular Interactions. *Molecules* **2020**, *25*, 5512. [CrossRef]
23. Jablónski, M.; Monaco, G. Different Zeroes of Interaction Energies as the Cause of Opposite Results on the Stabilizing Nature of C-H . . . O Intramolecular Interactions. *J. Chem. Inf. Modeling* **2013**, *53*, 1661–1675. [CrossRef]
24. Jablónski, M. Bond paths between distant atoms do not necessarily indicate dominant interactions. *J. Comput. Chem.* **2018**, *39*, 2183–2195. [CrossRef]
25. Jablónski, M. On the Uselessness of Bond Paths Linking Distant Atoms and on the Violation of the Concept of Privileged Exchange Channels. *Chem. Open* **2019**, *8*, 497–507.
26. Jablónski, M. Counterintuitive bond paths: An intriguing case of the $C(NO_2)_3^-$ ion. *Chem. Phys. Lett.* **2020**, *759*, 137946. [CrossRef]
27. Bader, R.F.W. Bond Paths are not chemical bonds. *J. Phys. Chem. A* **2009**, *113*, 10391–10396. [CrossRef] [PubMed]
28. Shahbazian, S. Why Bond Critical Points Are Not "Bond" Critical Points. *Chem. Eur. J.* **2018**, *24*, 5401–5405. [CrossRef] [PubMed]
29. Steiner, T.; Desiraju, G.R. Distinction between the weak hydrogen bond and the van der Waals interaction. *Chem. Commun.* **1998**, 891–892. [CrossRef]
30. Bauer, J.S.H.; Wilcox, C.F. On malonaldehyde and acetylacetone: Are theory and experiment compatible? *Chem. Phys. Lett.* **1997**, *279*, 122–128. [CrossRef]
31. Afonin, A.V.; Vashchenko, A.V.J. Quantitative decomposition of resonance-assisted hydrogen bond energy in β-diketones into resonance and hydrogen bonding (π- and σ-) components using molecular tailoring and function-based approaches. *Comput. Chem.* **2020**, *41*, 1285–1298. [CrossRef] [PubMed]

32. Belova, N.V.; Sliznev, V.V.; Oberhammer, H.; Girichev, G.V.J. Tautomeric and conformational properties of β-diketones. *Mol. Struct.* **2010**, *978*, 282–293. [CrossRef]
33. Ochterski, J.W.; Peterson, G.A. Complete basis set model chemistry. V. Extensions to six or more heavy atoms. *J. Chem. Phys.* **1996**, *104*, 2598–2619. [CrossRef]
34. Cremer, D.; Kraka, E. Chemical Bonds without Bonding Electron Density–Does the Difference Electron Density Analysis Suffice for a Description of the Chemical Bond? *Angew. Chem. Int. Ed.* **1984**, *23*, 627–628. [CrossRef]
35. Varandas, A.J.C. Straightening the Hierarchical Staircase for Basis Set Extrapolations: A Low-Cost Approach to High-Accuracy Computational Chemistry. *Annu. Rev. Phys. Chem.* **2018**, *69*, 177–203. [CrossRef]
36. Dunning, T.H. Gaussian basis sets for use in correlated molecular calculations. I. The atoms boron through neon and hydrogen. *J. Chem. Phys.* **1989**, *90*, 1007–1023. [CrossRef]
37. Wilson, A.K.; Mourik, T.; Dunning, J.T.H. Gaussian Basis Sets for use in Correlated Molecular Calculations. VI. Sextuple zeta correlation consistent basis sets for boron through neon. *J. Mol. Struct. (Theochem.)* **1996**, *388*, 339–349. [CrossRef]
38. Helgaker, T.; Klopper, W.; Koch, H.; Noga, J. Basis-set convergence of correlated calculations on water. *J. Chem. Phys.* **1997**, *106*, 9639–9646. [CrossRef]
39. Lesiuk, M.; Jeziorski, B.J. Complete Basis Set Extrapolation of Electronic Correlation Energies Using the Riemann Zeta Function. *Chem. Theory Comput.* **2019**, *15*, 5298–5403. [CrossRef]
40. Montgomery, J.; Frisch, M.J.; Ochterski, J.W.; Petersson, G.A. A complete basis set model chemistry. VI. Use of density functional geometries and frequencies. *J. Chem. Phys.* **1999**, *110*, 2822–2827. [CrossRef]
41. Curtiss, L.A.; Redfern, P.C.; Raghavachari, K. Gaussian-4 theory. *J. Chem. Phys.* **2007**, *126*, 084108. [CrossRef]
42. Martin, J.M.L.; de Oliveira, G. Toward standard methods for benchmark quality thermochemistry—W1 and W2 theory. *J. Chem. Phys.* **1999**, *111*, 1843–1856. [CrossRef]
43. Parthiban, J.; Martin, J.M.L. Assessment of W1 and W2 theories for the computation of electron affinities, heats of formation, and proton affinities. *J. Chem. Phys.* **2001**, *114*, 6014–6029. [CrossRef]
44. Tajti, A.; Szalay, P.G.; Császár, A.G.; Kállay, M.; Gauss, J.; Valeev, E.F.; Flowers, B.A.; Vázquez, J.; Stanton, J.F. HEAT: High accuracy extrapolated *ab. initio. thermochemistry. J. Chem. Phys.* **2004**, *121*, 11599. [CrossRef]
45. Bader, R.F.W. A Quantum Theory of Molecular Structure and Its Applications. *Chem. Rev.* **1991**, *91*, 893–928. [CrossRef]
46. Koch, U.; Popelier, P.L.A. Characterization of C-H-O Hydrogen Bonds on the Basis of the Charge Density. *J. Phys. Chem.* **1995**, *99*, 9747–9754. [CrossRef]
47. Fradera, X.; Austen, M.A.; Bader, R.F.W. The Lewis Model and Beyond. *J. Phys. Chem. A* **1999**, *103*, 304–314. [CrossRef]
48. Fradera, X.; Poater, J.; Simon, S.; Duran, M. Electron-pairing analysis from localization and delocalization indices in the framework of atoms-in-molecules theory. *Theor. Chem. Acc.* **2002**, *108*, 214–224. [CrossRef]
49. Johnson, E.R.; Keinan, S.; Mori-Sanchez, P.; Contreras-García, J.; Cohen, A.J.; Yang, W. Noncovalent Interactions. *J. Am. Chem. Soc.* **2010**, *132*, 6498–6506. [CrossRef] [PubMed]
50. Contreras-García, J.; Johnson, E.R.; Keinan, S.; Chaudret, R.; Piquemal, J.-P.; Beratan, D.N.; Yang, W. NCIPLOT: A Program for Plotting Noncovalent Interaction Regions. *J. Chem. Theory Comput.* **2011**, *7*, 625–632. [CrossRef] [PubMed]
51. Contreras-García, J.; Yang, W.; Johnson, E.R. Analysis of Hydrogen-Bond Interaction Potentials from the Electron Density: Integration of Noncovalent Interaction Regions. *J. Phys. Chem. A* **2011**, *115*, 12983–12990. [CrossRef] [PubMed]
52. Narth, C.; Maroun, Z.; Boto, C.R.; Bonnet, M.L.; Piquemal, J.P.; Contreras-Garcia, J. A Complete NCI Perspective from new bonds to reactivity. In *Applications of Topological Methods in Molecular Chemistry*; Springer: Cham, Switzerland, 2016; Volume 22, pp. 491–527.
53. Wilson, E.B.; Decius, J.C.; Cross, P.G. *Molecular Vibrations: The Theory of Infrared and Raman Vibratioal Spectra*; McGraw-Hill Book Co.: New York, NY, USA, 1955; Dover reissue, Courier Corporation 1980.
54. Konkoli, J.Z.; Cremer, D. A new way of analyzing vibrational spectra. I. Derivation of adiabatic internal modes. *Int. J. Quantum Chem.* **1998**, *67*, 1–9. [CrossRef]
55. Kraka, E.; Zou, W.; Tao, Y. Decoding chemical information from vibrational spectroscopy data: Local vibrational mode theory. *WIREs Comput. Mol. Sci.* **2020**, *34*, 1480. [CrossRef]
56. Freindorf, M.; Kraka, E.; Cremer, D. A comprehensive analysis of hydrogen bond interactions based on local vibrational modes. *Int. J. Quantum Chem.* **2012**, *112*, 3174–3187. [CrossRef]
57. Kalescky, R.; Zou, W.; Kraka, E. Cremer D Local vibrational modes of the water dimer—Comparison of theory and experiment. *Chem. Phys. Lett.* **2012**, *554*, 243–247. [CrossRef]
58. Kalescky, R.; Kraka, E.; Cremer, D. Local vibrational modes of the formic acid dimer—The strength of the double H-bond. *Mol. Phys.* **2013**, *111*, 1497–1510. [CrossRef]
59. Frisch, M.J.; Trucks, G.W.; Schlegel, H.B.; Scuseria, G.E.; Robb, M.A.; Cheeseman, J.R.; Scalmani, G.; Barone, V.; Mennucci, B.; Petersson, G.A.; et al. Gaussian 09, Revision, E.01. 2009. Available online: https://gaussian.com (accessed on 29 April 2021).
60. Frisch, M.J.; Trucks, G.W.; Schlegel, H.B.; Scuseria, G.E.; Robb, M.A.; Cheeseman, J.R.; Scalmani, G.; Barone, V.; Petersson, G.A.; Nakatsuji, H.; et al. Gaussian 16, Revision, A.03. 2016. Available online: https://gaussian.com (accessed on 29 April 2021).
61. Keith, T.A.; Todd, A.; Keith, T.K. Gristmill Software. AIMAll (Version 19.10.12). 2019. Available online: http://aim.tkgristmill.com (accessed on 29 April 2021).

62. Boto, R.A.; Peccati, F.; Laplaza, R.; Quan, C.; Carbone, A.; Piquemal, J.P.; Maday, Y.; Contreras-Garcia, J. NCIPLOT4: Fast, Robust, and Quantitative Analysis of Noncovalent Interactions. *J. Chem. Theory Comput.* **2020**, *16*, 4150–4158. Available online: http://www.lct.jussieu.fr/pagesperso/contrera/index-nci.html and https://github.com/juliacontrerasgarcia/nciplot (accessed on 29 April 2021). [CrossRef]

63. Zou, W.; Tao, Y.; Freindorf, M.; Makos, M.; Verma, N.; Kraka, E. *LMODEA2020*; Southern Methodist University: Dallas, TX, USA, 2020.

MDPI

St. Alban-Anlage 66

4052 Basel

Switzerland

Tel. +41 61 683 77 34

Fax +41 61 302 89 18

www.mdpi.com

Molecules Editorial Office

E-mail: molecules@mdpi.com

www.mdpi.com/journal/molecules